Sustainable Manufacturing Systems

Sustainable Manufacturing Systems

An Energy Perspective

Lin Li
University of Illinois at Chicago
Chicago, IL, USA

MengChu Zhou
University Heights
Newark, NJ, USA

IEEE Press Series on Systems Science and Engineering
MengChu Zhou, Series Editor

IEEE PRESS

WILEY

Published by John Wiley & Sons, Inc., Hoboken, New Jersey.
Published simultaneously in Canada.

For general information on our other products and services or for technical support, please contact our Customer Care Department within the United States at (800) 762-2974, outside the United States at (317) 572-3993 or fax (317) 572-4002.

Wiley also publishes its books in a variety of electronic formats. Some content that appears in print may not be available in electronic formats. For more information about Wiley products, visit our web site at www.wiley.com.

Library of Congress Cataloging-in-Publication Data applied for
Hardback ISBN: 9781119578246

Cover Design: Wiley
Cover Images: © Aleksey Popov

Set in 9.5/12.5pt STIXTwoText by Straive, Pondicherry, India

Contents

Author Biography

Dr. Lin Li joined the Department of Mechanical and Industrial Engineering, University of Illinois Chicago in 2011, and is now a Professor in Mechanical and Industrial Engineering. He also serves as the Director of U.S. Department of Energy Industrial Assessment Center and the founding Director of the Sustainable Manufacturing Systems Research Laboratory at the University of Illinois Chicago. He received a B.E. degree in Mechanical Engineering from Shanghai Jiao Tong University in 2001, and an M.S.E. degree in Mechanical Engineering, an M.S.E. degree in Industrial and Operations Engineering, and a Ph.D. degree in Mechanical Engineering from the University of Michigan, Ann Arbor, in 2003, 2005, and 2007, respectively. His research interests include energy control and electricity demand response of manufacturing systems, environmental sustainability of additive manufacturing processes, cost-effective cellulosic biofuel manufacturing system, lithium-ion electric vehicle battery remanufacturing and reliability assessment, multi-machine system modeling and throughput estimation, and intelligent maintenance of manufacturing systems. He is a recipient of Harold A. Simon Award and University of Illinois Chicago Teaching Recognition Program Award. He is a founding member of the technical committee of Sustainable Production and Service Automation in the IEEE Robotics and Automation Society, an academic Editor for journal Sustainability, and was Chair of quality and reliability technical committee, ASME Manufacturing Engineering Division.

Dr. MengChu Zhou joined the Department of Electrical and Computer Engineering, New Jersey Institute of Technology in 1990, and is now a Distinguished Professor in Electrical and Computer Engineering. His interests are in intelligent automation, semiconductor manufacturing, AI, Petri nets, Internet of Things, edge/cloud computing, and big data analytics. He has over 1000 publications including 12 books, over 700 journal papers including over 600 IEEE transactions/journal/magazine papers, 31 patents and 30 book-chapters. He is the founding Editor of IEEE Press Book Series on Systems Science and Engineering and Editor-in-Chief of IEEE/CAA Journal of Automatica Sinica. He is founding Chair/Co-chair of Technical Committee on AI-based Smart Manufacturing Systems of IEEE Systems, Man, and Cybernetics Society, Technical Committee on Semiconductor Manufacturing Automation and Technical Committee on Digital Manufacturing and Human-Centered Automation of IEEE Robotics and Automation Society. He is a recipient of Excellence in Research Prize and Medal from NJIT, Humboldt Research Award for US Senior Scientists from Alexander von Humboldt Foundation, and Franklin V. Taylor Memorial Award and the Norbert Wiener Award from IEEE Systems, Man, and Cybernetics Society, and Edison Patent Award from the Research & Development Council of New Jersey. He is Fellow of IEEE, International Federation of Automatic Control, American Association for the Advancement of Science, Chinese Association of Automation and National Academy of Inventors.

Preface

Sustainable Manufacturing Systems are one of modern technologies and have played a significant role in economic growth worldwide. Currently, the total value added by the global manufacturing industry reaches USD 13.5 trillion, accounting for nearly 16% of the global economy. Despite the continued strength of manufacturing industry, it also faces a pressing concern over energy consumption and environmental sustainability. Approximately, the industry sector possesses near one-quarter of the total energy consumption in the U.S., where over 75% of energy use is primarily attributed to manufacturing activities.

The issues of resource scarcity and environmental impacts are becoming vital due to the constantly rising demand for energy in the manufacturing sector. Several critical questions arise in proposing energy management strategies in manufacturing and evoke different aspects of energy efficiency studies, including (i) improving the energy efficiency of manufacturing systems considering the complex manufacturing conditions, (ii) reducing the energy cost with no sacrifice of manufacturing productivity, and (iii) generating policies or incentives to promote energy efficiency in the manufacturing industry and encourage the manufacturers' transition to environmentally conscious manufacturing. All these questions lead to the joint modeling and analysis of production and energy for manufacturing systems.

This book provides a holistic view of energy efficiency assessment and improvement measures for sustainable manufacturing systems, delivered through the state-of-the-art on sustainable manufacturing and energy efficiency issues, fundamentals and mathematical tools for manufacturing system modeling, and energy management methodologies for different manufacturing systems. Meanwhile, this book transfers the recent academic research results into various representative examples and case studies, which provide insights into the current sustainable practices and energy management strategies in manufacturing systems at different scales and levels. From the application aspect, this book is expected to help

(i) energy consumers, participants and administrators in energy efficiency programs, and (ii) research participants embrace the opportunities for advanced energy management. Furthermore, this book is intended to bring about learning initiatives for students in mechanical, industrial, environmental, and electrical engineering programs by effectively integrating concepts in academic research into real-world problem solving, which helps cultivate the student's enthusiasm for energy conservation and green manufacturing.

Organization of the Book

Part I: Introductions to Energy Efficiency in Manufacturing Systems

Chapter 1 provides an overview of this book and introduces background knowledge about manufacturing systems and concepts of sustainable manufacturing. First, it reviews the current status and development of the manufacturing industry and demonstrates a series of representative manufacturing systems. Then, it presents the key concepts of sustainable manufacturing and discusses the existing challenges that may impede sustainable development in manufacturing industries. Finally, it generalizes the problem statements and scopes of research in the context of sustainable manufacturing systems.

Chapter 2 provides more detailed background information on energy efficiency in manufacturing systems. The overall energy consumption and major energy end-users in manufacturing facilities are first introduced, followed by the discussions on the energy-saving potentials and energy management strategies at the machine, system, and plant levels. In addition, the significance of demand-side energy management is illustrated with the detailed explanations of associated techniques.

Part II: Mathematical Tools and Modeling Basics

Chapter 3 introduces the necessary mathematical tools used in the following chapters of this book. Specifically, the fundamentals of probability theory and application scenarios of several common probability distributions used in manufacturing system modeling are introduced, followed by the demonstration of Petri nets for the visual representation of manufacturing systems as discrete event systems and discussions on the optimization problems with metaheuristics algorithms, specifically a particle swarm optimizer.

Chapter 4 presents the mathematical modeling techniques for manufacturing systems, which play a critical role in sustainable manufacturing system design and analysis. This chapter introduces the basics of manufacturing system modeling, followed by detailed discussions on some typical modeling approaches to simple two-machine production lines and complex multi-machine ones.

Chapter 5 extends the modeling and analysis techniques discussed in the previous chapter into energy efficiency characterization in manufacturing systems. First, the energy consumption modeling approaches are discussed based on the inter-process dependency or the machines' operation schemes. Then, the energy cost models of manufacturing systems under different electricity tariffs are demonstrated with illustrative examples.

Part III: Energy Management in Typical Manufacturing Systems

Chapter 6 presents the electricity demand response (DR) strategies for manufacturing systems. The instant high demand can hinder the stability of a power grid, and thus the utility providers charge industrial customers specifically for their electricity demand in addition to the total energy consumption. In this chapter, the time-of-use (TOU) and critical peak pricing (CPP) tariffs are first introduced. The production scheduling methods that can respond to electricity price signals based on the system models are then discussed. Finally, case studies are presented to compare the peak demand and energy costs under TOU, CPP, and traditional flat-rate tariffs.

Chapter 7 extends the DR scheduling methods presented in the previous chapter by integrating a combined heat and power (CHP) system with manufacturing systems. As an on-site energy generation method, a CHP system can provide electricity and heat to the manufacturing plant, leading to a reduction in the grid power demand of the manufacturing plant. In this chapter, the key concepts of a CHP system are first reviewed, followed by the formulation of an energy cost optimization model for a combined CHP and manufacturing systems. The case studies are presented to demonstrate the effectiveness of the combined system in demand and energy cost reduction.

Chapter 8 addresses an energy management problem in manufacturing systems considering the heating, ventilation, and air conditioning (HVAC) system, which is one of the primary contributors to the direct non-process end use energy consumption in manufacturing plants. The heat emissions from manufacturing operations can significantly affect the thermal load of an HVAC system, and the relationships between manufacturing and HVAC systems are discussed in this chapter. Specifically, the formulation of an energy cost optimization problem for the integrated systems is first introduced, and then the metaheuristic algorithm used to solve the problem is discussed in detail. Finally, case studies demonstrate the optimal DR strategy for the integrated system.

Part IV: Energy Management in Advanced Manufacturing Systems

Chapter 9 specifically focuses on the energy analysis of additive manufacturing (AM) systems. In this chapter, stereolithography (SL), one of the most commonly used AM technologies, is adopted to demonstrate the energy modeling and analysis methods for an AM process. This chapter starts with the introduction of the technical advantages of AM technologies and a detailed description of an SL process. Then, it presents the energy consumption model of such SL process and its experimental validation results. The impacts of different parameters on the overall energy consumption are revealed through a Design-of-experiments (DOE) methodology. Finally, it gives case studies to illustrate the optimal combination of control parameters.

Chapter 10 presents the energy efficiency modeling and optimization of cellulosic biofuel manufacturing systems. The background knowledge and major processes of cellulosic biofuel manufacturing are first introduced. Then, the formulation of the energy consumption model for cellulosic biofuel manufacturing is illustrated by considering the intra-process and inter-process variables. Afterward, the optimization problem is solved through a metaheuristic algorithm, and the energy efficiency improvement under optimal process variables is presented at the end of this chapter.

Chapter 11 demonstrates the energy consumption modeling using Petri nets (PN) and production scheduling optimization for flexible manufacturing systems (FMS). In this chapter, the formulation of a place-timed PN model for FMS is first introduced, followed by a discussion of a dynamic programming (DP) algorithm to find production schedules that can minimize the energy consumption of small-size FMS. Next, a Modified DP (MDP) algorithm is presented to solve large-scale problems by addressing the state explosion issue. Finally, experimental results on FMS are presented to show the effectiveness of MDP.

Part V: Summaries and Conclusions

Chapter 12 summarizes the contribution of this book and highlights several important future research directions. The following figure illustrates the organization of the contents in this book.

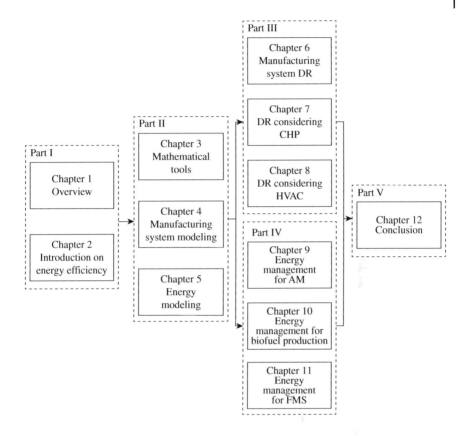

How to Read

This book can be used as a reference or a text book for senior and graduate students in mechanical, industrial, environmental, and electrical engineering programs as well as researchers, engineering professionals and policymakers in the areas of energy management and sustainable manufacturing.

The chapters in Parts I and II provide background knowledge and a mathematical foundation for the later chapters and are especially recommended to be read by students and new researchers. The chapters in Part III discuss the energy efficiency and power demand response in typical manufacturing systems and are encouraged to be read in order, as each chapter builds on the concepts in the previous chapter. The chapters in Part IV present the energy management in advanced manufacturing systems and can more or less be approached in any order, as each chapter discusses a different type of manufacturing systems. The last chapter as Part V summarizes this book and is recommended to be read in the end.

Acknowledgments

From the first author of this book:

I would like to thank all the people who have contributed to this book and the research team at the Sustainable Manufacturing Systems Research Laboratory at the University of Illinois at Chicago for their full dedication and quality research. In particular, I would like to acknowledge the following individuals.

First, I would like to express my great appreciation to this book's co-author, Professor MengChu Zhou from the New Jersey Institute of Technology, for his inspirational advices and insightful suggestions to help strengthen the visions and concepts of this book.

I would like to thank the significant help from my doctoral students Lingxiang Yun and Muyue Han for content and material preparations, as well as the research outcomes from my former doctoral students, especially Dr. Yong Wang from Binghamton University, Dr. Zeyi Sun from Missouri University of Science and Technology, Dr. Fadwa Dababneh from German Jordanian University, and Dr. Yiran (Emma) Yang from University of Texas at Arlington.

I would like to appreciate the Wiley-IEEE Press for providing the opportunity to publish this book and the esteemed editor and anonymous reviewers for reviewing our work. Special thanks are given to Ms. Teresa Netzler, Senior Managing Editor of Wiley-IEEE Press at United Kingdom, who kindly and patiently helped us move smoothly during our book writing and preparation period.

I would like to acknowledge the funding support for the research contents partially covered in Chapters 6, 7, 9, and 10 from the U.S. National Science Foundation, under Grants CMMI-1131537, CMMI-1434392, and CBET-1604825.

Finally, I truly appreciate the continuous support and endless love from my family, especially from my parents, who taught me to never give up anything that you feel deserves putting efforts wholeheartedly.

From the second author of this book:

Numerous collaborations were behind this book and its related work. It would be impossible to reach this status without the following collaborators, some of whom were already mentioned in the first author's message. In particular, I would like to acknowledge the following individuals:

Professor Keyi Xing, the State Key Laboratory for Manufacturing Systems Engineering, Systems Engineering Institute, Xi'an Jiaotong University, China, and his group members, e.g. Dr. Yanxiang Feng (Department of Computer Science and Technology and the State Key Laboratory for Manufacturing Systems Engineering, Xi'an Jiaotong University, China), Dr. Xiaoling Li (now with the School of Electronic and Control Engineering, Chang'an University, Xi'an, China), and Dr. Jianchao Luo (now with the Research and Development Institute and the School of Software, Northwestern Polytechnical University, Xi'an, China), have collaborated with me for many years in the areas of Petri net theory and applications to automated manufacturing systems. Specifically, we have developed several scheduling methods based on timed-place Petri nets. Some of our collaborated contributions are reflected in Chapter 11 of this book. I have enjoyed much collaboration with many outstanding researchers in the area of intelligent automation, transportation and sustainable manufacturing, e.g. Professors Naiqi Wu (Fellow of IEEE, Macau Institute of Systems Engineering, Macau University of Science and Technology, China), Zhiwu Li (Fellow of IEEE, Macau Institute of Systems Engineering, Macau University of Science and Technology, China), and Maria Pia Fanti (Fellow of IEEE, Dipartimento di Elettrotecnica ed Elettronica, Polytechnic of Bari, Italy).

I have enjoyed the full support and love from my family (my wife, Fang Chen, my two sons, Albert and Benjamin) for long. It would be impossible to accomplish this book and many other achievements without their support and love.

The work presented in this book was in part supported by FDCT (Fundo para o Desenvolvimento das Ciencias e da Tecnologia) under Grant No. 0047/2021/A1, and Lam Research Corporation through its Unlock Ideas program.

Lin Li, University of Illinois, Chicago, IL, USA
MengChu Zhou, New Jersey Institute of Technology, Newark, NJ, USA
Macau University of Science and Technology, Macao, China
20 August 2022

List of Figures

Part I

Introductions to Energy Efficiency in Manufacturing Systems

1

Introduction

In this chapter, the background knowledge about manufacturing systems and concept of sustainable manufacturing are demonstrated. In Section 1.1, an overview of the current status of manufacturing industry development is given, followed by a discussion on existing challenges that need to be addressed in order to sustain the continuous growth in manufacturing sectors. More specifically, the significant obstacles that may impede the sustainable development of manufacturing industries are discussed, and the implications for sustainability and energy efficiency in manufacturing systems are depicted. In addition, the definition of sustainable manufacturing and associated essential factors are demonstrated in Section 1.1.2. To better illustrate the significance of the industrial transition to sustainable manufacturing, several industrial paradigms and representative case studies are presented to strengthen the connections between the concepts of sustainable manufacturing and real-world problems. In Section 1.2, the key components of manufacturing systems are discussed from the perspective of a product life cycle. A series of representative manufacturing systems are demonstrated, which are associated with the discussions on system configurations, component functionality, and respective system performances. Section 1.3 is the overview of the problem statement and scope, which are facilitated by the hierarchical categorization of research expertise under the context of sustainable manufacturing.

1.1 Definitions and Practices of Sustainable Manufacturing

1.1.1 Current Status of Manufacturing Industry

Ever since the conception of industrialization, manufacturing production, as an indispensable corner stone, has been of decisive significance to the development of the world economy. The concept of the manufacturing value added (MVA) is

Sustainable Manufacturing Systems, First Edition. Lin Li and MengChu Zhou.
© 2023 The Institute of Electrical and Electronics Engineers, Inc.
Published 2023 by John Wiley & Sons, Inc.

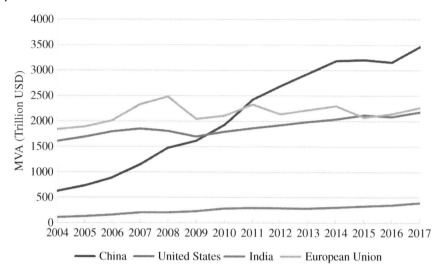

Figure 1.1 Changes in MVA among countries of different regions over the time from 2004 to 2017. *Source:* Adapted from [4].

proposed as one of the main indicators to measure the growth rate of manufacturing industry. By definition, it represents the total estimate of net output of all resident manufacturing activity units obtained by adding outputs and subtracting intermediate consumption [1]. According to the data released from the World Bank, in 2018, the total MVA has reached USD 13.976 trillion worldwide, which corresponds to approximately 16.82% of the global gross domestic product (GDP). Although the trend of declining manufacturing has been reported in developed regions due to the increased wage share of skilled worker and the competition from the service industry and other tertiary sectors [2, 3], manufacturing industries still maintain a good growth momentum and sustain their market value, especially in some developing countries and emerging economies. In reference to the World Bank report [4], Figure 1.1 illustrates the variations of MVA among countries of different regions over the period from 2004 to 2017. In particular, United States, European Union, China, and India are selected as the demonstrative regions.

As demonstrated in Figure 1.1, in the developed regions, such as United States and European Union, a steady increase in MVA can be observed. For example, the MVA in the United States was USD 1.61 trillion in 2004, and it increased by 34.8% to USD 2.17 trillion in 2017; a similar increment in MVA was also reported in the European Union. It is worth noting that fluctuations accompany the rise in MVA after 2008, which can be attributed to the relatively slower growth in manufacturing sectors after the nations survive during the post-economic-crisis period. Distinctively, in terms of the developing region, rapidly industrializing countries such as China and India possessed a significant increase in MVA over the past decades. In particular, the value added of the China manufacturing sector was USD 0.63 trillion in 2004, which was approximately 2.5 times less than the counterpart in the United States during the same period. The MVA in China was continuously increasing during the study period and reached USD 3.46 trillion in 2017, which indicates a total increase of 449.2% since 2004. The continuous rise of MVA also suggests that the manufacturing sectors remain one of the main propellers of economic growth. It can be depicted that the manufacturing industry has a pivotal position in the world economy and should sustain its pace of change in the foreseeable future.

1.1.2 Sustainability in the Manufacturing Sector and Associated Impacts

Despite the rapid development in a manufacturing sector, the proliferation of manufacturing systems also brings about sustainability concerns. More specifically, manufacturing in the traditional sense refers to processing raw materials to make useful products. In such manufacturing systems, production activities are built upon the consumption of feedstock materials, energy, and other resources. Meanwhile, production processes are often coupled with manufacturing emissions, waste generation, and residual heat. With the increasing consensus on resource scarcity and environmental sustainability, nowadays, the manufacturing industries face more and more challenges, such as efficient energy management, greenhouse gas (GHG) emissions, waste material, and resource reclamation. In general, the main challenges in the manufacturing industry can be examined from the following three aspects: economy, environment, and society, as demonstrated in Figure 1.2.

In terms of economic challenges, considering the high dependence of manufacturing industries on energy resources, the cost fluctuations in the energy market can significantly affect the manufacturing output and overall production cost. For example, the price of crude oil reached about USD 166 per barrel in 2008, which dealt a severe blow to the manufacturing industry during the 2008 international financial crisis [5]. On the other hand, from the product life cycle perspective, due to the rapid technological advancements and the increased diversity and functionality in products, the life cycle of products is continuously shortening. The faster evolution of manufactured products also brings about additional capital investment in manufacturing facility upgrading and employment of well-trained personnel and skilled workers.

Apart from the economic challenges, the environmental burdens caused by manufacturing productions drew increasing attention around the world in the past decades. Energy consumption and emissions are regarded as two major environmental

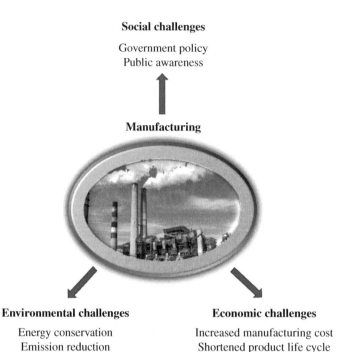

Social challenges

Government policy
Public awareness

Manufacturing

Environmental challenges

Energy conservation
Emission reduction

Economic challenges

Increased manufacturing cost
Shortened product life cycle

Figure 1.2 Challenges in the manufacturing sector. *Source:* cwizner/Pixabay.

concerns from production activities and other associated auxiliary processes. According to the Annual Energy Outlook 2020 published by the U.S. Energy Information Administration (EIA), the US energy consumption by sector in 2019 is illustrated in Figure 1.3. The overall energy consumption is subdivided into the following five primary sectors: industrial manufacturing, industrial non-manufacturing, residential, commercial, and transportation. As shown in Figure 1.3, industrial, electric power, and transportation sectors are the largest contributors to the total energy consumption in the United States, with the contributions of 23%, 37%, and 28%, respectively. The proportions of residential and commercial sectors are similar, which correspond to 7% and 5% of the total energy consumption, respectively.

In addition to the significance of energy consumption in the manufacturing sector, emission-related environmental issues may also hinder the development of the manufacturing industry. Considering the severity of global warming and resource scarcity, the environmental issues originated from manufacturing industries have gained increasing attention from the public, manufacturing enterprises, and government agencies. Based on the inventory report of US GHG emissions released by the U.S. Environmental Protection Agency (EPA) in 2018 [7], the sources of GHG emissions can be attributed to five major economic sectors, including transportation,

Figure 1.3 The US energy consumption by sector in 2019. *Source:* Adapted from [6].

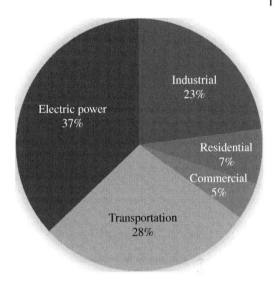

Figure 1.4 The total US GHG emissions by sector in 2018. *Source:* Adapted from [7].

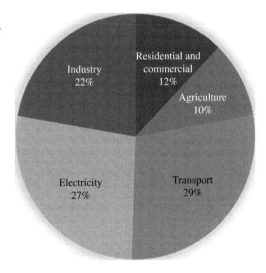

electricity, industry, commercial and residential buildings, as well as agriculture. The total US GHG emissions by sector in 2018 are illustrated in Figure 1.4. More specifically, the total emissions in 2018 are approximately 6677 million metric tons of carbon dioxide (CO_2) equivalent. As shown in Figure 1.4, transportation, electricity, and industry sectors are the major contributors to the total GHG emissions. Among the five economic sectors, the industry sector possesses the third-largest proportion (22%) of the total GHG emissions in the United States. The GHG emissions from the industrial sector can be further decomposed into two categories: direct emissions and indirect ones. The former are mainly contributed by the consumption of fossil fuels for power or heat at the manufacturing facilities, while the latter are primarily associated with the overall use of electricity.

From a societal perspective, the manufacturing industry is also facing challenges from the government and the public. On the government side, energy conservation, emission reduction, and environmental protection have increasingly become the guide for policy making. Internationally, the signing of the Kyoto Protocol and the Paris Agreement reflects the joint efforts of various governments on climate change and global warming. On the public side, the public consensus press is increasingly concerning energy conservation and emission reduction in the manufacturing sector. From the perspective of public awareness, consumers are more environmentally conscious as environmental degradation and pollution become more severe, and people are more eager for environmental-friendly products. At the same time, many organizations continue to call for restrictions on the development of high-polluting manufacturing industries. These factors have forced the manufacturing enterprises to brunch new paths and adjust their development strategies.

Due to these existing challenges in the development of the manufacturing sector, traditional manufacturing strategies need to be urgently changed. The emergence of the sustainable development concept provides a direction for the sustainable transformation of the manufacturing industry. More specifically, the concept of sustainable development was first raised by the United Nations General Assembly in 1987 as "the development that meets the needs of the present without compromising the ability of future generations to meet their own needs." The concept has been widely applied to all walks of life, including manufacturing. In the United Nations Conference on Environment and Development (UNCED) in 1992, sustainable manufacturing was formally proposed and used to guide future manufacturing development.

Ever since the conception of sustainable manufacturing, many institutions and researchers have put forward new discussions and definitions of manufacturing sustainability. For example, the Lowell Center for Sustainable Production at the University of Massachusetts Lowell (UMASS) defines sustainable production as "the creation of goods and services using processes and systems that are non-polluting, conserving of energy and natural resources, economically viable, safe and healthful for workers, communities, and consumers, socially and creatively rewarding for all working people" [8]. To date, the most widely accepted definition of sustainable manufacturing comes from the International Trade Administration (ITA) under the U.S. Department of Commerce (DOC). Sustainable manufacturing is defined as "the creation of manufactured products that use processes that are non-polluting, conserve energy and natural resources, and are economically sound and safe for employees, communities, and consumers." To reach the criteria mentioned above in manufacturing practices, a series of changes are required, such as eliminating the use of nonrenewable resources, switching to clean energy, and implementing energy-efficient production processes.

In general, sustainable manufacturing mainly consists of three pillars: economic, environmental, and social ones. The interconnections among them are demonstrated in Figure 1.5. More specifically, sustainable manufacturing can realize the protection of the environment and natural resources, which is conducive to mitigating the associated environmental impacts. On one hand, sustainable manufacturing responds to public concerns about environmental protection. For example, it can enhance environmental performances in terms of the consumption of raw materials and energy, which contributes positively to efficient resource utilization and emission reduction. On the other hand, sustainable manufacturing also contributes to business ethics and social responsibility, which ensures that companies do not disregard the common interests of the whole society for commercial interests. The implications of sustainability in the manufacturing enterprises are twofold. First, it facilitates the establishment of a sense of social responsibility in the manufacturing industry. Second, it reinforces the company's image, which increases customer's purchase intention and overall profits.

In addition, government agencies have paid more and more attention to the implementation of sustainable manufacturing. Companies that practice sustainable manufacturing have the opportunity to receive incentives such as government

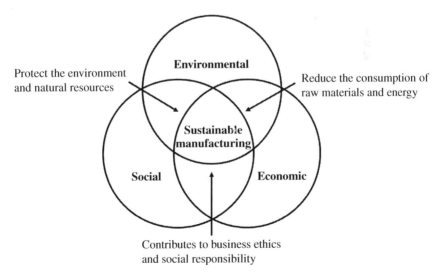

Figure 1.5 The three key pillars in sustainability. *Source:* Adapted from [9].

funding and tax relief. With the joint efforts from both government agencies and indigenous manufacturing enterprises, from 2005 to 2017, the United States' net GHG emissions were reduced by 13% [10]. The European Union also proposes the 2030 Climate Target Plan aiming at reducing the GHG emissions by at least 55% by 2030 [11]. In addition, the government works closely with universities and research institutions to provide additional technical support to local manufacturing business enterprises. For example, there are many Industrial Assessment Centers (IAC) located at universities across the United States, which are funded by the U.S. Department of Energy (DOE) to provide free industrial assessments for small- and medium-sized companies and facilitate the sustainable transitions of local manufacturing companies under the context of green manufacturing.

1.1.3 Sustainable Manufacturing Practices

To date, an increasing number of manufacturing enterprises have taken "sustainability" as an important goal of their development strategy and operation management to increase their market share and global competitiveness in the future. This trend has become more evident in recent years, and many well-known companies across different industry sectors have also begun taking actions. There are a number of reasons why manufacturing enterprises are pursuing sustainability [12]:

1) Increasing operational efficiency by reducing costs and wastes.
2) Responding to or reach new customers and increasing competitive advantages.
3) Protecting and strengthening brand and reputation and building public trust.
4) Building long-term business viability and success.
5) Responding to regulatory constraints and opportunities.

Sustainable manufacturing can be achieved from many perspectives, such as lean manufacturing, energy management, waste reduction, and re-/de-manufacturing. These efforts have led to remarkable results. In the following sections, a series of sustainable manufacturing practices are illustrated [13]:

- Lean manufacturing refers to the application of lean practices, principles, and tools to the development and manufacture of physical products to reduce waste. In terms of lean manufacturing, New Mexico Manufacturing Extension Partnership, a statewide assistance center, has been dedicated to increasing the competitiveness of the indigenous small- and medium-sized businesses by providing analysis and assessment services with respect to the production process arrangement and the production area layout planning. Through the implementation of lean manufacturing, companies can reduce the excessive transportation and excessive use of tools and materials. Through the practice of sustainable manufacturing, the company can reduce costs by 65%, increase production from

20 to 45 units per shift, reduce the scale of production facilities by 73%, and reduce the scrap rate from 24 to 1.8% according to [13].

- Energy management includes the planning and operation of energy production and energy consumption units. As an example of the implementation of energy management, Besam, a manufacturer of automatic doors, collaborated with the Carolina State University Industrial Evaluation Center and North Carolina State Industrial Extension Services' E3 (Economy, Energy and Environment) to improve energy efficiency in production processes. Several recommendations have been presented in the energy management survey, including replacing fluorescent lamp fixtures with metal halide lighting and installing high-efficiency lamps with occupancy sensors and electronic ballasts to reduce compressor air pressure and repair compressed air leakage. Through these efforts, a total reduction of 233,555 kWh in electricity consumption can be achieved per year, which corresponds to an annual saving of approximately USD 25,776.
- Waste reduction, also known as source reduction, is the practice of using less material and energy to minimize waste generation and preserve natural resources. In terms of manufacturing practice in waste reduction, Guardian Automobile's Ligonier plant takes measures to reduce waste generation by finding ways to recycle and reuse materials. The candidate recyclable materials include unused glass scrap, glass fiber, and waste polyvinyl chloride. In 2005, the Ligonier plant recycled more than 13,000 tons of waste, saving USD 360,000.

De-manufacturing and remanufacturing include a set of tools, knowledge-based methods, and technical solutions to systematically recover, reuse, and upgrade functions and materials from waste and end-of-life products [14–17]. Specifically, de-manufacturing liberates target materials and components [18, 19], while remanufacturing restores or upgrades their functions [20], as shown in Figure 1.6. The total 2022 US market size for automotive parts remanufacturing reaches USD 5.3 billion.

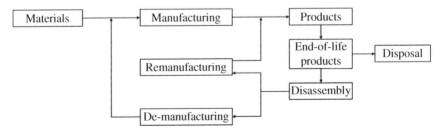

Figure 1.6 Illustration of the connections among manufacturing, remanufacturing, and de-manufacturing.

1.2 Fundamental of Manufacturing Systems

1.2.1 Stages of Product Manufacturing

In general, product manufacturing comprises four major stages: design, planning, control, and manufacturing. It has material and energy exchanges with the outside world, as demonstrated in Figure 1.7.

- Input, which represents the basic resources required in product development and manufacturing. The input resources can be categorized into the following types [22]:
  1) Raw materials: the primary feedstock materials required for a manufacturing system.
  2) Auxiliary and operating materials: the additional materials and supporting parts required by a manufacturing system.
  3) Energy: the energy and power required for system operation, such as electricity, gas, and fuel.
  4) Labor: the workforce required to maintain the operation of a system.
  5) Technical equipment: the equipment for supporting a main production process and various secondary processes, such as transport and storage.
  6) Information: the information obtained from the external of a system, such as market demand and price.

- Product design refers to the design of products with the joint consideration of market demand, availability of production resources, capability of machines and tools, and production cost.
- System planning, which aims to determine the optimal manufacturing system configuration, takes into account the capital investment, overall production cost, system reliability, and overall performance. In addition, the operation sequences of machines and resource allocations, as well as the processing parameters, are specified during this stage.
- Manufacturing system, which takes the feedstock materials as inputs, adds value, and eventually transforms the inputs into final products. Two major

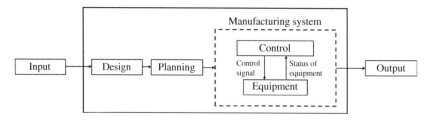

Figure 1.7 Illustrations of product manufacturing stages. *Source:* Adapted from [21].

factors that can affect the sustainable performance of a manufacturing system are equipment and control.

- Equipment is a fundamental component of a manufacturing system. The properties of the equipment can significantly affect the sustainable performance of the system. For example, the electric power of a machine determines the total energy consumed during a production period. The machine's processing accuracy, which indicates the percentage of non-defective items in all produced items, determines the amount of extra energy and materials need to be used to meet the production order quantity. The equipment properties provide a baseline of the system performance, and the control system determines whether the manufacturing system can perform its intended functions as expected.
- Control system analyzes the status of the equipment, develops the optimal operation strategies, and manages the equipment operations. The control system affects the system performance by adjusting the equipment behaviors, such as machine scheduling, resource assignment, and material handling route planning. An efficient control strategy can improve system sustainability without sacrificing product quality or system productivity [23].
- Output, which can be subdivided into final products and production wastes. In particular, the wastes generated during the manufacturing activities may include waste water, exhaust gas, solid and liquid wastes, excess heat, etc.

1.2.2 Classification of Manufacturing Systems

In general, manufacturing systems are classified into five categories based on the configuration of their material processing area, including job shop, project shop, cellular system, flow line, and continuous system [21].

1.2.2.1 Job Shop

In a job shop, machines are placed on the shop floor by functionality, and machines with the same or similar material processing capabilities are grouped together to form a work center. In this system, parts and materials need to move around different work centers to fulfill the processing requirements, as demonstrated in Figure 1.8.

This functionality-oriented job shop system layout has the following advantages: (i) when a specific machine breaks down or is under maintenance, its job can be easily taken over by other machines located in the same work center; and (ii) the system is capable of handling different types of parts with various process sequences, such as the Part 1 and Part 2 in Figure 1.8. However, in order to benefit from its internal flexibility, some challenges must be addressed. The first challenge

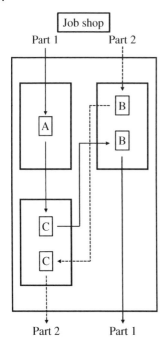

Figure 1.8 Schematic diagram of a job shop, where A, B, and C represent three types of machines.

is the material handling scheduling. Different types of parts have different process routings; hence, flexible material handling equipment, such as handcarts, forklifts, and automated ground vehicles, are more suitable for this system. Precise, dynamic, and real-time material handling equipment assignment and delivery route planning strategies are also indispensable to this system. In addition, since each work center works individually, comprehensive machine scheduling is required by considering the workflows of different types of parts. A lack of decision-making coordination among different work centers can increase the semifinished product inventory levels and eventually affect overall production throughput [24].

1.2.2.2 Project Shop

In a project shop, the position of a product is fixed during its manufacturing process. Different machines and workers are transported in and out of this production zone during the operation, as shown in Figure 1.9. Project shops are often used to manufacture products of large size and high weight, such as airplanes and ships. This

Figure 1.9 Schematic diagram of a project shop, where A, B, and C represent three types of machines.

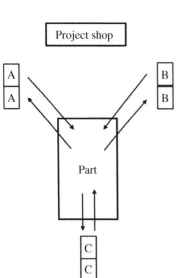

configuration is also widely used in bridges or building constructions.

1.2.2.3 Cellular System

In a cellular system, the machines are arranged and divided into different cells according to the workflow of parts, as shown in Figure 1.10. In each cell, all the necessary devices are sequentially placed to process a single type of part. Unlike the job shop, where the material handling system needs to ship parts among different work centers, machines in one single cell can perform all the necessary work for a given type of part. This property eliminates the need for intercell material handling. The cellular system can significantly simplify the material handling system scheduling and reduce transportation time. Despite these advantages, this system is more fragile to machine failure. A failed machine can break down the entire cell, and it is hard to maintain productivity during the machine maintenance period [25].

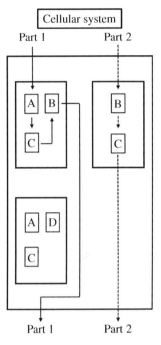

Figure 1.10 Schematic diagram of a cellular system, where A, B, C, and D represent four types of machines.

1.2.2.4 Flow Line

In a flow line, the machines are arranged and connected according to the part's process sequence, as illustrated in Figure 1.11. Generally, a flow line is designed for a specific type of part or several parts with the same workflow. Since the machine sequence is predefined and fixed, automated material handling methods, such as belt conveyors, powered roller conveyors, and overhead trolleys, are usually implemented to improve the system stability and production efficiency. The system configuration, machine processing rates, and inventories of semifinished products can be optimized in advance. Therefore, the flow line has a significant advantage in the mass production of a particular part. On the contrary, it may take hours or even days to reconfigure and optimize the flow line when the system is switched from the current part production to another type of parts.

1.2.2.5 Continuous System

Apart from the four aforementioned discrete manufacturing systems, a continuous system is designed to process materials without interruption. An example of a continuous system is shown in Figure 1.12. Different tanks or reactors are generally

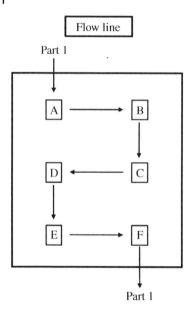

Figure 1.11 Schematic diagram of a flow line, where A–F represent six types of machines.

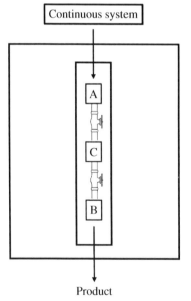

Figure 1.12 Schematic diagram of a continuous system, where A, B, and C represent three types of machines.

connected by pipelines. Unlike in a discrete system, where parts are processed from one machine to another until the manufacturing process of the product is finished, the materials pass through the continuous system as a whole and undergo chemical reactions and mechanical or thermal treatments without pauses. Continuous systems are commonly adopted in industries dealing with liquid or gas materials, such as oil, chemicals, and natural gas [26].

The selection of types of manufacturing systems depends on many factors, such as the lot size of products, the variety of processing workflows, the properties of the feedstock materials, and the responsiveness to market changes and customers' demands. For example, when the quantity of a product ordered by customers is relatively small, job shops and project shops are preferred over the other system configurations. A cellular system is more suitable for medium lot-size production since the different types of parts can be processed simultaneously in separate cells [28]. Besides, attributed to the implementation of the automatic material handling equipment, the flow line system provides the highest productivity and can meet the need for mass production. It should be noted that if the priority consideration is the system flexibility or ability to process various parts, the system selection could be different. More specifically, attributed to the flexible material handling systems and general-purpose machines, job shops are more suitable to manufacture various products from multiple product orders. In terms of a cellular system, each cell acts as an individual unit, which is designed for a specific workflow. The cellular system can handle different types of parts, but the total number of cells limits the part types. Additionally, compared to the other systems, the flow line is relatively more stable and harder to reconfigure; hence, it is only suitable for manufacturing of a single product family. The comparisons among different manufacturing systems are demonstrated in Figure 1.13. In the field applications, these

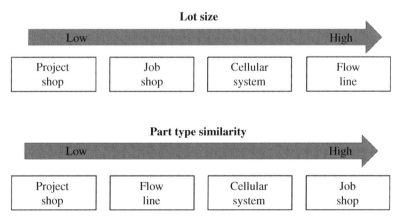

Figure 1.13 Selections of manufacturing systems.

standard systems may be combined together or adapted with necessary adjustments depending on the real production scenarios.

1.3 Problem Statement and Scope

To date, sustainable manufacturing has drawn wide attention and research interests. Current research on sustainable manufacturing can be mainly categorized into four layers, including manufacturing technologies, product life cycles, value creation networks, and global manufacturing impacts [27], as shown in Figure 1.14.

- Manufacturing technology and system performance. Research in this layer is mainly focused on "how to manufacture", and the studies usually are system or process oriented. One research direction aims to systematically model or simulate the operational conditions of the major components in manufacturing systems and propose solutions to optimize the resource allocation and energy consumption without sacrificing the overall system performances. The other research direction is the advancement in manufacturing technologies and specialized apparatus toward low-cost and energy-efficient product manufacturing.
- Product life cycle. Research in this layer mainly addresses "what is to be produced," and the research focuses on integrating sustainability criteria into product development processes. The main research directions include the life cycle management of products, the development of intelligent products, and the product sustainability assessment. This layer mainly integrates sustainability into several stages of a product life cycle, ranging from product design to manufacturing, and end-of-life management, considering the interactions among environmental, social, and economic

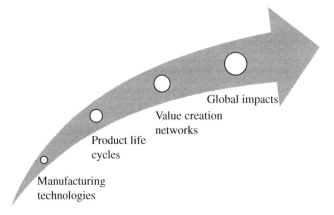

Figure 1.14 Four layers of sustainable manufacturing problems. *Source:* Adapted from [27].

factors. Due to the complex relationship among different life cycle stages, the research scope at this layer is broader in comparison with that of the first layer.

- Value creation network. Research in this layer is built upon the organizational context of manufacturing activities, and the main research objects are organizations such as companies and manufacturing networks. The main research directions involve resource-efficient supply chain planning and the development of industrial ecology. This layer is mainly focused on sustainability at the organizational level to make it meet the requirements of sustainable manufacturing. Compared with the first two layers, it has a broader research scope as the sustainable concerns grow beyond the products or production systems and involve the resource allocations and collaborations in different manufacturing networks.

- Global manufacturing impact. Research in this layer is mainly dedicated to the transition mechanism toward sustainable manufacturing. Different from the previous layer of manufacturing networks, the main research directions are the development of sustainability assessment methods, establishment of sustainability goals, and promotion of consensus for a sustainable future. Research in this layer mainly focuses on applying sustainable manufacturing on a global scale and the applicability of sustainable decision-making methods in the field. This layer of research sheds light on the standards in the implementation of sustainable manufacturing and potential directions for future development in manufacturing.

Problems

1.1 Why do we need to consider sustainability in the manufacturing industry? Presents some manufacturing challenges to the industry.

1.2 How to choose an appropriate manufacturing system?

1.3 State some commonly faced problems in the development of sustainable manufacturing.

References

1 United Nations Industrial Development Organization (2022). What is manufacturing value added? https://stat.unido.org/content/learning-center/what-is-manufacturing-value-added%253f (accessed 5 December 2022).

2 Dauth, W., Findeisen, S., and Suedekum, J. (2017). Trade and manufacturing jobs in Germany. *Am. Econ. Rev.* 107 (5): 337–342. https://doi.org/10.1257/aer.p20171025.

3 Stanford, J. (2016). Manufacturing (still) matters: why the decline of Australian manufacturing is not inevitable, and what government can do about it. *Brief. Pap. Cent. Futur. Work Aust. Inst.*

4 The World Bank (2020). Manufacturing, value added (current US$). https://data. worldbank.org/indicator/NV.IND.MANF.CD (accessed 5 December 2020).

5 Macrotrends (2020). Crude oil prices: 70 year historical chart. https://www. macrotrends.net/1369/crude-oil-price-history-chart (accessed 5 December 2020).

6 Sugawara, E. and Nikaido, H. (2019). EIA energy outlook 2020. *Antimicrob. Agents Chemother.* 58 (12): 7250–7257.

7 United States Environmental Protection Agency (2020). Sources of greenhouse gas emissions. https://www.epa.gov/ghgemissions/sources-greenhouse-gas-emissions (accessed 5 December 2020).

8 Lowell Centre for Sustainable Production (2015). What is sustainable production? https://www.uml.edu/Research/Lowell-Center/About/Sustainable-Production-Defined.aspx (accessed 5 December 2020).

9 Rosen, M.A. and Kishawy, H.A. (2012). Sustainable manufacturing and design: concepts, practices and needs. *Sustainability* 4 (2): 154–174. https://doi.org/ 10.3390/su4020154.

10 Pompeo, M.R. (2019). On the U.S. withdrawal from the Paris agreement - United States Department of State. *U.S Department of State.* https://www.state.gov/on-the-u-s-withdrawal-from-the-paris-agreement/ (accessed 5 December 2020).

11 European Commission (2020). 2030 Climate target plan. https://ec.europa.eu/ clima/eu-action/european-green-deal/2030-climate-target-plan_en (accessed 5 December 2020).

12 United States Environmental Protection Agency (2020). Sustainable manufacturing. https://www.epa.gov/sustainability/sustainable-manufacturing (accessed 5 December 2020).

13 United States Environmental Protection Agency (2020). Featured sustainable manufacturing case studies. https://archive.epa.gov/sustainablemanufacturing/ web/html/case-studies.html (accessed 5 December 2020).

14 Ying, T., Zhou, M., and Caudill, R.J. (2001). An integrated approach to disassembly planning and demanufacturing operation. *IEEE Trans. Robot. Autom.* 17 (6): 773–784. https://doi.org/10.1109/70.975899.

15 Tang, Y., Zhou, M., Zussman, E., and Caudill, R. (2002). Disassembly modeling, planning, and application. *J. Manuf. Syst.* 21 (3): 200–217. https://doi.org/10.1016/ S0278-6125(02)80162-5.

16 Guo, X., Zhou, M., Liu, S., and Qi, L. (2021). Multiresource-constrained selective disassembly with maximal profit and minimal energy consumption. *IEEE Trans. Autom. Sci. Eng.* 18 (2): 804–816. https://doi.org/10.1109/TASE.2020.2992220.

17 Guo, X., Zhang, Z., Qi, L. et al. (2021). Stochastic hybrid discrete grey wolf optimizer for multi-objective disassembly sequencing and line balancing planning in disassembling multiple products. *IEEE Trans. Autom. Sci. Eng.* 1–13. https://doi. org/10.1109/TASE.2021.3133601.

18 Guo, X., Zhou, M., Abusorrah, A. et al. (2021). Disassembly sequence planning: a survey. *IEEE/CAA J. Autom. Sin.* 8 (7): 1308–1324. https://doi.org/10.1109/ JAS.2020.1003515.

19 Tang, Y. (2009). Learning-based disassembly process planner for uncertainty management. *IEEE Trans. Syst. Man, Cybern. Part A Syst. Humans* 39 (1): 134–143. https://doi.org/10.1109/TSMCA.2008.2007990.

20 Tolio, T., Bernard, A., Colledani, M. et al. (2017). Design, management and control of demanufacturing and remanufacturing systems. *CIRP Ann.* 66 (2): 585–609. https://doi.org/10.1016/j.cirp.2017.05.001.

21 Chryssolouris, G. (2013). *Manufacturing Systems: Theory and Practice*. Springer Science & Business Media.

22 Thiede, S. (2012). *Energy Efficiency in Manufacturing Systems*. Springer Science & Business Media.

23 Mehrabi, M., Ulsoy, A., and Koren, Y. (2002). *Manufacturing Systems and Their Design Principles*. CRC Press.

24 Applegate, D. and Cook, W. (1991). A computational study of the job-shop scheduling problem. *ORSA J. Compu.* 3 (2): 149–156.

25 Irani, S.A., Subramanian, S., and Allam, Y.S. (1999). *Handbook of Cellular Manufacturing Systems*. Wiley.

26 Konstantinov, K.B. and Cooney, C.L. (2015). White paper on continuous bioprocessing May 20–21 2014 continuous manufacturing symposium. *J. Pharm. Sci.* 104 (3): 813–820. https://doi.org/10.1002/jps.24268.

27 Bonvoisin, J., Stark, R., and Seliger, G. (2017). Field of research in sustainable manufacturing. In: *Sustainable Manufacturing: Challenges, Solutions and Implementation* (ed. R. Stark, G. Seliger and J. Bonvoisin), 3–20. Cham: Springer International Publishing. Sustainable Production, Life Cycle Engineering and Management. https://doi.org/10.1007/978-3-319-48514-0.

28 Wang, J., Liu, C., and Zhou. (2020). Improved Bacterial Foraging Algorithm for Cell Formationand Product Scheduling Considering Learning and Forgetting Factors in CellularManufacturing Systems. *IEEE Systems Journal.* 14 (2): 3047–3056.

2

Energy Efficiency in Manufacturing Systems

In this chapter, background information about energy consumption in manufacturing systems is presented, followed by the discussions on the energy-saving potentials, energy management strategies, and demand-side management (DSM) programs. In Section 2.1, the physical definitions of energy and power are demonstrated, followed by the illustration of the energy generation, distribution, and consumption. More specifically, a variety of energy sources are categorized into primary and secondary energy, and the energy flows from offsite facilities and onsite energy generation are illustrated. The overall energy consumption in manufacturing facilities is decomposed into direct and indirect end uses, and the major components in each category are demonstrated. Section 2.2 explains the necessity of energy management in manufacturing systems. The energy-saving potentials and energy management strategies for the manufacturing industry are discussed at three levels, i.e. machine level, system level, and plant level. In addition, in Section 2.3, the significance of demand-side energy management on both the manufacturing industry and power grid is illustrated. In particular, two typical DSM programs, i.e. the energy efficiency programs and the demand response programs, are presented with detailed explanations on associated tools and techniques.

2.1 Energy Consumption in Manufacturing Systems

2.1.1 Energy and Power Basics

Energy (E), in physics, is defined as the ability to do work [1]. Energy exists in various forms, such as light energy, thermal energy, mechanical energy, chemical energy, electrical energy, and potential energy. Energy conversion and transformation can occur among different forms of energy. For example, the combustion of fuels allows the conversion of the chemical energy in the fuel to thermal one; the rotation of the rotor in an electric motor converts electrical energy to mechanical

Sustainable Manufacturing Systems, First Edition. Lin Li and MengChu Zhou.
© 2023 The Institute of Electrical and Electronics Engineers, Inc.
Published 2023 by John Wiley & Sons, Inc.

one. Different types of energy can be transformed into specific forms depending on the real-world application scenarios and actual demand in practice through the conversion of energy.

The conversion of energy is conducted in the form of work (W). It is generally regarded as a process variable, which reflects the state change of energy. In addition, the rate of generating, transferring, or consuming energy is commonly referred to as power (P) [2], which reflects the ability of a system to do work per unit of time. Mathematically, the relationship between work and power can be expressed by the following equations, where W is the integral of P over the time period t, and P is the time derivative of W.

$$W = \int_0^t P \cdot dt \tag{2.1}$$

$$P = \frac{dW}{dt} \tag{2.2}$$

In the International System of Units (SI), commonly known as the metric system, energy is measured in kilogram, meters squared, per second squared (denoted as $kg \cdot m^2 \cdot s^{-2}$), which is called joule (J). One joule equals the work done by a force of one newton (N) when its point of application moves through a distance of one meter in the direction of the force [2], expressed as

$$1\,J = 1\,kg \cdot m^2/s^2 = 1\,N \cdot m \tag{2.3}$$

In the metric system, the SI derived unit of power is watt (W), which is equivalent to joule per second (J/s). In terms of electric power, one watt also equals one ampere (A) under a pressure of one volt (V). In addition, another commonly used unit of power is horsepower (hp), which is often used to represent the output of motors or engines. One electrical horsepower is equal to 746 W. The relationships among the units mentioned above are illustrated as

$$1\,W = 1\,J/s = 1\,A \cdot V \tag{2.4}$$

In particular, the unit of energy may change depending on the specific scenarios of energy generation and applications. For example, in the calculation of electrical power, kilowatt-hour (kWh) is commonly used; the British thermal unit (BTU) and kilocalorie (kcal) are often used in thermodynamics. Table 2.1 demonstrates the conversions between different energy units.

2.1.2 Energy Generation

In reference to the conservation of the energy principle (commonly referred to as the first law of thermodynamics), energy can be neither created nor destroyed but rather changed into various forms. Therefore, the term energy generation is

Table 2.1 Energy units and conversions.

Common energy units	Symbol	Unit conversions
Kilowatt-hour	kWh	3.6×0^6 J
British thermal unit	BTU	1.055×10^3 J
Kilocalorie	kcal	4.184×10^3 J

defined as the process of converting energy from one form into another. An energy generation process is often regarded as a manifestation of energy transformation or energy conversion. A few examples of energy generation include the processes of fossil fuel combustion to produce heat and electricity in thermal power plants, and the energy conversions in hydroelectric power plants by harnessing the kinetic energy of moving water to turn turbines, which converts mechanical energy to electricity.

In energy generation, energy can be categorized into two types based on the nature of energy transformation, i.e. primary energy and secondary one.

2.1.2.1 Primary Energy

Primary energy is defined as "the energy sources that only involve extraction or capture, with or without separation from contiguous material, cleaning or grading, before the energy embodied in that source can be converted into heat or mechanical work" [3]. The distinguishing characteristic of primary energy is the extraction or capture of energy sources found in the natural environment without human-engineered energy transformation, i.e. the physical and chemical properties of the energy sources are not altered. In particular, based on energy renewability, primary energy can be further divided into the following two categories:

- Nonrenewable energy: The energy source that is exhaustible when used as a major energy source. The nonrenewability of energy is originated from the restrictions on the storage duration and cycling time of energy. More specifically, the supplies of these energy sources are limited to the amount extracted or captured from the earth. The consumption rate of non-renewable energy far exceeds their replenishment through natural processes [4]. There are many types of nonrenewable energy, such as petroleum, coal, natural gas, and nuclear energy. In particular, petroleum, natural gas, and coal are commonly referred to as fossil fuels as they are formed from the remains of long-dead organisms buried millions of years ago.
- Renewable energy: The energy source that can be continuously replenished through natural processes [4]. In contrast to the finite stock of depletable energy

such as petroleum and natural gas, the supply of renewable energy can be restored over a short period of time after consumption. Renewable energy mainly includes hydropower, wind energy, solar energy, biomass energy, geothermal energy, tidal energy, etc. Considering the possible natural resource scarcity resulting from current energy consumption, there is an increasing consensus in reducing fossil fuel consumption. The replenishable nature of renewable energy also brings the assurance of seeking promising energy alternatives for clean and green energy consumption in the future.

According to the U.S. Energy Information Administration (EIA)'s monthly energy review [5], we use Figure 2.1 to illustrate the US primary energy consumption by energy source in 2019. Notably, the total US primary energy consumption is approximately 100.2 quadrillion BTU. As shown in Figure 2.1, nonrenewable energy sources are the major energy contributors in the US, which account for a total of more than 80% of the total energy consumption. Among non-renewable sources, fossil fuels still dominate the US energy consumption. The respective proportions of petroleum, natural gas, and coal are 36.7%, 32.0%, and 11.3%. In addition, it can be observed that, in comparison with fossil fuels, the share of energy from renewable energy sources is relatively small, which accounts for a total of 11.4%. In particular, biomass energy possesses a proportion of 43.8% within renewable energy, with the percentage of wind energy and hydropower being similar, which correspond to 21.9% and 23.7%, respectively.

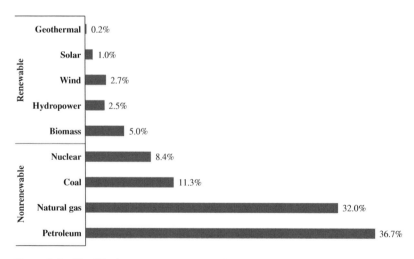

Figure 2.1 The US primary energy consumption by energy source in 2019. *Source:* Adapted from [5].

2.1.2.2 Secondary Energy

Secondary energy is defined as "all sources of energy that results from the transformation of primary sources" [3]. Its distinguishing characteristic is the energy conversion of primary energy sources. For example, primary energy such as fossil fuels and nuclear fuels can be used to produce heat and raise steam, which is subsequently transformed into secondary energy such as electricity. Unlike the activities of extraction or capture in primary energy harvesting directly from nature, the generation of secondary energy involves human-induced energy transformation.

Different types of energy sources have been used in electricity generation. According to the inventory of the US electricity generation released by the U.S. EIA in 2020 [6], the domestic electricity generation can be contributed by three major categories of energy, including fossil fuels, nuclear energy, and renewable energy. Figure 2.2 demonstrates the detailed breakdown of the US electricity generation by energy source. As depicted in Figure 2.2, fossil fuels possess the largest share of electricity generation, with the dominance of natural gas accounting for 40% of the total electricity generation, and coal as the third-largest energy source with a contribution of 19%. Less than 1% of the domestic electric power is produced by petroleum. Besides, one-fifth of US electricity generation is contributed by nuclear energy, and the share of renewable energy in electricity generation is about 20%, which is mainly contributed by wind power (8.4%) and hydropower (7.3%).

2.1.3 Energy Distribution

The flow of energy involves both the energy transmissions from energy generation sites to industrial end users, and the internal energy distribution inside the industrial facilities. In general, the energy flow in a manufacturing sector can be categorized into two major pathways: (i) energy transmitted from offsite facilities, such

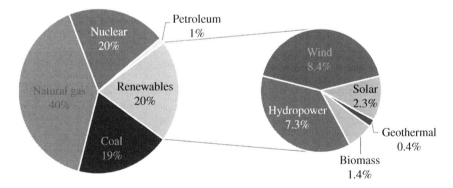

Figure 2.2 The US electricity generation by energy source in 2020. *Source:* Adapted from [6].

as the electric power transmitted through power lines from a power generation site to manufacturing plant, and other primary energy sources (such as oil, natural gas, and coal) delivered directly to industrial sites through railway, waterway, pipeline, etc.; and (ii) energy generated onsite for use in the operations of industrial facilities, such as steam and compressed air generated inside the facility and then passes through distribution pipework to individual pieces of equipment for onsite applications [7]. In particular, electricity, steam, and compressed air are selected as the representative energy flows in the manufacturing sectors. The respective advantages and energy distribution pathways are demonstrated in the following sections.

2.1.3.1 Electricity

Electricity is widely used in manufacturing systems in terms of energy supply, electronic control systems, electric space heating, and lighting, etc. It has many advantages in the manufacturing systems, including [7]

- As a secondary energy source, the electricity generation capacity is dependent on the availability of primary energy sources, and the delivery of these resources to electricity generation facilities is supported by various infrastructure networks.
- Electricity is suitable for long-distance transportation, and energy loss during transmission is relatively small.
- Electricity can be used directly to power electronic control systems and other electric equipment in the manufacturing facilities.

In particular, the supply chain of electricity involves electricity generation in power plants, transmission, and distribution. Generally, a power plant is distant from the demand centers (i.e. electricity consumers). The electricity needs to be transmitted and distributed through a series of substations and local distribution networks with the consideration of inevitable power loss over long distances, and the utilization levels of various end users. The combination of transmission and distribution networks is commonly referred to as the power grid, which is mainly composed of generator step-up transformers, transmission lines, substation step-down transformers, and end users (including commercial and industrial users, as well as residential customers) [8].

Figure 2.3 shows a schematic diagram of an electricity supply chain. As demonstrated in Figure 2.3, the entire supply chain can be divided into three major components, i.e. power generation, distribution, and consumption. A series of substations and transformers are involved in this complex network as critical nodes to ensure the crucial connections among power generation facilities, transmission and distribution networks, and end users. More specifically, the voltage of electricity generated in a power plant is generally 5–34.5 kilovolts (kV); however, considering the power loss induced by the wire resistance over long-distance

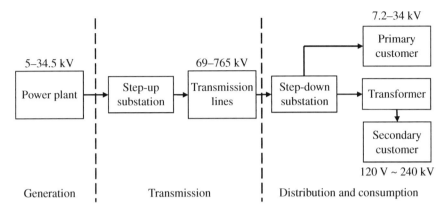

Figure 2.3 Illustration of the supply chain of electricity. *Source:* Adapted from [8].

transmissions, the voltage boosting is required prior to the bulk power transmissions through transmission lines. Hence, the "step-up" substations usually are first employed to significantly increase the power voltage in the range of 69–765 kV depending on the normal operating voltage of transmission lines. The typical transmission lines operate at 138 kV, 230 kV, 345 kV, 500 kV, and currently, the highest voltage lines in power grid systems of North America operate at 765 kV [9]. It should be noted that the transmission lines are different from the distribution lines. In particular, once the high-voltage power reaches the load center, the distribution substations are employed to reduce the power voltage and pass the bulk power to medium-voltage distribution networks. In addition, a series of modular and smaller distribution transformers are used to further step down the power voltage in medium-voltage distribution lines to local distribution lines with low utilization levels required by different types of end users. Normally, the power voltage is between 7.2 and 34 kV for primary distribution lines, and the common secondary distribution lines operate at 120 and 240 V for secondary customers [8].

In addition, from the perspective of reliability and stability in a power grid, multiple small power grids can be interconnected to form larger power grids. For example, the bulk power system in the United States consists of the following three major power grids, which are also referred to as interconnections. It should be noted that the power grid systems in Alaska and Hawaii are not connected to the grids in the lower 48 states [10].

- The Eastern Interconnection encompasses the area east of the Rocky Mountains and a portion of the Texas panhandle.
- The Western Interconnection encompasses the area from the Rockies west.
- The Electric Reliability Council of Texas (ERCOT) covers most of the State of Texas.

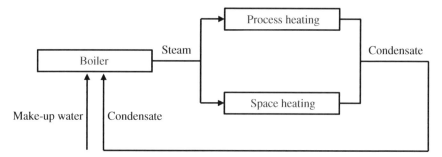

Figure 2.4 The schematic diagram of a typical steam generation and distribution system. *Source:* Adapted from [12].

2.1.3.2 Steam

In manufacturing systems, steam is mainly generated through boilers or steam generators, which primarily involve the combustion of fossil fuel or electricity consumption. It is one of the widely used energy to provide power and heat in manufacturing facilities and has many advantages in the manufacturing systems, including [11]

- Steam can supply heat to industrial facilities, given its high heat capacity.
- Steam temperature can remain constant once the pressure is determined, which enables a constant temperature environment.
- Steam is recyclable through efficient condensate recovery systems.

Figure 2.4 illustrates the typical boiler-based steam generation and distribution processes in a manufacturing site. More specifically, the boiler feedwater is comprised of freshwater (also referred to as make-up water) and recovered condensate (return water). The feedwater is heated in the boiler, and the generated steam is conveyed through the pipework inside the manufacturing site. There are multiple branches of steam pipes that connect with different end-use equipment. The steam distribution pathways in a manufacturing system can be generally categorized into process heating and spacing heating. In addition, as steam condenses in a distribution system, a localized pressure drop can be induced due to the relatively small volume of condensate compared to the steam. The pressure difference pushes the flow of steam in a supply pipe. The condensate formed as the steam travels along the pipe can be recycled through condensate return lines and returned to the boiler. Besides, additional make-up water is loaded into the boiler to compensate for water loss in the system.

2.1.3.3 Compressed Air

Compressed air is generally generated by air compressors in the manufacturing systems. The generation of compressed air can be regarded as a process of converting electrical energy into mechanical energy. Compressed air has many applications in

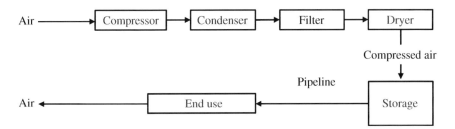

Figure 2.5 The schematic diagram of a typical compressed air distribution system. *Source:* Adapted from [8].

manufacturing systems, such as pneumatic transportation, cleaning dust, sandblasting, and vehicle propulsion [7]. The main benefits of compressed air in manufacturing include [13]

- Compressed air can be efficiently transported and distributed over long distances through pipelines. Meanwhile, the compressed air distribution system does not require the return lines as there is almost no residual compressed air after use.
- In comparison with steam, compressed air has lower storage requirements. Compressed air can be easily stored onsite in low-cost gas storage tanks.
- Compressed air can work safely in harsh environments, and the pneumatic system has low wear and tear, which is accompanied by a low failure rate.
- Pneumatic components are generally more cost-effective and have a longer service life.
- Compressed air is considered to be overload proof. Overloading can only cause the work to stop without affecting the equipment.

A typical workflow of compressed air generation and distribution is illustrated in Figure 2.5. As shown in Figure 2.5, to generate compressed air, the air compressor inducts air and increases the inlet air pressure by reducing its volume. The compressed air then passes through the condenser, filter, and dryer to remove moisture and other impurities in the air. The compressed air can be stored in the compressed air storage tank (also referred to as air receiver tank) and used at a later time. The pneumatic regulating valves are employed to modulate the steam flow in response to the demand and maintain a constant steam outlet pressure. After the compressed air is delivered to the end-use equipment, it is released to the surrounding environment during operating activities.

2.1.4 Energy Consumption

Energy is one of the crucial components in the operation of manufacturing systems. According to the manufacturing energy consumption survey (MECS) published by

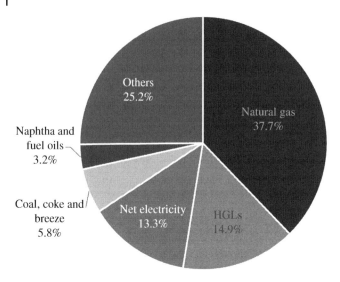

Figure 2.6 Illustration of energy consumption in the US manufacturing sector by energy. *Source:* Adapted from [14].

the U.S. EIA [14], the US manufacturing sector consumed 19 436 trillion BTU of energy in 2018, which can be primarily attributed to the energy consumption of fossil fuels, electricity, and biomass. A detailed breakdown of the energy consumption by energy source is demonstrated in Figure 2.6. It can be observed that natural gas and hydrocarbon gas liquids (HGLs) contribute 53% of the total energy consumption in the US manufacturing sector. The net electricity accounts for 13.3% of total energy consumption. Near 9% of the energy consumption originates from the use of coal, coke, and breeze, as well as naphtha and fuel oils. Other energy consumption, such as still gas, petroleum gas, black liquor and other petroleum products, contributes 25.2% of total energy consumption.

Apart from the overview of energy consumption in the US manufacturing sector, the energy consumption involved in manufacturing systems can be further decomposed into two large categories, i.e. direct end use and indirect one. More specifically, the former refers to the energy directly used for the operations of manufacturing facilities. The energy consumption estimations for direct end use take into account only the quantities of electricity or fossil fuels used in their original states (i.e. not transformed) [15]. The latter, also referred to as fuel for onsite energy generation, reflects the amount of fuel used in the boiler to produce secondary energy sources [15]. For example, a boiler can convert the chemical energy of fuel into secondary energy sources, such as electricity and steam. According to the reported energy consumption data released in MECS [15], the former contributes 70.2% of manufacturing end-use energy consumption; while the latter contributes

29.8% of manufacturing end-use energy consumption. Moreover, based on the end-use purpose, the former can be subdivided into direct non-process end use (10.9%) and direct process end use (59.3%).

2.1.4.1 Indirect End Use

Based on the type of boilers, the indirect end use can be mainly categorized into two groups: conventional boiler use and combined heat and power (CHP) cogeneration process use. More specifically, almost all of the heat generated by conventional boilers is mainly used for facility and process heating. In comparison, the CHP cogeneration process enables the concurrent production of electricity or mechanical power and thermal energy from a single energy source [16]. Compared to the traditional process with the separated boiler and the power generation, the waste heat in the boiler can be reused, which leads to an improvement in the overall thermal efficiency of a CHP cogeneration process [17].

2.1.4.2 Direct Process End Use

The direct process end use refers to the end uses specific to the manufacturing operations, which mainly involves the following five categories: process heating, process cooling and refrigeration, machine drive, electrochemical processes, and other process use [15].

- Process heating: The direct end use energy consumption, which is used to raise the temperature of substances involved in a manufacturing process. Process heating is normally involved in the manufacturing activities that need to be performed at high temperature, such as the use of the blast furnace for iron smelting in iron production and the use of heat to dry paint in automobile manufacturing.
- Process cooling and refrigeration: The direct end-use energy consumption, which is used to lower the temperature of substances involved in a manufacturing process. For example, in the chemical industry, specific reactions can only occur at a temperature below the ambient temperature, and energy is required for low-temperature maintenance.
- Machine drive: The direct process end use in which thermal or electrical energy is converted into mechanical energy. Generally, at a manufacturing site, the energy consumption for machine drive can be mainly attributed to the operation of motors. For example, in electric motor-based metal processing, the torque generated through the interactions between the motor's magnetic field and the electric current is applied on the main shaft of electrical motors, which drives the drill bit to rotate and conduct the drilling operations [18].
- Electrochemical process: The direct process end use in which electricity is used to cause a chemical transformation, where the electrical energy is converted into chemical energy. For example, in active metal fabrication, aluminum can be

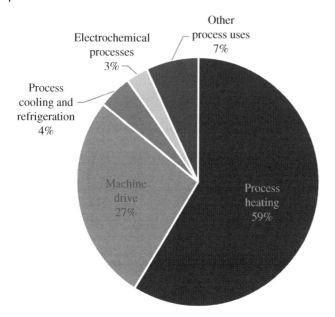

Figure 2.7 The direct process end use energy consumption in the US manufacturing sector. *Source:* Adapted from [19].

extracted using an aluminum smelter. Particularly, the aluminum oxide is dissolved in molten cryolite, and the aluminum is produced through the electrolytic reduction of aluminum oxide.

According to the data published in MECS [19], the direct-process end-use energy consumption in the US manufacturing sector in 2018 is demonstrated in Figure 2.7. In particular, among the aforementioned five process end-use categories, process heating is the most dominant component of direct process end-use energy consumption in manufacturing systems, which possess a proportion of 59% in the total energy consumption. The machine drive is the second-largest energy consumer with a share of 27%. In addition, the energy consumed in process cooling and refrigeration, and electrochemical processes account for 4% and 3%, respectively.

2.1.4.3 Direct Non-process End Use

The direct non-process end use refers to the end uses that may be found on commercial, residential, or other sites, as well as at manufacturing establishments. The major contributors to direct non-process end use include facility heating, ventilation, and air conditioning (HVAC), facility lighting, facility support, onsite transportation, and conventional electricity generation [15].

- Facility heating, ventilation, and air conditioning (HVAC): The direct non-process end use includes energy use in systems that condition air in a building [15]. HVAC mainly consists of the following two components: (i) heating and air conditioning, which are used to regulate the working environment temperature in the office area and provide suitable temperature for manufacturing establishment sites; and (ii) ventilation, which is used to maintain indoor air quality. The HVAC system is considered to be one of the major contributors to the total electricity consumption in a manufacturing plant [20].
- Facility lighting: The direct non-process end use takes into account the energy consumed in equipment that illuminates buildings and other areas on the establishment sites [15].
- Facility support: The direct non-process end use of energy in diverse applications that are usually associated with the office or building operations [15]. For example, the facility support includes office equipment, such as computers and printers, building cleaning service, cafeteria, and foodservice facilities.
- Onsite transportation: The direct non-process end-use energy consumption in vehicles and other transportation equipment that primarily consume energy within the boundaries of the manufacturing establishment [15]. The energy consumed by trucks out of a factory to transport finished products for sale does not belong to onsite transportation.

The direct non-process end-use energy consumption in the US manufacturing sector in 2018 is demonstrated in Figure 2.8. In particular, among the aforementioned six non-process end-use categories, facility HVAC is the major contributor, which accounts for a proportion of 61% of the total energy consumption. Facility lighting is the second-largest energy consumer with a share of 16%. In addition, the energy consumed in facility support and onsite transportation accounts for 10% and 7%, respectively. Besides, the rest 6% of the total energy consumption can be attributed to conventional electricity generation and other non-process miscellaneous uses.

2.2 Energy Saving Potentials and Energy Management Strategies for Manufacturing Systems

The industrial sector is regarded as one of the main engines of economic growth, and the development of manufacturing industries contributes positively to technological innovations and the development of associated infrastructure services. Despite the social-economic benefits, the tendency to increased energy consumption is more pronounced in the industrial sector. In reference to the Annual Energy Outlook 2020 (AEO2020) published by the U.S. EIA, the modeled projections (i.e. the AEO2020 reference case) of long-term trends of the energy

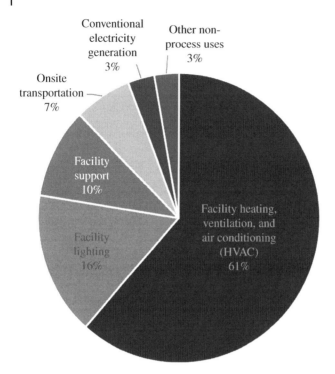

Figure 2.8 The direct non-process end use energy consumption in the US manufacturing sector. *Source:* Adapted from [19].

consumption in the US industrial sector is demonstrated in Figure 2.9, with the projections to the year 2050 [21]. In particular, the projected annual growth rate of the total industrial energy consumption is 1.0% per year on average during the projection period in the AEO2020 reference case. As shown in Figure 2.9, in the past two decades, the total energy consumption in the US industrial sector remained at a relatively stable rate. In the AEO2020 reference case, the total energy consumption in the industrial sector increases from 26 quadrillion BTU in 2019 to 36 quadrillion BTU by the end of the year 2050, which indicates an increment of 36% during the projection period.

Meanwhile, the manufacturing activities are also accompanied by energy-related greenhouse gas (GHG) emissions. According to the US manufacturing energy use and GHG emissions analysis published by the U.S. Department of Energy (DOE), the total U.S. manufacturing GHG emissions equaled 1261 million metric tons of CO_2 equivalent (MMT CO_2e) in 2006 [22]. Figure 2.10 illustrates a detailed breakdown of GHG emissions by energy end-use type. In particular, the GHG emissions associated with the electricity consumption are contributed by both

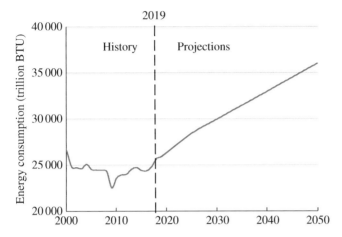

Figure 2.9 The US industrial energy consumption in the AEO 2020 reference case. *Source:* Adapted from [21].

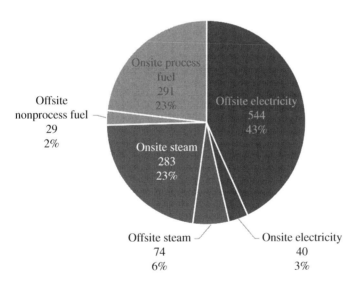

Figure 2.10 Total GHG combustion emissions in US manufacturing by energy end-use type. *Source:* Adapted from [22].

offsite (i.e. the local electrical power grids) and onsite electricity sources (i.e. the electric power generated at the manufacturing site), and the fuel combustion-related GHG emissions take into account both the process and non-process end uses in the manufacturing sites. As shown in Figure 2.10, a total of 584 MMT CO_2e emissions can be assigned to electricity consumption (both offsite and onsite), which

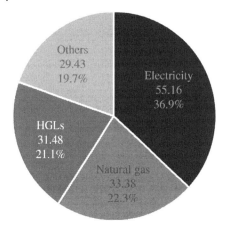

Figure 2.11 The cost of energy consumption in US manufacturing by energy type. *Source:* Adapted from [23].

corresponds to 46% of the total GHG emissions. A further 29% of the GHG emissions can be attributed to the steam production (with a total of 357 MMT CO_2e emissions), while the remaining 25% of emissions are contributed by the fuel combustion for process and non-process end uses.

Apart from the energy-related GHG emissions in the manufacturing industries, from an economic perspective, the energy cost in the manufacturing sector is also determined by the energy consumption and the corresponding energy price. Based on the U.S. EIA's manufacturing energy consumption survey in 2014 (MECS 2014) [23], the majority of energy cost is originated from carbon-based energy sources. Particularly, the energy cost in the US manufacturing sector is decomposed into five categories, and the respective cost estimations are demonstrated in Figure 2.11 based on the average energy prices for manufacturers summarized in MECS 2014. As shown in Figure 2.11, electricity expenditure accounts for the largest proportion (USD 55.16 billion, 36.9%) of the total energy cost in the manufacturing sector, with the cost shares of natural gas (22.3%) and HGL (21.1%) being similar. A further 19.7% of the energy cost is contributed by all other energy sources.

As the manufacturing sector continues to grow, the increasing concerns about climate change, resources security, and restricted energy supply have elevated the importance of energy management in manufacturing. Given the environmental awareness of manufacturing activity associated carbon footprints, the growing energy price, and stricter environmental regulations and standards, effective energy management has been highlighted as a crucial element for sustainable development in manufacturing industries [24]. In the following three sections, the energy-saving potentials in the manufacturing industry and representative energy management strategies are discussed in detail. Due to the significant electricity-related GHG emissions and energy costs in manufacturing systems, the following sections are mainly presented from the aspect of electricity use. More specifically, energy management in manufacturing systems is divided into three different levels. Figure 2.12 illustrates the relationships among these three levels, as well as the essential contributors that may affect the overall energy efficiency at

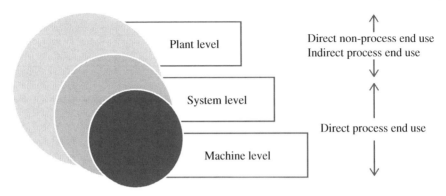

Figure 2.12 Three levels of energy management in manufacturing.

each level. As shown in Figure 2.12, the energy management in machine level and system level emphasizes the energy efficiency issues related to the direct process energy end use. While moving upward to the plant level, the energy saving potentials and energy management strategies are discussed along with the aspects of both direct non-process end use and indirect process one.

2.2.1 Machine Level

Machines are the crucial components in the manufacturing systems, and machine drive is regarded as one of the major contributors to process energy consumption [22]. The machine energy consumption rate is usually dynamic rather than static, which is closely related to the specified production activities and machine states. Five typical machine states are demonstrated in Table 2.2, including off, startup, idle, ready for processing, and processing states [7].

To better illustrate the relationships between the energy consumption and machine states, a typical electrical load of a machining center is demonstrated in Figure 2.13. When the machine is in the off state, the electrical load of the machine is zero. At the instant of machine startup, an electrical load spike is commonly observed as a result of machine homing, machine tool initialization, pre-warming, etc. When the machine is in the idle state, a constant base energy consumption is still required to maintain the functioning of machine components such as control units and cooling modules. Under certain circumstances, some machines may still run at a low speed during the idle state with relatively lower energy consumption. For example, the spindle speed at idle state could be one half of its normal operating speed [25]. Furthermore, when the machine is in the ready for processing state, additional energy is required for machine tool positioning to the designated workpiece, which is normally associated with an increase in the

Table 2.2 Illustrations of typical machine states.

State	Description
Off	The machine is turned off, and it does not consume energy
Startup	Certain machines require a startup state before normal operations. It can be visualized as spikes in the energy load of machines upon turning them on. The possible factors that can lead to a startup state include the initial startup check, homing of machine components, prewarming process, etc.
Idle	The idle state indicates that the machine is currently available and in a standby mode. The possible reasons for entering it include waiting for a task assignment and material delivery. The energy consumption during the idle state is relatively constant
Ready for processing	In this state, positioning and loading are performed prior to the actual processing, e.g. moving the spindle toward the workpiece before the milling process
Processing	The processing state of a machine indicates the ongoing actual production activities

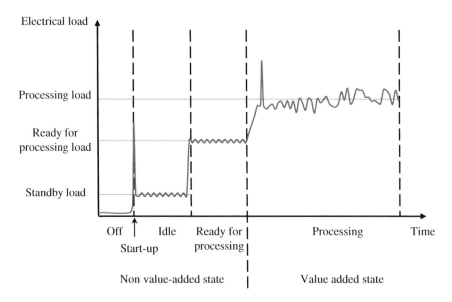

Figure 2.13 A typical electrical load profile of a machining center.

electrical load. It should be noted that the machine does not add value to the product during the aforementioned states. When a machine is in the processing state, it runs at the designated operating speed, which is accompanied by significant increases in electrical load. The processing state is often referred to as the value-added state.

At the machine level, the intrinsic characteristics of machine tools and processing conditions can affect total energy consumption [26]. In general, the intrinsic characteristics of machine tools are mainly associated with energy loss issues, such as electric motor efficiency loss, and hydraulic system loss. More specifically, some common sources of motor loss include (i) friction loss, which can be primarily attributed to the force to overcome the friction associated with rotor rotation; (ii) windage loss, which can be caused by the turbulence against the rotation of a rotor in an air-cooled motor; (iii) core loss, which is related to the magnetic paths of a motor; and (iv) ohmic loss, which is induced by the current flowing through the conductors. In addition, motor loss is usually influenced by motor topologies, power factors, and operating conditions (such as running speed, and lubrication situation of transmission parts). Hydraulic system loss is commonly referred to as the energy loss in the form of pressure (which can be due to friction, flow deflection, and throttling) and leakage loss. It is usually related to the system structure and properties of hydraulic fluid. From the perspective of machine processing conditions, the energy efficiency of a single machine can be primarily affected by the respective proportions of energy consumption during value-added periods, as well as non-value-added periods. Particularly, the selection of machining parameters can affect the effective operation time, which gives rise to variations in energy consumption. Meanwhile, the energy use profile during the non-value-added periods can be influenced by the duration of machine idling, energy consumed in machine tool allocation, acceleration-deceleration control, and toolpath planning issues, etc. Given the energy loss induced by the intrinsic characteristics of machine tools and the effects of processing conditions on the energy efficiency of machines, the representative energy management can be conducted along the following two directions:

2.2.1.1 Intrinsic Characteristics of Machine Tools

To address the energy loss issues of machine tools, it is necessary to achieve low energy consumption without sacrificing machine productivity and product quality in the early design stage of machine tools. As a result, the concept of lightweight design in machine tools has emerged. In particular, a lightweight factor is proposed to assess the lightweight effectiveness of certain machine tool designs, which is commonly referred to as the ratio of elastic modulus to density for materials in machine tools or the ratio of structure stiffness over mass with the consideration of machine geometry design [27]. The formulation of the lightweight factor is illustrated as follows:

$$\text{Lightweight factor} = \frac{\delta}{\rho} \text{ or } = \frac{k}{m} \tag{2.5}$$

where δ denotes the elastic modulus; ρ represents the density of the material; k signifies the structure stiffness; and m is the mass. A high lightweight factor in the design of the machine components can be achieved by jointly considering the material selection and topology design of the key mechanical components. The implementation of lightweight design can facilitate the reduction in the energy consumption of machine tools. On the one hand, a lighter mechanical component is associated with smaller inertia force, which enables the machine tool allocation and acceleration–deceleration with less energy. In addition, a lightweight design can potentially make it easier to move the machine tool through a series of motions, which ultimately reduces the amount of time the machine spends in non-value-added states, such as idle and ready for processing states. Moreover, lighter components can induce less gravity force at contact points in bearings and guides, which reduces the friction loss during the machine tool movement. On the other hand, lightweight design can improve the process stability and contribute positively to the increase of the yield rate, which can be regarded as indirect energy savings. Meanwhile, the improved process stability can also lead to shortened primary processing time and eventually reduces the overall energy consumption of machines [27].

Particularly, the strategies for lightweight design include structure optimization and material selection. The structure optimization emphasizes the topology design of a machine tool toward improved structural stiffness and reduction in component weight. For example, a machine tool column can be redesigned based on the configuration principles of biological skeletons and sandwich stems [28]. The column weight is reduced by 6.13% with improved static and dynamic performances. In addition, new materials are proposed for the lightweight design of machine tools to overcome the restrictions of traditional materials. It has been demonstrated that materials such as carbon/epoxy composites and resin concrete can reduce the weight of a table-top machining center by 36.8% and increase the structure stiffness by 16% [29].

2.2.1.2 Processing Conditions

Energy management strategies associated with the intrinsic characteristics of machine tools necessitate design innovations in machine tools, and the implementation of these strategies can be cost-prohibitive in some small-scale manufacturing enterprises due to the additional investments in capital equipment upgrades and replacement. A more cost-friendly alternative energy management strategy is the processing condition optimization of existing machines [30]. Some common energy-saving-oriented processing condition adjustment strategies are introduced as follows:

a) Optimization of process parameters. In practice, the same working procedure and product quality can be delivered by various combinations of machining parameters. However, the associated energy consumption profile may differ depending on the adjustment of parameters. For example, previous research investigates the energy consumption of drilling and face/end milling processes [31]. The energy performance of machine tools is evaluated by selecting different levels of cutting speed, feed rate, and cutting depth within the recommended operating ranges indicated in the equipment manuals. The results reveal that an energy consumption reduction of up to 39% can be achieved through the optimization of process parameters.

b) Machine tool trajectory planning. Machine tool movements are often involved in machine operations, such as the transition from the standby position to working one, and the motions of machining head relative to the surface of workpieces. The machine tool trajectory planning aims to minimize energy consumption by determining the optimal toolpath and velocity/acceleration profiles. For example, previous research has demonstrated the trajectory planning on an Ortho-glide system, which is three degrees of freedom translating Parallel Kinematics Machine with a cubic workspace [32]. The study investigates the relationships between the energy consumption and the position and orientation of the end-effectors of a manipulator, and the optimal path placement can lead to an up to 60% energy saving.

2.2.2 System Level

A modern manufacturing system is a dynamic system consisting of multiple independent machines, buffers, and other devices. The productivity performance of various machines may vary depending on the machine-specific intrinsic characteristics and operating conditions. However, due to the complexity of the interconnection among different machines, it is challenging to maintain the maximum productivity for every single machine during the entire production horizon [33]. As a result, in a mass production environment, more than 50% of energy is consumed during the non-value-added states, and the machine utilization rate is relatively low [26].

At the system level, energy management is mainly originated from how the interconnections among different machines affect the total energy consumption. Hence, different from the considerations of detailed characteristics of a single machine (such as the electrical load profiles and machine tool path), machines

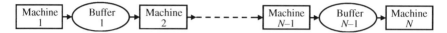

Figure 2.14 Schematic diagram of a typical serial production line with N machines and $N-1$ WIP buffers.

are often described by some general features at the system level. In particular, a schematic diagram of a typical serial production line is presented in Figure 2.14. The rectangles represent the machines, and the circles are work-in-process (WIP) buffers. Each machine is characterized by its own cycle time (i.e. the time required to process one part), time between failures, and time to repair. In particular, the WIP buffers represent the material handling devices, such as conveyors, forklifts, which transfer WIP (i.e. semi-finished products) from the upstream machines to downstream ones. From the system-level perspective, the WIP buffer is characterized by its storage capacity.

In the ideal situation, upon the processing operations in the upstream machine being completed, the WIP can be transferred to the downstream machine by the WIP buffer, and the downstream processing can be initiated immediately. In this case, machines can stay in the processing state through the entire production horizon, which in turn minimizes the idle time and associated energy consumption. However, it is quite difficult to reach this ideal production scenario due to the existence of inhomogeneous machines and the occurrence of machine failures and maintenance operations in practice.

2.2.2.1 Inhomogeneous System

In this book, the term "inhomogeneous machines" refers to machines with non-identical features, such as different cycle time, different time between failures, and different time to repair. Attributed to the deviations in machine performance, some machines can become the bottleneck of a manufacturing system. More specifically, a bottleneck is defined as the machine whose performance impedes the overall system performance in the strongest manner [34]. In the case where the bottleneck exists in a production line, the upstream WIP buffer of the bottleneck machine becomes eventually full, and the downstream WIP buffer becomes empty. As a result, the upstream machine can be blocked by the full buffer, and the blockage can propagate upstream; additionally, the downstream machine can be starved by the empty buffer and the starvation can propagate downstream. Such blockage and starvation increase the nonworking time of the non-bottleneck machines and lead to increased energy consumption in the idle state. To address this energy waste issue, two methods can be implemented: bottleneck identification and production management.

- Bottleneck identification. Simulation-based methods are widely implemented for bottleneck detection in the manufacturing industries, such as automotive or aircraft assembly lines. For example, the General Motors develops a simulation system (C-MORE) to analyze the manufacturing system performance and identify the bottlenecks [35]. Besides the simulation-based methods, recent "big data"-based manufacturing environment and the digital transformation of manufacturing provide a foundation for the establishment of online data-driven bottleneck detection methods, where the system performance can be continuously monitored, and the energy management decisions can be made in real time [34, 36].

- Production management. The production management in inhomogeneous systems can be generally categorized into two strategies. One strategy is to improve the system productivity such that the share in the total energy consumption of a single part can be reduced. A previous study demonstrates that through the initial WIP buffer adjustment and maintenance task prioritization, the system productivity can be increased by 8–32% [37]. Another strategy is to turn off the starved or blocked machines temporally, and up to 30% of energy savings can be achieved without compromising the production throughput [38].

2.2.2.2 Machine Maintenance

Machine maintenance is critical for a smooth operation in the manufacturing systems. Once a machine breaks down, it can cause blockage and starvation, which eventually negatively impacts system productivity and energy efficiency. Depending on timing for maintenance, the machine maintenance strategies can be categorized into two types: preventive maintenance and corrective maintenance.

- Preventive maintenance is normally carried out prior to the occurrence of a machine breakdown. The preventive maintenance can be scheduled in advance, and the major purposes of this type of maintenance are to reduce the probability of machine breakdown and mitigate the degradation of equipment. A dynamic maintenance scheduling based on the real-time system operation status can minimize energy cost with the consideration of the time-dependent electricity price. A previous study indicates that up to 32% of energy-related cost savings can be achieved with the implementation of dynamic maintenance control [39].

- Corrective maintenance is usually performed when a machine failure has occurred. Due to the suddenness of machine failure, the timing for corrective maintenance cannot be pre-scheduled, and it has to be carried out in response to the failure to resume the normal operations of the machine. Perfect corrective maintenance can restore the failed machine to a brand-new condition. However, it is also associated with more maintenance time and cost. On the contrary, imperfect corrective maintenance can be quicker and less costly, but it can only recover the machine to a certain state, and additional maintenance task may be required in the near future. It has been reported that the energy-related cost can be reduced by 20% through proper selection of the maintenance level [40].

2.2.3 Plant Level

The energy saving potentials and energy management strategies discussed at the machine level and the system level are mainly related to the direct process end use. At the manufacturing plant level, apart from the main production components, some auxiliary systems also exist to facilitate the proper operation of the manufacturing systems, such as the HVAC system, and lighting system. Although these systems may not directly affect the manufacturing productivity, they are necessary components at manufacturing plants, and their associated energy consumption is not negligible. Hence, the energy management strategies related to indirect end use and direct non-process end use need to be studied and discussed.

2.2.3.1 Indirect End Use

The indirect end-use energy consumption refers to the fuel used for onsite energy generation, such as electricity and steam. According to the 2018 customer survey conducted by the S&C Electric Company (which specializes in consulting and engineering services related to electrical power systems), approximately 25% of their client companies experience power outages at least once per month [41]. Given the unplanned downtime and costly interruption of production lines during the power outage, there is an increasing demand for onsite energy generation for continuous production at manufacturing sites. Meanwhile, steam is widely used in the manufacturing facilities for process and space heating, and steam is mainly generated by onsite boilers [22]. Traditionally, the generation of electricity and steam is carried out in separate systems. In recent years, a combined heat and power system has emerged to provide simultaneous generation of electricity and steam with significantly higher energy efficiency.

In a typical CHP system, the excess heat that would be normally lost during the operation of electricity generators can be recovered and functions as a thermal energy source for the plant facilities and buildings. The cogeneration of heat and electricity from a CHP system delivers higher energy efficiency [42]. A demonstrative energy flow in a typical CHP system is presented in Figure 2.15 to further illustrate the energy saving potential of this cogeneration system. In particular, in reference to the conventional power plant and fuel combustion-based boiler, the corresponding energy efficiency is 33% and 80%, respectively [17]. Through back-calculation, in order to generate 30 units of electricity and 45 units of steam, a total of 175 units of fuel are required. On the contrary,

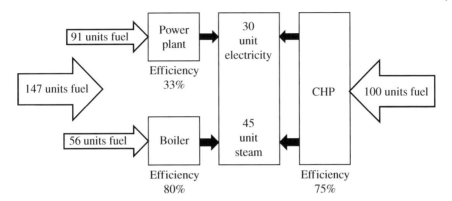

Figure 2.15 Energy efficiency comparison between CHP system and separated heat and power generation systems. *Source:* Adapted from [17].

to produce the same amount of electricity and steam, only 100 units of fuel is required using a CHP system. In general, CHP systems can achieve the overall system efficiency of 60–80% in comparison to around 45–50% in separated power plant and boiler systems under the similar operating condition [17].

2.2.3.2 Direct Non-process End Use

Facility HVAC System An HVAC system is considered to be one of the major contributors to the total electricity consumption in a manufacturing plant [20]. In some cases, to simplify the process management in complex manufacturing systems, the interactions between the HVAC system and the main production lines are often ignored. However, from the energy-efficiency aspect, it is necessary to jointly consider the heat generated during the manufacturing operations and its associated impacts on the HVAC load modulations. More specifically, by integrating the heat transfer characteristics of machines and the corresponding load variations in the facility HVAC systems, the concurrent determination of a production plan and HVAC system control scheme can be performed. The integrated control approach contributes positively to the power demand reduction during peak hours of energy consumption without compromising the designated production target.

A schematic chart of the heat flow in a manufacturing plant is illustrated in Figure 2.16. In a manufacturing system, the heat produced by machines is transferred to the surrounding environment by convection and radiation processes. More specifically, the convective heat transfer is regarded as an instantaneous heat gain of the surrounding environment as the heat is directly released in the production area. On the contrary, radiant heat is absorbed by the indoor surfaces and then dissipated into the surrounding air over time [44]. The radiant heat imposes a delayed response in cooling load modulation of the HVAC systems. To tackle this problem, a radiation

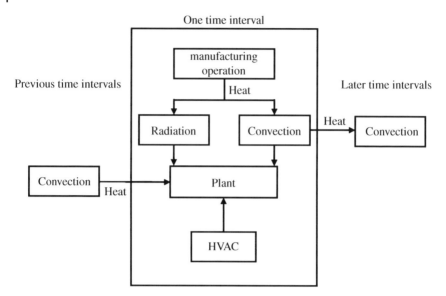

Figure 2.16 Illustration of convective and radiant heat transfer due to manufacturing operations. *Source:* Adapted from [43].

time series method is proposed to simulate the delay of radiant heat and determine the percentage of radiant heat dissipated in each time interval [45].

Facility Lighting According to the Annual Energy Review published by the U.S. EIA [46], facility lighting accounts for 3% of the total direct end-use energy consumption in the US manufacturing sector. In terms of facility lighting, the most common energy management method is to change the types of lighting. Table 2.3 summarizes the energy consumption associated with different lighting methods under similar lighting levels. Compared with traditional incandescent light bulbs, compact fluorescent light (CFL) bulbs and light-emitting diode (LED) lamps can deliver improved energy savings of up to 75–80%. Meanwhile, the increased lifetime of light bulbs can reduce lighting costs in manufacturing sites.

Table 2.3 The energy consumption of different lighting methods.

	Incandescent	CFL	LED
Power (W)	60	15	12
Energy saved (%)	0	75	80
Bulb lifetime (hour)	1 000	10 000	25 000

2.3 Demand-side Energy Management

In the previous section, the energy saving potentials and energy management strategies are mainly demonstrated from the manufacturing side. Indeed, these energy management strategies are not only beneficial for manufacturing enterprises to lower their utility bills, but also of great significance for the stable operation of the power grids. Traditionally, the operation of power grids is unidirectional from the supply side, i.e. only a limited number of power plants feed into the grid and help balance the demand and supply in an electricity system. In order to fulfill the rising electricity demand, new electricity facilities are needed, such as electricity generation station upgrades or expansion to transmission lines, which require significant investments in electrical infrastructures. More specifically, it has been estimated that the investment of more than USD 1.5 trillion is required for the construction of new electricity generation, transmission, and distribution networks by 2030 [47].

Considering this background, there has been increasing attention placed on the impacts of DSM on electricity networks. DSM is a portfolio of measures aiming to lower the electricity demand and improve the performance of electrical energy systems from the consumption side [48]. In comparison with the construction of new electricity infrastructure, it is far less cost-prohibitive to influence the electricity demand through DSM. In general, DSM programs are designed to modify the patterns of electricity use, which are mainly composed of energy efficiency programs and demand response programs. A demonstrative diagram of the impacts of energy efficiency and demand response on electricity demand is shown in Figure 2.17.

In particular, the special aim of energy efficiency programs is to permanently improve the energy efficiency of manufacturing systems by upgrading energy-consuming equipment. Therefore, as indicated by the energy efficiency curve (the dash line) in Figure 2.17, with the implementation of an energy efficiency program, both power demand (i.e. the height of the demand curve) and total energy consumption (i.e. the area under the demand curve) are reduced in comparison with the original energy consumption scenario (the solid line). As shown in Figure 2.17, the consumers' energy consumption patterns may remain unchanged under an energy efficiency program. On the contrary, a demand response program does not require equipment upgrade. It aims at mitigating the burden on the power grid caused by peak demand through load shifting and load shedding. As indicated by the demand response curve (dash-dotted line) in Figure 2.17, with the implementation of a demand response program, the total energy consumption may remain unchanged, but the peak demand is significantly reduced. Hence, a demand response program can help reduce the GHG emission from dirty backup

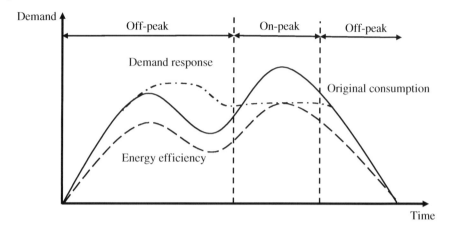

Figure 2.17 Impacts of energy efficiency and demand response programs on electricity demand.

electricity generators in order to meet the high peak demand. In general, compared to energy efficiency programs, demand response programs are more feasible for energy consumers who are unwilling to upgrade equipment or already have high energy-efficient equipment.

In addition, financial incentives are commonly offered for the adoption of DSM programs, which encourage consumers to modify their energy use behaviors to some extent or switch production to off-peak periods [49]. Considering the close relations of energy prices to the energy consumption and load control in energy management, in the following sections, the components of electricity bill are introduced, followed by the detailed discussions on the energy efficiency programs and demand response programs.

2.3.1 Electricity Bill Components

For manufacturers, the cost of electricity (C) mainly consists of three parts: electricity consumption cost (C_E), demand cost (C_D), and fixed cost (C_F). The total electricity cost is equal to the sum of these three parts, i.e.

$$C = C_E + C_D + C_F \tag{2.6}$$

Figure 2.18 demonstrates the decomposition of the electricity bill from an LED bulb manufacturing plant. As shown in Figure 2.18, the electricity cost dominates a manufacturing company's utility bill, followed by the demand cost. The fixed cost only accounts for a relatively small part of the total electricity bill.

2.3.1.1 Electricity Cost

The electricity cost (C_E) is the electricity fee charged by the utility companies or energy providers for the total energy consumption during the billing period. As a secondary energy source, the electricity cost is related to the cost of the primary energy sources used in electricity generation. C_E is calculated by multiplying the amount of energy consumed (denoted as e_T, with the unit of kWh) by the relevant energy price (denoted as c_E, with the unit of $/kWh), i.e.

$$C_E = e_T c_E \qquad (2.7)$$

In addition, electricity prices may vary in response to the changes in supply and demand, as well as the tariff rates. There are several types of electricity plans and energy rates offered by energy providers, such as flat electricity price, tiered electricity price, time-of-use electricity price, and real-time electricity price.

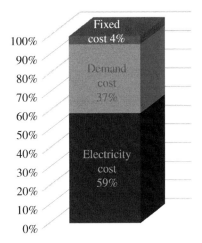

Figure 2.18 Illustration of cost compositions of electricity in a manufacturing plant.

a) Flat utility rate: The price of electricity remains constant over a long period of time.
b) Tiered utility rate: When the electricity consumption is within a certain range (also referred to as tier), the electricity price remains flat; when the electricity consumption exceeds the amount in the tier, a higher electricity price applies.
c) Time-of-use electricity price: During peak periods, such as from 4:00 p.m. to 9:00 p.m., the price of electricity is much higher than the price during off-peak hours [50].
d) Real-time electricity price: The electricity price varies over a short time interval, typically hourly.

2.3.1.2 Demand Cost

Demand cost (C_D) is the electricity fee paid based on the maximum electricity demand in a certain time period. As a major electricity consumer, the electricity consumption in the manufacturing industry is incomparable with that of the residential uses. Hence, the utility bill of manufacturing companies also includes the charges based on electricity demand [7]. More specifically, to ensure the stable operation of the power grid and avoid power failure, there must be a balance between power supply and demand [48]. To maintain this electricity balance, large

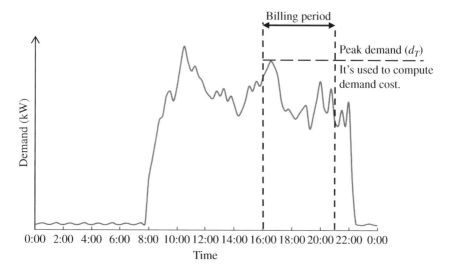

Figure 2.19 The peak demand for time-of-use rate.

quantities of backup equipment are required in the power grid to meet the peak power demand. Therefore, for users with higher power demand, such as the manufacturing industry, the demand cost is usually charged as a cost share of the total cost of the power grid. C_D is based on the peak demand of a manufacturing system (denoted as d_T, with the unit of kW) and the corresponding rate (denoted as c_D, with the unit of \$/kW) during the billing period, i.e.

$$C_D = d_T c_D \tag{2.8}$$

In general, the utility company usually records the average power usage in a demand interval (usually 15 minutes). As shown in Figure 2.19, the interval with the highest 15-minute electrical power usage is set as the peak demand during the billing period, which is used to determine the corresponding demand cost.

2.3.1.3 Fixed Cost

Fixed cost (C_F) is commonly referred to as a fixed charge set by the utility company to cover the expenditures in other auxiliary systems during electricity transmission from the generation site to the consumption site. This part of the cost is fixed and not related to electricity consumption and demand on the user side.

2.3.2 Energy Efficiency Programs

In physics, efficiency is defined as the ratio of input to output energy. Therefore, the formula for energy efficiency can be written as follows:

$$\text{Efficiency} = \frac{\text{Output}}{\text{Input}} \qquad (2.9)$$

In terms of energy utilization efficiency, the U.S. EIA defines energy efficiency as "using technology that requires less energy to perform the same function" [51]. From the aspect of energy demand and GHG emissions in the manufacturing industry, energy efficiency is considered to be a key success factor for sustainable production. Hence, the formula for energy efficiency in the manufacturing systems can be further generalized as Eq. (2.10), where the "input" is referred to as the total energy input to the manufacturing system, and the "output" is substituted by the total production output [7].

$$\text{Energy efficiency} = \frac{\text{Production output}}{\text{Total energy input}} \qquad (2.10)$$

The necessity of improving the energy efficiency in the manufacturing industry has led to a variety of government subsidies and energy efficiency programs for manufacturers, such as energy efficiency loans and tax incentives for energy efficiency upgrades [52]. Among the proposed programs, the ENERGY STAR® program is one of the representative energy efficiency projects. The ENERGY STAR® is a joint program of the U.S. Environmental Protection Agency (EPA) and the U.S. Department of Energy. More specifically, to certify a manufacturing facility as the ENERGY STAR®, an energy performance indicator (EPI), also referred to as the industry-specific benchmarking tool by the U.S. EPA, is used to measure the energy performance of a plant. By comparing its EPI with similar plants across the country, an ENERGY STAR score can be assigned in the range of 1–100. The manufacturing facility with an ENERGY STAR score over 75 is eligible for ENERGY STAR certification [53]. In addition, ENERGY STAR® also provides guidance for energy management in the manufacturing enterprises. Figure 2.20 illustrates the seven-step guidelines to achieve energy management.

Step 1: Make a commitment.
Only when the organization has made a commitment to improving energy efficiency and provides necessary human and economic supports, energy management has the possibility of success.
Step 2: Evaluate performance.
Through the assessment of energy usage in the organization, the goals of energy management can be effectively discovered, and no future actions have been established.
Step 3: Set goals.
Clear goals can be set to guide daily behavior and used to determine the progress of energy management projects.

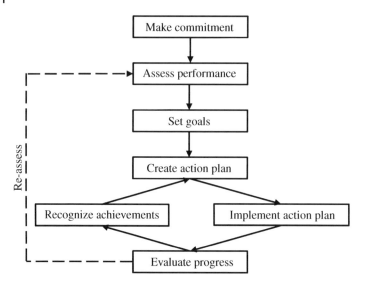

Figure 2.20 Illustration of seven-step guidelines for energy management. *Source:* Adapted from [54].

Step 4: Develop an action plan.

With the goals being determined, a detailed step-by-step action plan can be specified. A proper action plan can help a manufacturer facilitate the success of an energy management project, and it needs to be regularly updated.

Step 5: Implement the action plan.

Take specific actions for energy management.

Step 6: Evaluate progress.

The energy management project needs to be regularly evaluated to determine its progress toward the set goals.

Step 7: Recognize achievements.

The recognition of current achievements is conducive to the implementation of future energy management projects.

In terms of government economic measures, a variety of tax credits, rebates, and energy saving policies has been proposed at both federal and state levels. A search engine has been built into the official website of the U.S. DOE website with the database of State Incentives for Renewables & Efficiency (DSIRE) [55]. For example, at the state level in the United States, starting in July 2008, a sales tax exemption is in effect in Kentucky, which allows manufacturers to apply for a reimbursement of expenses on equipment replacement for renewable energy or energy efficiency projects. To apply the sales tax exemption, the energy efficiency projects in the manufacturing companies are required to reduce the measurable

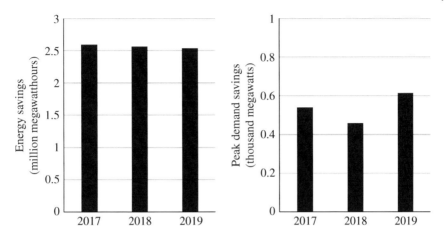

Figure 2.21 Industrial sector annual incremental savings resulting from energy efficiency programs.

amount of energy used by the facility by at least 15% while maintaining the regular manufacturing activities [56]. In addition, according to a survey from the U.S. EIA [57], the industrial sector annual incremental savings resulting from energy efficiency programs are illustrated in Figure 2.21. From 2017 to 2019, the average annual energy savings are above 2.5 million mega watthours, and peak demand savings are around 0.5 thousand megawatts.

2.3.3 Demand Response Programs

Electricity is a form of energy that cannot be effectively stored in bulk like fossil fuels. Therefore, from the generation side, electric utilities and power network companies have always tried to adjust their operations to maintain a balance between supply and demand [48]. However, this is not an easy task as the demand levels frequently change in different seasons and even at different hours of a day. Figure 2.22 shows the US hourly electricity consumption in December 2020 [58]. The historical data indicates an apparent oscillation in electricity consumption on a daily basis. The peak consumption is more than 125% of the trough value.

The unpredicted high electricity demand can impose negative impacts on the power grids. For example, high power demand can cause overburdened electric cables, transformers, and melt electrical equipment, which can eventually lead to power outages. To deal with this problem, as an alternative DSM approach to the investment in backup equipment, demand response programs are specifically designed to reduce the peak power demand. The U.S. Federal Energy Regulatory Commission (FERC) defines the demand response as "the changes in

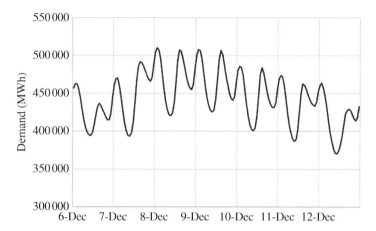

Figure 2.22 The US hourly electric demand in December 2020.

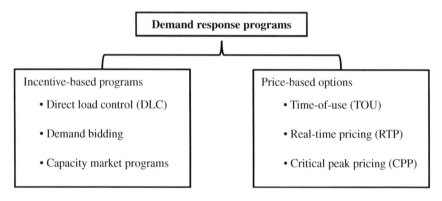

Figure 2.23 Classification of demand response programs.

electric use by demand-side resources from their normal consumption patterns in response to changes in the price of electricity, or to incentive payments designed to induce lower electricity use at times of high wholesale market prices or when system reliability is jeopardized" [59]. According to how load changes are brought about, the U.S. DOE classifies the demand response programs into two main categories: incentive-based programs and price-based options [60]. Figure 2.23 demonstrates the detailed classification of demand response programs.

2.3.3.1 Incentive-based Programs

The incentive-based programs provide incentives to participated customers, which are separate from or additional to their original electricity rates if customers successfully respond or fulfill the demand reduction per the requirements of the utilities. If the customers failed to reduce the amount of load that they previously agreed, they might face penalties in some programs.

a) In direct load control (DLC), the program operator has the authority to remotely shut down the customer's electrical equipment on short notice. This type of program is mainly offered to residential or small commercial customers. For example, in the State of Oregon in the United States, the Portland General Electric offers a DLC program to residential and business customers in exchange for a lower rate structure [61].

b) The demand bidding programs are mainly designed for large customers. The program encourages customers to either (i) bid into the electricity market and offer a price at which they are willing to curtail their energy load or (ii) identify how much load reduction they can accept with a utility-posted price. The customers whose offer is accepted can be offered with a lower electricity price, and a penalty may apply if the customers fail to reduce the energy load as contracted.

c) Capacity market programs can be viewed as a form of insurance [62]. Customers who participate in this program are committed to providing prespecified load reductions when system contingencies happen. For example, the PJM Interconnection provides a capacity market program to help maintain the long-term grid stability [63].

2.3.3.2 Price Base Options

The price-based programs aim at flattening the demand curve by offering dynamic pricing rates. During high-demand (on-peak) periods, the electricity price is higher than that during the low-demand (off-peak) periods. In some programs, the utilities may charge for power demand during on-peak periods in addition to energy consumption fees. If the price difference is significant, customers may consider adjusting the energy usage pattern according to the price structure to reduce the utility bills. It should be noted that the load modifications from the customer sides are completely voluntary, and there is no penalty for high electricity consumption during the peak periods.

a) Time-of-use (TOU) is a basic type of price-based programs. The pricing rates are predefined and differ during a day. The utilities may also provide different TOU rates depending on the seasons or days of the week. For example, Southern California Edison (SCE) offers different TOU rates in one program, as shown in Figure 2.24 [50]. In general, summer rates are slightly higher than winter rates, and even higher rates can apply during summer weekdays.

b) Real-time pricing programs (RTP) provide customers hourly fluctuating prices, which reflect the changes in the wholesale price of electricity. The hourly price

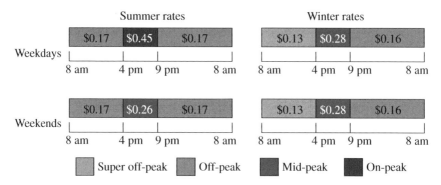

Figure 2.24 TOU rate plans offered by SCE with the price per kilowatt-hour.

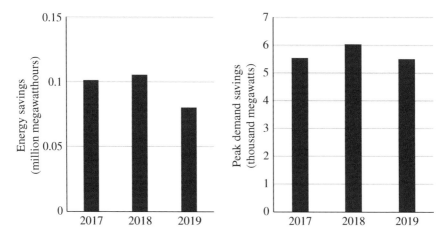

Figure 2.25 Industrial sector total annual savings resulting from demand response programs.

may vary based on the season, time of the day, and temperature [64]. RTP customers are typically informed of the price one day ahead. For example, SCE decides the daily pricing schedule according to the previous day's temperature in downtown Los Angeles, USA. Customers who participate in this program receive an email notification one day ahead to adjust their operations.

c) Critical peak pricing (CPP) rates are a hybrid design of TOU and RTP. The basic price structure of CPP is built upon TOU. However, a much higher CPP price applies if some specific CPP event is triggered, and the CPP event trigger is determined by utilities. For example, SCE defines it as (i) forecasts of extreme or unusual temperature conditions, (ii) SCE system emergency, and (iii) California independent system operator warning or emergency [64]. The event notification is sent to customers one weekday in advance.

According to the Assessment of Demand Response and Advanced Metering Report published by the U.S. EIA [57], Figure 2.25 illustrates the recent annual energy savings and peak demand savings in the industrial sector achieved by demand response programs. As shown in Figure 2.25, although the energy savings are less significant compared with the outcomes induced by energy efficiency programs (as referred to Figure 2.21), demand response programs can effectively help the industry reduce peak demand cost by approximately 5.5 thousand megawatts per year.

Problems

2.1 What are the major sources and users of energy in the US manufacturing sector?

2.2 State the major principles for energy management in the manufacturing system.

2.3 What are the benefits of demand-side management?

2.4 What is the difference between energy efficiency programs and demand response programs?

2.5 Provide examples of common demand response practices and programs.

References

1 U.S. Energy Information Administration (EIA). What is energy? explained. https://www.eia.gov/energyexplained/what-is-energy/ (accessed 9 December 2020).

2 U.S. Energy Information Administration (EIA). Glossary. https://www.eia.gov/tools/glossary/ (accessed 9 December 2020).

3 United Nathions (1982). Concepts and Methods in Energy Statistics, with Special Reference to Energy Accounts and Balances: A Technical Report.

4 U.S. Energy Information Administration (EIA). Sources of energy. https://www.eia.gov/energyexplained/what-is-energy/sources-of-energy.php (accessed 9 December 2020).

5 U.S. Energy Information Administration (2020). U.S. Energy Information Administration, Monthly Energy Review, April 2020, preliminary data. https://www.eia.gov/totalenergy/data/monthly/pdf/mer.pdf.

6 U.S. Energy Information Administration (2021). Electricity in the United States. https://www.eia.gov/energyexplained/electricity/electricity-in-the-us.php.

7 Thiede, S. (2012). *Energy Efficiency in Manufacturing Systems*. Springer Science & Business Media.

8 U.S. Department of Energy (2015). United States Electricity Industry Primer. https://www.energy.gov/sites/prod/files/2015/12/f28/united-states-electricity-industry-primer.pdf.

9 Molburg, J. C., Kavicky, J. A., and Picel, K. C. (2008). The design, construction, and operation of long-distance high-voltage electricity transmission technologies. Argonne, IL. https://doi.org/10.2172/929262.

10 U.S. Energy Information Administration (EIA). Delivery to consumers. https://www.eia.gov/energyexplained/electricity/delivery-to-consumers.php (accessed 14 December 2020).

11 U.S. Department of Energy (2004). *Improving Steam System Performance: A Sourcebook for Industry Second Edition* (ed. 3), 2. U.S. Department of Energy, The Office of Energy Efficiency and Renewable Energy (EERE) https://energy.gov/sites/prod/files/2014/05/f15/steamsourcebook.pdf.

12 Einstein, D., Worrell, E., and Khrushch, M. (2001). Steam systems in industry: energy use and energy efficiency improvement potentials publication date steam systems in industry: energy use and energy efficiency improvement potentials. https://escholarship.org/uc/item/3m1781f1 (accessed 18 December 2020).

13 BOGE KOMPRESSOREN (2020). 2.1. The advantages of compressed air. http://www.drucklufttechnik.de/www/temp/e/drucklfte.nsf/70479b556c5dc3b4c125662900334cdf/43f985f4de9f66c7c12566250057deec?OpenDocument (accessed 18 December 2020).

14 EIA (2021). 2018 Manufacturing Energy Consumption Survey (MECS). https://www.eia.gov/consumption/manufacturing/pdf/MECS%202018%20Results%20Flipbook.pdf (accessed 18 December 2020).

15 U.S. Energy Information Administration (2017). Manufacturing Energy Consumption Survey (MECS). https://www.eia.gov/consumption/manufacturing/reports/2014/enduse_intensity/ (accessed 17 December 2020).

16 Department of Energy. Combined heat and power basics. https://www.energy.gov/eere/amo/combined-heat-and-power-basics (accessed 23 December 2020).

17 U.S. Environmental Protection Agency (2015). Fuel and carbon dioxide emissions savings calculation methodology for combined heat and power systems. https://www.epa.gov/sites/production/files/2015s-07/documents/fuel_and_carbon_dioxide_emissions_savings_calculation_methodology_for_combined_heat_and_power_systems.pdf.

18 Andrew J. Gellman Group S.O.P. Chapter 4 Drilling Machines. http://uhv.cheme.cmu.edu/procedures/machining/ch4.pdf (accessed 23 December 2020).

19 U.S. Energy Information Administration (2021). Manufacturing Energy Consumption Survey (MECS), Table 5.2 End Uses of Fuel Consumption. https://www.eia.gov/consumption/manufacturing/data/2018/pdf/Table5_2.pdf.

20 Brundage, M.P., Chang, Q., Li, Y., et al. (2013). Energy efficiency management of an integrated serial production line and HVAC system. *IEEE International Conference on Automation Science and Engineering*, 2013, 634–639. https://doi.org/10.1109/CoASE.2013.6653896.

21 U.S. Energy Information Administration (2020). Annual Energy Outlook 2020: with projections to 2050. https://www.eia.gov/outlooks/aeo/pdf/aeo2020.pdf.

22 U.S. Department of Energy (2012). U.S. manufacturing energy use and greenhouse gas emissions analysis. https://www.energy.gov/sites/default/files/2013/11/f4/energy_use_and_loss_and_emissions.pdf.

23 EIA (2017). 2014 Manufacturing Energy Consumption Survey (MECS) - Expenditures for Purchased Energy Sources. https://www.eia.gov/consumption/manufacturing/data/2014/pdf/table7_9.pdf.

24 Apostolos, F., Alexios, P., Georgios, P. et al. (2013). Energy efficiency of manufacturing processes: a critical review. *Procedia CIRP* 7: 628–633. https://doi.org/10.1016/j.procir.2013.06.044.

25 Vijayaraghavan, A. and Dornfeld, D. (2010). Automated energy monitoring of machine tools. *CIRP Ann. Meanuf. Technol.* https://doi.org/10.1016/j.cirp.2010.03.042.

26 Zhou, L., Li, J., Li, F. et al. (2016). Energy consumption model and energy efficiency of machine tools: a comprehensive literature review. *J. Clean. Prod.* https://doi.org/10.1016/j.jclepro.2015.05.093.

27 Kroll, L., Blau, P., Wabner, M. et al. (2011). Lightweight components for energy-efficient machine tools. *CIRP J. Manuf. Sci. Technol.* https://doi.org/10.1016/j.cirpj.2011.04.002.

28 Zhao, L., Chen, W., Ma, J., and Yang, Y. (2008). Structural bionic design and experimental verification of a machine tool column. *J. Bionic Eng.* 5: 46–52. https://doi.org/10.1016/S1672-6529(08)60071-2.

29 Cho, S.K., Kim, H.J., and Chang, S.H. (2011). The application of polymer composites to the table-top machine tool components for higher stiffness and reduced weight. *Compos. Struct.* https://doi.org/10.1016/j.compstruct.2010.08.030.

30 Bi, Z.M. and Wang, L. (2012). Optimization of machining processes from the perspective of energy consumption: a case study. *J. Manuf. Syst.* 31 (4): 420–428. https://doi.org/10.1016/j.jmsy.2012.07.002.

31 Mori, M., Fujishima, M., Inamasu, Y., and Oda, Y. (2011). A study on energy efficiency improvement for machine tools. *CIRP Ann. Manuf. Technol.* https://doi.org/10.1016/j.cirp.2011.03.099.

32 Ur-Rehman, R., Caro, S., Chablat, D., and Wenger, P. (2010). Multi-objective path placement optimization of parallel kinematics machines based on energy consumption, shaking forces and maximum actuator torques: application to the Orthoglide. *Mech. Mach. Theory* https://doi.org/10.1016/j.mechmachtheory. 2010.03.008.

33 Chang, Q., Xiao, G., Biller, S., and Li, L. (2013). Energy saving opportunity analysis of automotive serial production systems (March 2012). *IEEE Trans. Autom. Sci. Eng.* 10 (2): 334–342. https://doi.org/10.1109/TASE.2012.2210874.

34 Li, L., Chang, Q., and Ni, J. (2009). Data driven bottleneck detection of manufacturing systems. *Int. J. Prod. Res.* 47 (18): 5019–5036.

35 Alden, J.M., Burns, L.D., Costy, T. et al. (2006). General motors increases its production throughput. *Interfaces (Providence).* 36 (1): 6–25. https://doi.org/ 10.1287/inte.1050.0181.

36 Subramaniyan, M., Skoogh, A., Salomonsson, H. et al. (2018). A data-driven algorithm to predict throughput bottlenecks in a production system based on active periods of the machines. *Comput. Ind. Eng.* 125: 533–544. https://doi.org/10.1016/ j.cie.2018.04.024.

37 L. Li, Q. Chang, J. Ni, and S. Biller, "Real time production improvement through bottleneck control," *Int. J. Prod. Res.*, 2009, https://doi.org/10.1080/ 002075408022 44240.

38 Brundage, M.P., Chang, Q., Li, Y., et al. (2014). Utilizing energy opportunity windows and energy profit bottlenecks to reduce energy consumption per part for a serial production line. *2014 IEEE International Conference on Automation Science and Engineering (CASE)*, August 2014, 461–466. https://doi.org/10.1109/ CoASE.2014.6899366.

39 Dababneh, F., Li, L., Shah, R., and Haefke, C. (2018). Demand response-driven production and maintenance decision-making for cost-effective manufacturing. *J. Manuf. Sci. Eng.* 140 (6): 061008. https://doi.org/10.1115/1.4039197.

40 T. Yu, C. Zhu, Q. Chang, and J. Wang, "Imperfect corrective maintenance scheduling for energy efficient manufacturing systems through online task allocation method," *J. Manuf. Syst.*, 2019, https://doi.org/10.1016/j. jmsy.2019.11.002.

41 S&C (2018). S&C's 2018 State of Commercial & Industrial Power Reliability Report. https://www.sandc.com/globalassets/sac-electric/documents/sharepoint/ documents---all-documents/technical-paper-100-t120.pdf?dt=637532366053266924 &utm_source=web&utm_medium=digital&utm_campaign=Reliability&utm_ content=2018report.

42 U.S. Environmental Protection Agency. What Is CHP?. https://www.epa.gov/chp/ what-chp (accessed 20 December 2020).

43 Dababneh, F., Li, L., and Sun, Z. (2016). Peak power demand reduction for combined manufacturing and HVAC system considering heat transfer characteristics. *Int. J. Prod. Econ.* 177: 44–52. https://www.sciencedirect.com/ science/article/pii/S0925527316300329?casa_token=ppindgHnO4UAAAAA:

x9t34cWES4PwRJfE2jbd9mrjKVyr8BuyeMSwAeC0NSkVyORk5Ob50s2nU3d44 KFUu21hny3kLA (accessed 20 December 2020).

44 Hosni, M., Jones, B., Xu, H. A. Transactions, and undefined (1999). Experimental results for heat gain and radiant/convective split from equipment in buildings. pdfs. semanticscholar.org, https://pdfs.semanticscholar.org/64fb/ 02ecb27509ca20e4c05a2e6eb468f97d3b05.pdf (accessed 20 December 2020).

45 Spitler, J., Fisher, D., and Pedersen, C. (1997). The radiant time series cooling load calculation procedure. https://www.osti.gov/biblio/349986 (accessed 20 December 2020).

46 U.S. Energy Information Administration (2012). Annual energy review 2011. https:// www.eia.gov/totalenergy/data/annual/pdf/aer.pdf (accessed 20 December 2020).

47 Chupka, M.W., Earle, R., Fox-Penner, P., and Hledik, R. (2008). Transforming America's power industry: the investment challenge 2010–2030. *Brattle Gr.*, p. 2.

48 Palensky, P. and Dietrich, D. (2011). Demand side management: demand response, intelligent energy systems, and smart loads. *IEEE Trans. Ind. Informatics* 7 (3): 381–388. https://doi.org/10.1109/TII.2011.2158841.

49 Chiu, W.-Y., Sun, J., and Poor, H.V. (2013). Energy imbalance management using a robust pricing scheme. *IEEE Trans. Smart Grid* 4 (2): 896–904. https://doi.org/ 10.1109/TSG.2012.2216554.

50 Southern California Edison. No 0000000392 schedule TOU-GS-1: time-of-use general service. https://library.sce.com/content/dam/sce-doclib/public/regulatory/ tariff/electric/schedules/general-service-&-industrial-rates/ ELECTRIC_SCHEDULES_TOUs-GS-1.pdf (accessed 22 December 2020).

51 U.S. Energy Information Administration. Energy efficiency and conservation. https://www.eia.gov/energyexplained/use-of-energy/efficiency-and-conservation. php (accessed 21 December 2020).

52 Department of Energy. Energy efficiency policies and programs. https://www. energy.gov/eere/slsc/energy-efficiency-policies-and-programs (accessed 21 December 2020).

53 Energy Star. ENERGY STAR plant certification. https://www.energystar.gov/ industrial_plants/earn-recognition/plant-certification (accessed 21 December 2020).

54 U.S. Environmental Protection Agency (2016). Guidelines for energy management. https://www.energystar.gov/sites/default/files/buildings/tools/Guidelines-fors-Energys-Managements-6_2013.pdf?07cfs-f522 (accessed 21 December 2020).

55 Department of Energy (2020). Tax credits, rebates & savings. https://www.energy. gov/savings/dsire-age (accessed 23 December 2020).

56 DSIRE (2020). Sales tax exemption for manufacturing facilities. https://programs. dsireusa.org/system/program/detail/2745/sales-tax-exemption-for-manufacturing-facilities (accessed 23 December 2020).

57 U.S. Energy Information Administration (2019). Annual Electric Power Industry Report. https://www.eia.gov/electricity/data/eia861/ (accessed 23 December 2020).

58 U.S. Energy Information Administration (EIA). Real-time operating grid. https://www.eia.gov/beta/electricity/gridmonitor/dashboard/custom/pending (accessed 13 December 2020).

59 U.S. Federal Energy Regulatory Commission (2017). Assessment of Demand Response and Advanced Metering Report. https://www.ferc.gov/sites/default/files/2020-04/DR-AM-Report2017_0.pdf (accessed 23 December 2020).

60 U.S. Department of Energy (2006). Benefits of demand response in electricity markets and recommendations for achieving them. https://www.energy.gov/sites/default/files/oeprod/DocumentsandMedia/DOE_Benefits_of_Demand_Response_in_Electricity_Markets_and_Recommendations_for_Achieving_Them_Report_to_Congress.pdf (accessed 23 December 2020).

61 Oregon Public Utility Commission. Utility regulation. https://www.oregon.gov/puc/utilities/Pages/Energy-Electric-Natural-Gas.aspx (accessed 23 December 2020).

62 Aalami, H.A., Moghaddam, M.P., and Yousefi, G.R. (2010). Demand response modeling considering interruptible/curtailable loads and capacity market programs. *Appl. Energy* 87 (1): 243–250. https://doi.org/10.1016/j.apenergy.2009.05.041.

63 U.S. Federal Energy Regulatory Commission (2019). Assessment of demand response and advanced metering. https://www.ferc.gov/sites/default/files/2020-04/DR-AM-Report2019_2.pdf (accessed 24 December 2020).

64 Southern California Edison. Demand response programs. https://www.sce.com/business/demand-response (accessed 24 December 2020).

Part II

Mathematical Tools and Modeling Basics

3

Mathematical Tools

In this chapter, the fundamentals of probability theory and application scenarios of several common probability distributions used in manufacturing system modeling are introduced, followed by the demonstration of a modeling tool, Petri net, for the visual representation of manufacturing systems and discussions on optimization models and representative algorithms. In Section 3.1, the basics of probability theory are presented, followed by the characteristics of commonly used distributions as well as the properties of Bernoulli and Markov processes. Section 3.2 presents a basic overview of the formal definition of the Petri net, the properties and modeling capabilities of the classical Petri nets. Meanwhile, examples are also provided to illustrate two-timed Petri nets, i.e. deterministic timed Petri net and stochastic one. Compared with the classical Petri nets, the concept of execution time is involved in the timed Petri net structures. In addition, in Section 3.3, the fundamentals of an optimization model are demonstrated, and the differences between local and global optima are discussed, followed by the basic concepts in optimization method classification. Introduction to optimization with metaheuristics algorithms, i.e. genetic algorithm and particle swarm optimization, are also given in this section.

3.1 Probability

3.1.1 Fundamentals of Probability Theory

3.1.1.1 Basics of Probability Theory

Probability theory is a branch of mathematics concerned with the analysis of random phenomena. In terms of a random event, a set of possible outcomes may be produced if a random event is repeated many times, and the event outcome cannot be determined before its occurrence. Hence, the actual outcome is classically determined by the chances that one observes the event at a certain time [1]. In a general

Sustainable Manufacturing Systems, First Edition. Lin Li and MengChu Zhou.
© 2023 The Institute of Electrical and Electronics Engineers, Inc.
Published 2023 by John Wiley & Sons, Inc.

sense, probability is the frequency of the occurrence of an event, and the range of probability is between 0 and 1. In particular, a fundamental concept of a probability model is the random experiment, which is a mechanism that produces a set of distinct possible outcomes [2]. A probability model mainly consists of three elements, i.e. sample space, event, and probability.

- The sample space (Ω) is a collection of all possible elementary outcomes $s_1, s_2, ...$, and s_n, where n (>1) is the number of outcomes. The elementary outcome denotes the occurrence of exactly one of the outcomes when the random experiment is performed [3]. Mathematically, this relationship can be written as

$$\Omega = \{ s_1, s_2, ..., s_n \} \tag{3.1}$$

  For example, in a production system with two machines, the sample space can be written as

$$\Omega = \{ s_1 = (\text{up, up}), s_2 = (\text{up, down}), s_3 = (\text{down, up}), s_4 = (\text{down, down}) \}$$

  where a machine is assumed to be up (working) or down (not working) only.
- An event (E) is the occurrence of either a single prescribed outcome or any number of possible favorable outcomes of an experiment; in other words, an event is a subset of the sample space. For example, as in the above example, one can define events A and B as "the second machine is up," and "at least one machine is up," which can be mathematically expressed as

$$A = \{\text{The second machine is up}\} = \{s_1, s_3\},$$

$$B = \{\text{At least one mahcine is up}\} = \{s_1, s_2, s_3\}.$$

- Probability (P) of an event is determined by the relative frequency of the event. Suppose that an experiment is conducted N times under the same condition and let N_A be the number of times event A occurs during experimental trials. Then the probability of A, denoted as $P(A)$, is defined as

$$P(A) = \lim_{N \to \infty} \frac{N_A}{N} \tag{3.2}$$

In mathematics, a limit (lim) is the value that a function "approaches" as the input "approaches" some value. Limits are essential to calculus and mathematical analysis and are used to define continuity, derivatives, and integrals.

Specifically, given a collection of k possible elementary outcomes $s_1, s_2, ...$, and s_k, in each of which event A occurs, if the probability of s_i is $P(s_i)$, we have the probability of event A as

$$P(A) = \sum_{i=1}^{k} P(s_i) \tag{3.3}$$

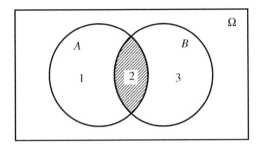

Figure 3.1 Illustration of Venn diagram.

In addition, for two events A and B in the same sample space, new events can be defined through set operations.

$A \cup B$ denotes the union of A and B, which contains all the elementary outcomes included either in event A or event B.

$A \cap B$ denotes the intersection of A and B, which contains all the elementary outcomes included in both events A and B.

$A - B$ denotes the difference of A and B, which contains all the elementary outcomes included in event A but not in event B.

Venn diagrams are usually used to visualize the relationships between a finite collection of different sets and sample space. In a Venn diagram, the sample space is represented by the outer rectangle; the elementary outcomes can be depicted as points in the plane, and the events can be represented as regions inside closed curves (as referred to as the circles inside the rectangle). Figure 3.1 shows an example of a typical Venn diagram.

As demonstrated in Figure 3.1, events A and B are represented by circles drawn inside the rectangle, and the region outside the circle represents the complement of the set. Particularly, $A \cup B$ is the addition of regions 1, 2, and 3, $A \cap B$ is the overlapping region of two circles (i.e. region 2), and $A - B$ is depicted as the region 1.

Specifically, if the intersection of A and B does not contain any elementary outcome, i.e. $A \cap B = \emptyset$, events A and B are mutually exclusive. Figure 3.2 shows a representative Venn diagram of mutually exclusive events A and B. Thus, the probability of $A \cap B$ is 0, i.e.

$$P(A \cap B) = 0 \tag{3.4}$$

3.1.1.2 Axioms of Probability Theory

The probability theory is built upon some axioms as the foundations. The axioms of probability include

1) For any event A belonging to the sample space, its probability is between 0 and 1.

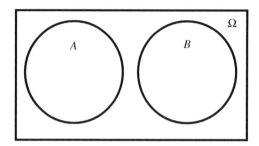

Figure 3.2 Venn diagram of mutually exclusive events.

$$0 \leq P(A) \leq 1 \tag{3.5}$$

2) For the sample space Ω, its probability equals 1.

$$P(\Omega) = 1 \tag{3.6}$$

3) For events A_1, A_2, ..., and A_n that are mutually exclusive in the same sample space, the probability of their union must be the summations of their probability.

$$P(A_1 \cup A_2 \cup \cdots \cup A_n) = P(A_1) + P(A_2) + \cdots + P(A_n) \tag{3.7}$$

According to the above three axioms and the definition of probability, some inferences can be drawn from probability theory.

For two events A_1 and A_2, which are mutually exclusive and their union is the sample space Ω, event A_2 is regarded as the complement of event A_1, which can be denoted as $\overline{A}_1$.

$$A_2 = \overline{A}_1 \tag{3.8}$$

Figure 3.3 demonstrates the complement rule for probability using a Venn diagram. The complement rule states that the summation of the probabilities of an event and its complement must equal 1. Mathematically, this relationship can be expressed as

$$P(A_2) = P(\overline{A}_1) = 1 - P(A_1) \tag{3.9}$$

4) For two events A_1 and A_2, if A_2 is included or equal to A_1, the relation between A_1 and A_2 can be mathematically expressed as

$$A_2 \subseteq A_1 \tag{3.10}$$

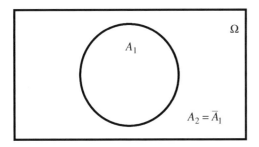

Figure 3.3 Venn diagram of complement.

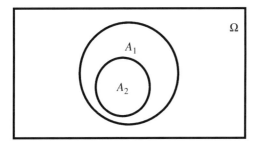

Figure 3.4 Venn diagram of inclusion.

Figure 3.4 demonstrates the Venn diagram of inclusion, and the corresponding probability relationship can be expressed as

$$P(A_1) \geq P(A_2) \tag{3.11}$$

In retrospect, the third axiom of probability addresses the probability of the union of mutually exclusive events. However, suppose that two events A and B are not mutually exclusive, the probability of their union follows the general probability addition rule, i.e.

$$P(A \cup B) = P(A) + P(B) - P(A \cap B) \tag{3.12}$$

where $A \cap B$ is the intersection between the two events. A demonstrative Venn diagram is illustrated in Figure 3.5.

As demonstrated in Figure 3.5, in the case of computing the probability of the union of two events A and B, which are not mutually exclusive (i.e. there is overlap between two events); when $P(A)$ and $P(B)$ are added, the probability of their intersection is counted twice. Hence, the $P(A \cap B)$ needs to be subtracted to compensate for that double addition.

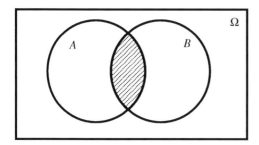

Figure 3.5 Venn diagram of addition.

For multiple events $A_1, A_2, ..., $ and A_n ($n \geq 3$) that are not mutually exclusive, the general formula for the probability of multiple non-mutually exclusive events can be expressed as

$$P\left(\bigcup_{i=1}^{n} A_k\right) = \sum_{i=1}^{n} P(A_i) - \sum_{1 \leq j < k \leq n} P(A_j \cap A_k) + \sum_{1 \leq j < k < l \leq n} P(A_j \cap A_k \cap A_l) - \cdots$$
$$+ (-1)^{n-1} P(A_1 \cap A_2 \cap \cdots \cap A_n)$$

$$(3.13)$$

3.1.1.3 Conditional Probability and Independence

In probability theory, the probability of event A occurring, given that another event B has already occurred, is referred to as conditional probability. The notation for the probability of A given B is written as $P(A \mid B)$. In this case, since the occurrence of event B is known (i.e. $P(B) > 0$), the sample space is narrowed down to event B instead of Ω, as shown in Figure 3.6. The conditional probability of A given B is defined by

$$P(A \mid B) = \frac{P(A \cap B)}{P(B)}$$

$$(3.14)$$

The conditional probability makes it possible to calculate the probability of the intersection of two events, which can be mathematically expressed as follows:

$$P(A \cap B) = P(A) \cdot P(B \mid A) = P(B) \cdot P(A \mid B)$$

$$(3.15)$$

If the occurrence of one event does not affect the probability that the other event occurs, these two events are independent. Mathematically, events A and B are independent if

$$P(A \mid B) = P(A)$$

$$(3.16)$$

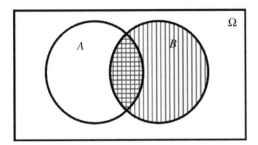

Figure 3.6 Venn diagram of conditional probability.

According to (3.15) and (3.16), the probability of the intersection of independent events A and B is

$$P(A \cap B) = P(A) \cdot P(B) \tag{3.17}$$

Equation (3.17) is another alternative definition of two independent events. Particularly, for multiple events $A_1, A_2, ..., A_n$ ($n \geq 3$), they are independent if any subsets of them are independent, which can be expressed as follows:

$$P(A_i \cap A_j) = P(A_i) \cdot P(A_j)$$
$$P(A_i \cap A_j \cap A_k) = P(A_i) \cdot P(A_j) \cdot P(A_k) \tag{3.18}$$
$$......$$
$$P(A_1 \cap A_2 \cap \cdots \cap A_n) = P(A_1) \cdot P(A_2) \cdot \cdots \cdot P(A_n)$$

where $1 \leq i < j < k \cdots \leq n$.

3.1.1.4 Total Probability Theorem

Sometimes in practice, one is often interested in calculating the probabilities based on certain available information. Hence, the probability calculation of an event is usually based on its conditional probabilities and the probabilities of those conditions. For example, the sample space Ω can be partitioned into n mutually exclusive events, $B_1, B_2, ..., $ and B_n, as shown in Figure 3.7. Suppose that another event A also belongs to the same sample space Ω, what is the probability that event A occurs?

According to the third axiom and (3.15), the probability of event A can be calculated as

$$
\begin{aligned}
P(A) &= P(A \cap B_1) + P(A \cap B_2) + \cdots + P(A \cap B_n) \\
&= P(B_1) \cdot P(A \mid B_1) + \cdots + P(B_n) \cdot P(A \mid B_n) \\
&= \sum_{i=1}^{n} P(B_i) \cdot P(A \mid B_i)
\end{aligned} \tag{3.19}
$$

Equation (3.19) is known as the total probability theorem.

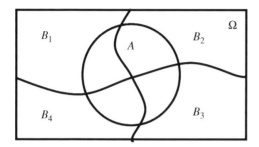

Figure 3.7 The Venn diagram of total probability.

3.1.1.5 Bayes' Law

Recall the sample space and events shown in Figure 3.7. If it is known that event A has already occurred, then what is the probability that event B_i occurs given the knowledge about event A, i.e. what is the conditional probability of $P(B_i \mid A)$?

According to (3.14), $P(B_i \mid A)$ can be written as

$$P(B_i \mid A) = \frac{P(A \cap B_i)}{P(A)} \tag{3.20}$$

Considering (3.15) and the total probability theorem in (3.19), we have

$$P(B_i \mid A) = \frac{P(B_i) \cdot P(A \mid B_i)}{\sum\limits_{k=1}^{n} P(B_k) \cdot P(A \mid B_k)} \tag{3.21}$$

Equation (3.21) is known as the Bayes' law.

3.1.2 Random Variables

A random variable X is a function that maps sample space Ω into the real number space R.

$$X(s) = x \subseteq \mathrm{R}, \quad s \in \Omega \tag{3.22}$$

where X is a random variable, usually represented by a capital letter, and its corresponding lowercase letter, x in this case, is one of its values. Particularly, x belongs to the real number space R, and s is a possible elementary outcome that belongs to Ω.

Random variables can usually be divided into two categories, discrete variables and continuous variables [2].

- A discrete random variable X is a function that maps events in a sample space into a finite or countably infinite set of real numbers.
- A continuous random variable X is a function that maps events in a sample space into an uncountably infinite set of real numbers.

3.1.2.1 Discrete Random Variables

The probabilities that a discrete random variable is equal to some values can be described by the probability mass function (PMF). Assume that the possible values for the discrete random variable X are $\{x_1, x_2, ..., x_n\}$. Then the PMF can be written as

$$P(X = x_k) = p(x_k) \quad k = 1, 2, ..., n \tag{3.23}$$

Recalling the second and third axioms of probability theory,

$$\sum_{k=1}^{n} p(x_k) = 1 \tag{3.24}$$

The cumulative distribution function (CDF) of a random variable X is the probability that the value of X is less or equal to a specific value x. In the discrete case, when the random variable X attains the value of x_1, ..., and x_s with the respective probabilities of $p(x_1)$, ..., and $p(x_s)$, its CDF of X is discontinuous at any x_s and it can be calculated as

$$F(x_s) = P(X \le x_s) = \sum_{k=1}^{s} p(x_k) \quad s = 1, 2, ..., n \tag{3.25}$$

Figure 3.8 shows the examples of PMF and CDF of a discrete random variable.

- **Expectation**

The weighted average of the possible values that a random variable X can take is measured by the expected value, denoted by $E[X]$ or μ. For a discrete random variable, the expected value can be calculated as

$$E[X] = \sum_{k=1}^{n} x_k \cdot p(x_k) \tag{3.26}$$

- **Variance**

The degree of dispersion of a random variable X, i.e. how far the possible values are spread out from the expected value, is measured by variance, denoted by Var (X) or σ^2. The variance is the expectation of the squared deviations of a random variable from its expected value, i.e.

$$\text{Var}(X) = E\left[(X - E[X])^2\right] \tag{3.27}$$

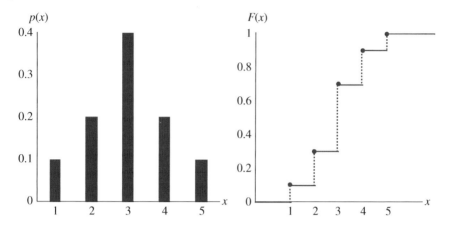

Figure 3.8 Examples of PMF (left) and CDF (right) of a discrete random variable.

For a discrete random variable X with expected value μ, the variance can be calculated as

$$\text{Var}(X) = \sum_{k=1}^{n} (x_k - \mu)^2 \cdot p(x_k) \tag{3.28}$$

In the following sections, some representative discrete random variables that are commonly used in manufacturing systems are introduced.

1) Bernoulli random variable

A Bernoulli random variable has two possible values only, i.e. it equals to 1 with the probability of p; and 0 with the probability of $1 - p$. The probability distribution of a Bernoulli random variable is called a Bernoulli distribution, and its PMF is shown as follows:

$$p(x) = \begin{cases} p, & x=1 \\ 1-p, & x=0 \end{cases} \tag{3.29}$$

The expected value and variance for a Bernoulli random variable can be calculated as

$$\begin{aligned} E(X) &= p \\ \text{Var}(X) &= p(1-p) \end{aligned} \tag{3.30}$$

In a discrete-time-based manufacturing system, such as production lines for automobiles or electronic devices, if the state of a machine in each cycle is independent of its state in all other cycles, the machine obeys the Bernoulli reliability model and it is called a Bernoulli machine. For example, the probability that machine states can be formulated as

$$p(1) = P(\text{machine is up during a cycle}) = p$$
$$p(0) = P(\text{machine is down during a cycle}) = 1 - p$$

(3.31)

Alternatively, if the quality of parts produced in a machine in each cycle is independent of the part quality in all other cycles, the machine obeys the Bernoulli quality model. The probability of the part quality can be formulated as

$$p(1) = P(\text{part quality is acceptable during a cycle}) = p$$
$$p(0) = P(\text{part is defective during a cycle}) = 1 - p$$

(3.32)

2) Binomial random variable

Consider a Bernoulli machine, which is up during a cycle with the probability of p, and operates in n cycles. Let X be the number of cycles that the machine is up, then X is called a binomial random variable with the parameters of (n, p). The probability distribution of this discrete random variable is called a binomial distribution. The PMF of a binomial random variable X can be formulated as

$$p(x) = \binom{n}{x} p^x (1-p)^{n-x} \quad x = 0, 1, ..., n$$

(3.33)

where $\binom{n}{x}$ is the binomial coefficient, which represents the number of combinations that the machine is up in x cycles out of n cycles, and it can be calculated as

$$\binom{n}{x} = \frac{n!}{x!(n-x)!} \quad x = 0, 1, ..., n$$

(3.34)

Figure 3.9 shows some demonstrative PMFs of a binomial distribution with different parameter settings.

The expected value and variance of a binomial random variable X can be calculated by

$$E(X) = np$$
$$\text{Var}(X) = np(1-p)$$

(3.35)

Example 3.1 Consider a Bernoulli machine and the corresponding part inventory with a maximum capacity of 10 parts. The probability that the machine consumes one part during each cycle is 0.9. Assume that the part inventory is initially fully loaded, and the part replenishment will be triggered if the probability that no more than three parts are in the inventory is greater than 90%. What time will the first part replenishment be triggered?

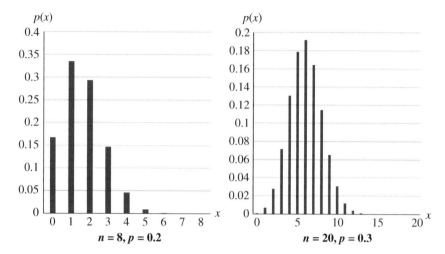

Figure 3.9 Two example PMFs of Bernoulli distribution.

Solution

Let X be the number of parts in the inventory. Since the inventory is initially fully loaded and its maximum capacity is 10, it takes at least 7 time steps for the machine to consume 7 parts. Thus, in this question, one needs to check if $P\{X \leq 3\}$ is greater than 90% when $t = 7$. If it is not, it is necessary to check the corresponding probabilities when $t = 8$, $t = 9$, …, until the minimal time is found which satisfies the part replenishment trigging condition, $P\{X \leq 3\} > 0.9$.

When $t = 7$, X is a binomial random variable with $n = 7$, and $p = 0.9$:

$$P\{X \leq 3\} = P\{X = 3\} + P\{X = 2\} + P\{X = 1\} + P\{X = 0\}$$

$$= \binom{7}{7} p^7 + 0 + 0 + 0$$

$$= 0.9^7$$

$$= 0.4783 < 0.9$$

When $t = 8$, X is a binomial random variable with $n = 8$, and $p = 0.9$:

$$P\{X \leq 3\} = P\{X = 3\} + P\{X = 2\} + P\{X = 1\} + P\{X = 0\}$$

$$= \binom{8}{7} p^7(1-p) + \binom{8}{8} p^8 + 0 + 0$$

$$= \frac{8!}{7!1!} 0.9^7 0.1 + 0.9^8$$

$$= 0.8131 < 0.9$$

When $t = 9$, X is a binomial random variable with $n = 9$, and $p = 0.9$:

$$P\{X \le 3\} = P\{X = 3\} + P\{X = 2\} + P\{X = 1\} + P\{X = 0\}$$

$$= \binom{9}{7} p^7 (1-p)^2 + \binom{9}{8} p^8 (1-p) + \binom{9}{9} p^9 + 0$$

$$= \frac{9!}{7!2!} 0.9^7 0.1^2 + \frac{9!}{8!1!} 0.9^8 0.1 + 0.9^9$$

$$= 0.9470 > 0.9$$

Hence, the first replenishment will be triggered at time $t = 9$.

3) Poisson random variable

Let X be a binomial random variable with parameters (n, p). Consider the number of trials is large ($n \to \infty$), and the probability of success in any given one trial is small ($p \to 0$), such that the expected value of np (denoted as λ) remains constant, and the binomial random variable can be approximated by a Poisson random variable with parameter λ. The PMF of a Poisson random variable is given as

$$p(x) = \frac{\lambda^x e^{-\lambda}}{x!} \quad x = 0, 1, 2, \dots \tag{3.36}$$

The probability distribution of a Poisson random variable is called the Poisson distribution. The Poisson distribution becomes more symmetric as the parameter λ increases, as illustrated in Figure 3.10.

The expected value and variance for a Poisson random variable are

$$\begin{aligned} E(X) &= \lambda \\ \text{Var}(X) &= \lambda \end{aligned} \tag{3.37}$$

Example 3.2 In a manufacturing plant, workplace accident occurs infrequently. According to the historical record, the probability of an accident on any given day is 0.003, and accidents are independent of each other.

a) What is the probability that exactly one accident happens in one year (approximately 365 days)?
b) What is the probability that more than three accidents occur in one year?

Solution

Let X be a binomial random variable with $n = 365$, and $p = 0.003$. Then, use Poisson distribution to approximate the binomial distribution.

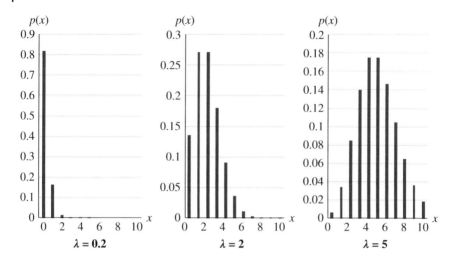

Figure 3.10 The PMFs of Poisson distribution with different value of λ.

a) Calculate the parameter of the Poisson distribution: $\lambda = np = 365 * 0.003 = 1.095$. Hence,

$$P\{X = 1\} = \frac{\lambda^x e^{-\lambda}}{x!} = 1.095 * e^{-1.095} = 0.3663$$

b) According to the complement rule and the third axiom of probability, one can obtain

$$P\{X > 3\} = 1 - P\{X \leq 3\} = 1 - \sum_{x=0}^{3} \frac{\lambda^x e^{-\lambda}}{x!} = 0.0254$$

4) Geometric random variable

Considering a Bernoulli machine with the probability p of breakdown during a cycle, what is the probability that this machine can be in normal operation during a certain period before it breaks down? To model such a situation, the geometric distribution can be used. The PMF and CDF of a geometric random variable X with parameter p are as follows:

$$p(x) = (1-p)^{x-1} \cdot p$$
$$F(x) = 1 - (1-p)^x \quad x = 1, 2, 3, \ldots \tag{3.38}$$

In addition, a survival function (also known as reliability function) is used to represent the probability that the event of interest has not yet occurred by

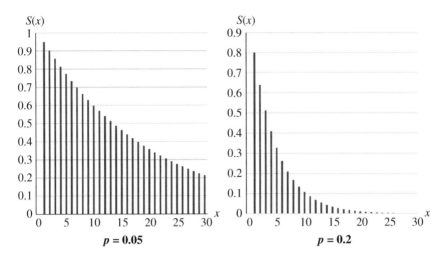

Figure 3.11 Examples of the survival function of geometric distribution.

time t. Let X denote the time until the occurrence of a machine breakdown, the survival function $S(x)$ denotes the probability that the machine can survive beyond any specific time x:

$$S(x) \equiv 1 - F(x) = P(X > x) \quad x > 0 \tag{3.39}$$

In the discrete case with a Bernoulli machine, the survival function can be written as

$$S(x) = 1 - F(x) = (1 - p)^x \quad x = 1, 2, 3, \ldots \tag{3.40}$$

Figure 3.11 shows examples of the survival functions of the geometric distributions. The x-axis is time (i.e. the number of cycles), and the y-axis is the probability that a Bernoulli machine can operate for a certain amount of time without breaking down.

The expected value and variance of a geometric random variable are

$$E(X) = \frac{1}{p}$$
$$\text{Var}(X) = \frac{1-p}{p^2} \tag{3.41}$$

Recall the definition of a Bernoulli machine. The probability of a Bernoulli machine being up during a cycle is constant and irrelevant to the machine state. In a general case, if the state transition probability of a machine is related to its state, one can model the machine with geometric distribution, as shown in

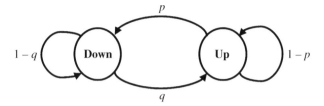

Figure 3.12 The state transition diagram of a geometric machine.

Figure 3.12. More specifically, when the machine was up during cycle $t-1$, the probabilities that it is down and up at cycle t are p and $1-p$, respectively. Similarly, when the machine was down during cycle $t-1$, it could have another pair of probabilities q (the machine is up) and $1-q$ (the machine is down) at cycle t. A machine with such a reliability model is called a geometric machine.

For a geometric machine, assuming that the machine is initially up, let the random variable X represent that its first breakdown occurs at the xth cycles. The probability of $X=x$ can be written as

$$p(x) = P(\{\text{machine is initially up and the first breakdown occurs at the } x\text{th cycle}\})$$
$$= P[\{\text{up at cycle 1}\}] \cdot P[\{\text{up at cycle 2}\}] \cdots P[\{\text{up at cycle } x-1\}]$$
$$\cdot P[\{\text{down at cycle } x\}]$$
$$= (1-p)^{x-1} \cdot p$$

$$(3.42)$$

Similarly, for a geometric machine, assuming that machine is down before the first cycle starts, and the machine operation resumes at the xth cycle, the probability of $X=x$ under this situation is

$$p(x) = (1-q)^{x-1} \cdot q \qquad (3.43)$$

Although the machine's state transition probability in each cycle is related to its state in one previous cycle, geometric random variables are considered memoryless. This is because the state transition probability does not depend on how long the machine has been in up (or down) state before the state transition occurs.

3.1.2.2 Continuous Random Variables

For a continuous random variable X, since it can take an infinite number of possible values, and the probability of any point over a continuous interval is zero. Therefore, a continuous random variable cannot be defined by a PMF. Instead, a function $f(x)$, also referred to as the probability density function (PDF), is used

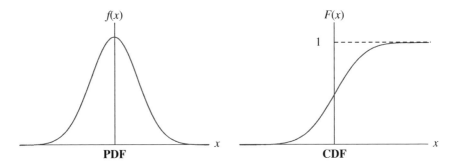

Figure 3.13 Examples of PDF and CDF for a continuous random variable.

to calculate the probability of the continuous random variable falling within a certain range of values. The PDF has the following properties:

1. $f(X) \geq 0, \quad$ for all $x \in R$
2. $\int_{-\infty}^{\infty} f(x)\mathrm{d}x = 1$
3. $P(x_1 < X \leq x_2) = \int_{x_1}^{x_2} f(x)\mathrm{d}x$ $\qquad$ (3.44)
4. $\lim_{x \to -\infty} f(x) = 0$ and $\lim_{x \to \infty} f(x) = 0$

The CDF of a continuous random variable X, i.e. $F(x)$, is a nonnegative and non-decreasing function, which can be formulated as

$$F(x) = P(X \leq x) = \int_{-\infty}^{x} f(t)\mathrm{d}x \qquad (3.45)$$

Examples of PDF and CDF of a continuous random variable are shown in Figure 3.13. For a continuous random variable X, the expected value and variance can be calculated as

$$E(X) = \int_{-\infty}^{\infty} xf(x)\mathrm{d}x \qquad (3.46)$$

$$\mathrm{Var}(X) = E\left[(X - E[X])^2\right] = \int_{-\infty}^{\infty} (x - \mu)^2 f(x)\mathrm{d}x \qquad (3.47)$$

Some continuous random variables that are commonly used in manufacturing systems are presented next.

1) Exponential random variable

An exponential random variable is often used to model the distribution of the amount of time until an event of interest, e.g. machine breakdown, occurs. The PDF and CDF of an exponential random variable X are

$$f(x) = \begin{cases} \lambda e^{-\lambda x}, & x > 0 \\ 0, & \text{otherwise} \end{cases} \qquad (3.48)$$

$$F(x) = \int_0^x \lambda e^{-\lambda x} = 1 - e^{-\lambda x} \qquad (3.49)$$

Note that sometimes geometric distribution is also used to model a similar distribution. Indeed, an exponential random variable can be regarded as a continuous version of a geometric random variable.

Example 3.3 Consider a Bernoulli machine that breaks down with the probability of λ/n. Assume that the machine is initially up and let a geometric random variable X represent the period of time until its first breakdown. Show that

$$\lim_{n \to \infty} P(X \le nt) = 1 - e^{-\lambda t}$$

Recall that in calculus, the limit of the following equation equals $e^{-\lambda x}$ as n approaches the infinity:

$$\lim_{n \to \infty} \left(1 - \frac{\lambda}{n}\right)^{nx} = e^{-\lambda x}$$

Solution

In the geometric distribution, the probability that a machine breaks down at or before the xth cycles is

$$P(X \le x) = \sum_{k=1}^{x} p \cdot (1 - p)^{k-1}$$

where p is the probability that the machine breaks down during a cycle.

Let $x = nt$ and $s = k/n$, such that

$$\lim_{n \to \infty} P(X \le nt) = \lim_{n \to \infty} \sum_{k=1}^{nt} \frac{\lambda}{n} \cdot \left(1 - \frac{\lambda}{n}\right)^{k-1}$$

$$= \lim_{n \to \infty} \frac{1}{n} \cdot \sum_{k=0}^{nt-1} \lambda \cdot \left(1 - \frac{\lambda}{n}\right)^{n \cdot \frac{k}{n}}$$

$$= \lim_{n \to \infty} \frac{1}{n} \cdot \sum_{k=0}^{nt-1} \lambda \cdot e^{-\lambda \frac{k}{n}}$$

$$= \int_0^t \lambda \cdot e^{-\lambda s} ds$$

$$= 1 - e^{-\lambda t}$$

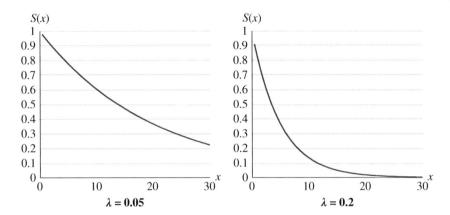

Figure 3.14 Example survival functions of the exponential distribution.

Example 3.3 implies that if X is a geometric random variable with parameter λ/n, then the random variable $T = X/n$ can be approximated by the exponential distribution when n is sufficiently large.

The survival function of an exponential distribution is

$$S(x) = P(X > x) = 1 - F(x) = e^{-\lambda x} \quad x > 0 \tag{3.50}$$

Figure 3.14 shows the examples of exponential distribution-based survival functions. Compared to the survival function of geometric distribution (as shown in Figure 3.11), the survival function pattern for exponential distribution is analogous to that of the geometric distribution.

The expected value and variance of an exponential random variable are

$$E(X) = \frac{1}{\lambda} \tag{3.51}$$

$$\text{Var}(X) = \frac{1}{\lambda^2} \tag{3.52}$$

Similar to the geometric distribution, the exponential distribution is also memoryless. In other words, the probability of an event that has not occurred by the time $t_0 + t$ given that it did not occur by time t_0 equals to the probability of an event that has not occurred during time t.

Example 3.4 Consider an exponential random variable X with the parameter λ. Prove that:

$$P(X > t_0 + t \mid X > t_0) = P(X > t)$$

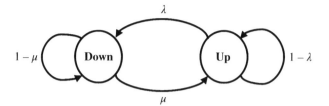

Figure 3.15 State transition diagram of an exponential machine.

Solution

$$P(X > t_0 + t \mid X > t_0) = \frac{S(t_0 + t)}{S(t_0)}$$

$$= \frac{e^{-\lambda(t_0 + t)}}{e^{-\lambda t_0}}$$

$$= e^{-\lambda t} = P(X > t_0)$$

The exponential distribution can be used to model the machine failures in manufacturing systems. The average time elapsed before the machine breaks down is generally different from the mean time to repair the failed machine. Hence, two exponential random variables with different parameters should be applied to model the probability distributions of these two types of time. As shown in Figure 3.15, an exponential random variable with parameter λ is used to model the time before the machine breaks down, and thus the average time between two machine failures is $1/\lambda$. Similarly, the mean time to repair a failed machine is $1/\mu$. A machine with such a reliability model is also referred to as an exponential machine.

2) Weibull random variable

Weibull random variables are widely used in engineering practice to approximate the lifetime distribution of a particular item. The PDF and CDF of a Weibull random variable are

$$f(x) = \begin{cases} \dfrac{k}{\zeta}\left(\dfrac{x}{\zeta}\right)^{k-1} e^{-\left(\frac{x}{\zeta}\right)^{k}}, & x > 0 \\ 0, & \text{otherwise} \end{cases} \tag{3.53}$$

$$F(x) = 1 - e^{-\left(\frac{x}{\zeta}\right)^{k}} \tag{3.54}$$

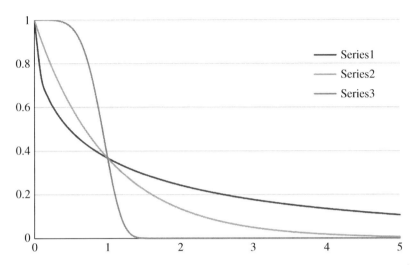

Figure 3.16 The survival function of Weibull distribution with parameter values.

where ζ is the scale parameter, and k is the shape parameter of the distribution. When $k = 1$, the Weibull distribution is an exponential distribution with the parameter of $\frac{1}{\zeta}$. Hence, the exponential distribution can be regarded as a special case of the Weibull distribution. Figure 3.16 shows the illustrative survival functions of Weibull distributions with different parameters and the effects of shape parameter k.

Example 3.5 Consider a certain type of machines in a manufacturing plant with an average operating time of T months before maintenance. Suppose that random variable T can be nicely modeled by the Weibull distribution with parameters $\zeta = 4$, and $k = 2$. If five machines are employed in the plant, what is the probability that at least four machines have been maintained in half a year?

Solution

The probability that a machine has been maintained in half a year is

$$P(T \le 6) = F(6) = 1 - e^{-\left(\frac{6}{4}\right)^2} = 0.8946$$

Let X represent the number of machines that have been maintained in half a year. Applying the binomial distribution gives

$$P(X \ge 4) = \sum_{x=4}^{5} \binom{5}{x} 0.8946^x (1 - 0.8946)^{5-x} = 0.9105$$

3.1.3 Random Process

A random process, also known as a stochastic process, is a time-varying function that maps sample space Ω to state space S_x, which is typically a set of real numbers:

$$X(s,t) = x(t) \quad s \in \Omega, \, x \in S_x, \, t \in T \tag{3.55}$$

where s is a possible outcome belonging to sample space Ω, and t is in an index of time belong to a set of time indices, i.e. T. Unlike a random variable, time t is involved in a random process. More specifically, at a specific time t', a random process becomes a random variable:

$$X(s, t = t') = X(s) \tag{3.56}$$

On the other hand, if the outcome is fixed as $s = s'$, then the random process is a deterministic function of t, which is called a realization of a random process $x(t)$:

$$X(s = s', t) = x(t) \tag{3.57}$$

In practice, a random process is usually denoted by a capital letter $X(t)$, and its realization is denoted by the corresponding lowercase letter $x(t)$. A simple example of a random process is the Bernoulli process. Consider a Bernoulli machine with probability p to be up and $(1 - p)$ to be down during each cycle. If this machine operates in multiple cycles, then this process is a Bernoulli process. In other words, a Bernoulli process is a sequence of independently and identically distributed Bernoulli random variables.

In particular, Markov processes are one of the most important classes of random processes and are widely used in the engineering field. Markov processes are used to model systems with limited memory of the past, and the underlying idea to build the model is the Markov property. Markov property indicates that the future state of the process depends only on the present state and is independent of the past. This means that given the present state of the process, one can make the best possible prediction of the future position, and no additional information from the past is required. This property induces the significant simplification and reduction of parameter quantities in system modeling [4].

Mathematically, the Markov property can be formulated as follows. Notably, for a random process $X(t)$ with $(t_0 < t_1 < \cdots < t_n) \in R$, the conditional CDF of $X(t_n)$ satisfies the following equation:

$$
\begin{aligned}
&P[X(t_n) \leq x(t_n) \,|\, X(t_{n-1}) \leq x(t_{n-1}), X(t_{n-2}) \leq x(t_{n-2}), \cdots, X(t_0) \leq x(t_0)] \\
&= P[X(t_n) \leq x(t_n) \,|\, X(t_{n-1}) \leq x(t_{n-1})]
\end{aligned}
\tag{3.58}
$$

Depending on whether the process has a discrete or continuous state space and parameter space (time), Markov processes can be classified into four categories, as

Table 3.1 The categories of Markov processes.

		State space	
		Discrete	**Continuous**
Time space	*Discrete*	Discrete-time Markov chain	Discrete-time Markov process
	Continuous	Continuous-time Markov chain	Continuous-time Markov process

shown in Table 3.1. The state space of the manufacturing systems discussed in this book is generally discrete, such as the machine states (up and down) and the number of products in the inventory. Therefore, only the discrete state Markov Chains (in both discrete time and continuous time) are introduced in detail in the following sections.

3.1.3.1 Discrete-time Markov Chain

Discrete-time Markov Chain $X(t)$ is a discrete-time, discrete-state random process whose state space S_x consists of n finite points and satisfies the Markov property:

$$P[X(t) = i|X(t-1) = j, X(t-2) = k, ..., X(1) = s]$$
$$= P[X(t) = i|X(t-1) = j]$$
$$= P_{ij} \tag{3.59}$$

$$i, j, k, ..., s \in S_x$$

where P_{ij} is called the state-transition probability from state j to state i, satisfying:

$$0 \leq P_{ij} \leq 1 \quad \forall i, j \tag{3.60}$$

$$\sum_{i=1}^{n} P_{ij} = 1 \quad \forall j \tag{3.61}$$

The state-transition probabilities can be formulated into an $n \times n$ matrix P, which is referred to as a transition matrix:

$$P = \begin{bmatrix} P_{11} & P_{12} & \cdots & P_{1n} \\ P_{21} & P_{22} & \cdots & P_{2n} \\ \vdots & \vdots & \ddots & \vdots \\ P_{n1} & P_{n2} & \cdots & P_{nn} \end{bmatrix} \tag{3.62}$$

The state-transition probabilities among different states can be visualized in a state-transition diagram, as illustrated in Figure 3.17.

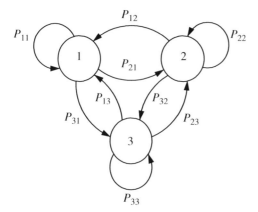

Figure 3.17 State transition diagram.

Particularly, if there is a nonnegative probability that the process can start from state i and reach state j after finite steps, j is said to be reachable from i, where $i, j \in S_x$. If any state j are reachable from any state i, the Markov chain can be called irreducible. In addition, suppose that the states are not partitioned into several sets such that all state transitions occur periodically from one set to another, then the Markov chain is aperiodic. If a Markov chain has the above two prosperities, it is said to be ergodic. For an ergodic Markov chain, there exists a unique steady state, which is independent of the initial state, and the steady-state probability can be calculated as follows:

First, based on the state transition probabilities and the probabilities of all states at $t - 1$, the probability of state i at time t can be calculated by employing the total probability theorem:

$$P[X(t) = i] = \sum_{j=1}^{n} \left\{ P_{ij} \cdot P[X(t-1) = j] \right\} \tag{3.63}$$

Once a Markov process reaches its steady state, the probability of the process at each state does not change, which gives

$$P_i = P[X(t) = i] = P[X(t+k) = i] \quad k = 1, 2, \dots \tag{3.64}$$

Let P_i denote the steady-state probability at state i. The steady-state probabilities can be obtained by solving the following linear equations:

$$P_i = \sum_{j=1}^{n} P_{ij} P_j \quad \forall i \tag{3.65}$$

$$\sum_{i=1}^{n} P_i = 1 \tag{3.66}$$

Note that (3.65) is referred to as balance equations, (3.66) as a normalization equation, and the corresponding solution to (3.65) and (3.66) describes the steady-state behavior of the discrete-time Markov chain.

Example 3.6 Consider a geometric machine with two states: up (denoted as state 1) and down (denoted as state 0). The state transition probability matrix is given as

$$P = \begin{bmatrix} 0.5 & 0.1 \\ 0.5 & 0.9 \end{bmatrix}$$

a) If the machine is up at the first step, what are the probabilities of states in the third step?

b) When the machine operates for a long period of time (reaches a steady state), what is the proportion of time spent in each state?

Solution

a) Let $P_i(t)$ denote the probability of state i at time t, and the state transition can be calculated in a matrix form:

$$\begin{bmatrix} P_{00} & P_{01} \\ P_{10} & P_{11} \end{bmatrix} \cdot \begin{bmatrix} P_0(t) \\ P_1(t) \end{bmatrix} = \begin{bmatrix} P_0(t+1) \\ P_1(t+1) \end{bmatrix}$$

Hence,

$$\begin{bmatrix} P_0(3) \\ P_1(3) \end{bmatrix} = \begin{bmatrix} P_{00} & P_{01} \\ P_{10} & P_{11} \end{bmatrix} \cdot \begin{bmatrix} P_0(2) \\ P_1(2) \end{bmatrix}$$

$$= \begin{bmatrix} P_{00} & P_{01} \\ P_{10} & P_{11} \end{bmatrix}^2 \cdot \begin{bmatrix} P_0(1) \\ P_1(1) \end{bmatrix}$$

$$= \begin{bmatrix} 0.5 & 0.1 \\ 0.5 & 0.9 \end{bmatrix}^2 \cdot \begin{bmatrix} 0 \\ 1 \end{bmatrix}$$

$$= \begin{bmatrix} 0.14 \\ 0.86 \end{bmatrix}$$

b) Denote the steady-state probability at states 0 and 1 as P_0 and P_1, respectively. Applying (3.65) and (3.66) gives

$$P_0 = 0.5P_0 + 0.1P_1$$
$$P_1 = 0.5P_0 + 0.9P_1$$
$$P_0 + P_1 = 1$$

Then solving them yields:

$$P_0 = \frac{1}{6}, \quad P_1 = \frac{5}{6}$$

3.1.3.2 Continuous-time Markov Chain

Continuous-time Markov chain is a continuous-time, discrete-space random process whose state space S_x consists of n finite points and satisfies the Markov property:

$$
\begin{aligned}
P[X(t + \delta t) = i | X(t) = j, X(l) = k, 0 \le l < t] \\
= P[X(t + \delta t) = i | X(t) = j] \\
= \nu_{ij}\delta t + o(\delta t) \\
= P_{ij}(\delta t)
\end{aligned}
\tag{3.67}
$$

$$i, j, k, \dots, s \in S_x$$

where δt is a short period of time, and $o(\delta t)$ is an infinitesimal of higher order than δt, i.e.

$$\lim_{\delta t \to 0} \frac{o(\delta t)}{\delta t} = 0 \tag{3.68}$$

The transition probability $P_{ij}(\delta t)$ represents the probability that a process is currently in state j and it will be in state i in a δt time later. $P_{ij}(\delta t)$ is a function of the time interval δt, and the coefficient ν_{ij} is a constant referred to as the transition rate from state j to state i.

$$0 \le P_{ij}(\delta t) \le 1 \quad \forall i, j \tag{3.69}$$

$$\sum_{i=1}^{n} P_{ij}(\delta t) = 1 \quad \forall j \tag{3.70}$$

The state transition rates can be written as an $n \times n$ matrix ν, which is referred to as a transition rate matrix, or infinitesimal generator:

$$
\nu = \begin{bmatrix}
\nu_{11} & \nu_{12} & \cdots & \nu_{1n} \\
\nu_{21} & \nu_{22} & \cdots & \nu_{2n} \\
\vdots & \vdots & \ddots & \vdots \\
\nu_{n1} & \nu_{n2} & \cdots & \nu_{nn}
\end{bmatrix}
\tag{3.71}
$$

The state transition rates satisfy the following conditions:

$$\nu_{ii} \leq 0 \quad \forall i \tag{3.72}$$

$$\nu_{ij} \geq 0 \quad \forall i \neq j \tag{3.73}$$

$$\sum_{i=1}^{n} \nu_{ij} = 0 \quad \forall i \tag{3.74}$$

The continuous-time Markov chain is ergodic if the infinitesimal generator admits a flow from each state to any other state either directly or indirectly. Similarly, an ergodic Markov chain has a unique steady state. The details of the steady state are illustrated as follows.

If the current probability of a state is known, the future probability of the state after time δt can be calculated by employing the total probability theorem:

$$P[X(t + \delta t) = i] = \sum_{j=1}^{n} \{P[X(t + \delta t) = i \mid X(t) = j] \cdot P[X(t) = j]\} \tag{3.75}$$

Denote $P_j(t) = P[X(t) = j]$ for simplicity, the preceding equation can be rewritten as

$$
\begin{aligned}
P_i(t + \delta t) &= \sum_{j=1}^{n} \left[\nu_{ij}\delta t + o(\delta t) \right] P_j(t) \\
&= P_i(t) \left[\nu_{ii}\delta t + o(\delta t) \right] + \sum_{j \neq i} \left[\nu_{ij}\delta t + o(\delta t) \right] P_j(t) \\
&= P_i(t) \left[1 - \sum_{j \neq i} \left[\nu_{ji}\delta t + o(\delta t) \right] \right] + \sum_{j \neq i} \left[\nu_{ij}\delta t + o(\delta t) \right] P_j(t)
\end{aligned}
\tag{3.76}
$$

Rearrange the equation:

$$\frac{P_i(t + \delta t) - P_i(t)}{\delta t} = -P_i(t) \sum_{j \neq i} \left[\nu_{ji} + \frac{o(\delta t)}{\delta t} \right] + \sum_{j \neq i} \left[\nu_{ij} + \frac{o(\delta t)}{\delta t} \right] P_j(t) \tag{3.77}$$

Note that at steady state ($t \to \infty$), the probability of state does not change. Denote P_i as the steady-state probability at state i, the left-hand side of (3.77) becomes

$$\lim_{t \to \infty, \, \delta t \to 0} \frac{P_i(t + \delta t) - P_i(t)}{\delta t} = \frac{P_i - P_i}{dt} = 0 \tag{3.78}$$

Hence, the right-hand side of (3.76) can be written as

$$\lim_{t \to \infty, \, \delta t \to 0} P_i(t) \sum_{j \neq i} \left[\nu_{ji} + \frac{o(\delta t)}{\delta t} \right] + \sum_{j \neq i} \left[\nu_{ij} + \frac{o(\delta t)}{\delta t} \right] P_j(t)$$

$$= -P_i \sum_{j \neq i} \nu_{ji} + \sum_{j \neq i} \nu_{ij} P_j \tag{3.79}$$

$$= 0$$

Thus, in the steady state, the flow into a state equals to the flow out of a state, as shown in (3.80). With the total probability (as shown in (3.81)), the steady-state probabilities can be calculated by solving the following equations:

$$\underbrace{P_i \sum_{j \neq i} \nu_{ji}}_{\text{Flow out of state } i} = \underbrace{\sum_{j \neq i} \nu_{ij} P_j}_{\text{Flow into state } i} \tag{3.80}$$

$$\sum_{i=1}^{n} P_i = 1 \tag{3.81}$$

Example 3.7   Consider an exponential machine with two states: up (denoted as state 1) and down (denoted as state 0). The breakdown rate and repair rate are λ and μ, respectively. What are the respective steady-state probabilities when the machine is up and down?

Solution

The infinitesimal generator is

$$\begin{bmatrix} \nu_{00} & \nu_{01} \\ \nu_{10} & \nu_{11} \end{bmatrix} = \begin{bmatrix} -\mu & \lambda \\ \mu & -\lambda \end{bmatrix}$$

At the steady state, $\nu_{10} P_0 = \nu_{01} P_1$, that is

$$\mu P_0 = \lambda P_1$$

Also, $P_0 + P_1 = 1$
Solving them yields

$$P_0 = \frac{\lambda}{\mu + \lambda}, \quad P_1 = \frac{\mu}{\mu + \lambda}$$

3.2   Petri Net

One of the main challenges of manufacturing system modeling is to identify appropriate components in the system and embed them with graphical notations for system representation, which depicts the interactions between the system of interest and its environment, or among the components. A proper system model can improve the visualizations of process synchronization, concurrent operations, inventory control, and specification of asynchronous events in various industrial automated systems such that the potential areas for system improvement can be identified and associated solutions can be proposed [5].

Petri nets, originally proposed by Carl Adam Petri in 1966 as a graphical tool and a symbolic method to describe concurrent and distributed systems, have been available as a system modeling tool and widely applied in many disciplines. Even since its conception, Petri nets have evolved into several forms to facilitate the design of discrete event systems concerning the formal description of system data and control, hierarchical structure abstraction, and physical system realization [6–11].

3.2.1 Formal Definition of Petri Net

3.2.1.1 Definition of Petri Net
A Petri net can be defined as a directed graph consisting of four fundamental components, as demonstrated in Figure 3.18, where

- *Places* are used to represent the states and conditions of system components. For example, the machine operation conditions, busy and idle, can be represented by two places. Places are depicted as circles in the directed graph.
- *Transitions* are used to represent operations, processes, activities, and/or events that may lead to state changes in system components. For example, a transition can be the initialization of the milling process. If this transition is executed, the milling machine's state is changed from idle to busy. Transitions are depicted by bars.
- *Tokens* are used to represent material, resources, information, or control. For example, tokens can depict the raw material waiting to be processed by a manufacturing system, or workers required to operate a machine. In addition, tokens can be simply treated as markers to indicate the current states of system components. Tokens are pictured by dots and deposited inside places. The number of tokens in each place may vary after the execution of a Petri net.

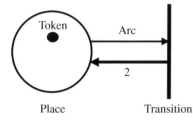

Figure 3.18 Illustration of fundamental components in Petri net.

- *Arcs* are used to represent the flow directions of material, resource, information, or control. Arcs are represented by directed line segments, which connect places to transitions or transitions to places. Note that arcs cannot be used to build connections between two places or two transitions. The number of arcs running from a place to a transition, or in the opposite direction, should be nonnegative integers. If there are multiple arcs connecting a place and a transition, it should be visualized using a bold arc with a number, where the number represents the number of arcs and is referred to as the arc weight. For example, the bold arrow in Figure 3.18 denotes two arcs running from a transition to a place.

Petri net can be mathematically defined as a five-tuple $Z = (P, T, I, O, M)$:

- $P = \{p_1, p_2, ..., p_n\}$ is the set of places, and n is the total number of places in Petri net.
- $T = \{t_1, t_2, ..., t_m\}$ is the set of transitions, and m denotes the total number of transitions.

I is an $n \times m$ matrix, called an input function, which represents the arcs connecting places to transitions. Its value at the ith row and jth column, denoted as $I(p_i, t_j)$, represents the number of arcs directly connecting place $p_i \in P$ to transition $t_j \in T$. If there is no direct connection from place p_i to transition t_j, then $I(p_i, t_j) = 0$. Otherwise, if $I(p_i, t_j) > 0$, p_i is called an input place of t_j.

O is an $n \times m$ matrix, called an output function, which represents the arcs connecting transitions to places. Similarly, the value of $O(p_i, t_j)$ represents the number of arcs that directly connect transition t_j to place p_i. If $O(p_i, t_j) > 0$, p_i is called an output place of t_j.

$M = \{m_1, m_2, ..., m_n\}$ is a vector of nonnegative integers and referred to as a marking, where m_i, also denoted as $M(p_i)$, represents the total number of tokens in $p_i \in P$.

The four-tuple (P, T, I, O) without the information about marking is called Petri net structure. With the addition of M, the five-tuple Z is called a marked Petri net. By introducing tokens into places and moving tokens through transitions, a Petri net can be used to model the discrete-event dynamic behaviors of its modeled manufacturing system. The initial marking of a Petri net is denoted as M_0, similarly, the kth marking is denoted as M_k.

Example 3.8   Given a Petri net in Figure 3.19 by using a five-tuple.

Solution

The marked Petri net can be formulated as $Z = (P, T, I, O, M_0)$, where

$$P = \{p_1, p_2, p_3, p_4\}, \quad T = \{t_1, t_2\},$$

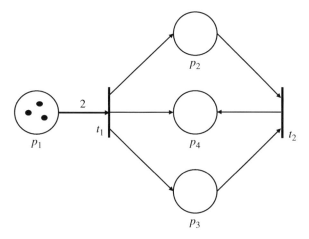

Figure 3.19 The Petri net for Example 3.8.

$$
I = \begin{bmatrix} 2 & 0 \\ 0 & 1 \\ 0 & 1 \\ 0 & 0 \end{bmatrix}, \quad O = \begin{bmatrix} 0 & 0 \\ 1 & 0 \\ 1 & 0 \\ 1 & 1 \end{bmatrix}, \quad M_0 = \begin{bmatrix} 3 \\ 0 \\ 0 \\ 0 \end{bmatrix}
$$

Execution Rules of Petri Net With the defined structure of a Petri net and the initial marking, the next question is how to move the tokens in the Petri net. The execution rules of Petri net can be divided into an enabling rule and firing rule.

Enabling rule: At a given marking M_k, transition $t_j \in T$ is enabled if for all $p_i \in P$, $M_k(p_i) \geq I(p_i, t_j)$.

The enabling rule indicates that transition t_j can be enabled only if all input places of t_j have enough tokens. For example, if a transition t_j represents an assembly operation, all its input places stand for the states of part inventories, and the number of tokens in each input place specifies the quantity of a particular part in that inventory. The enabling rule implies that an operation can occur (e.g. the starting of assembly) only if all necessary conditions are satisfied (e.g. enough parts stored in inventory).

Firing rule: If transition $t_j \in T$ is enabled at a given marking M_k, t_j fires and results in a new marking M_{k+1}, where

$$
M_{k+1}(p_i) = M_k(p_i) - I(p_i, t_j) + O(p_i, t_j), \quad \forall p_i \in P
$$

The firing rule specifies how a marking should be changed after a transition fires. The marking changes involve two steps. The first step is removing tokens from all transition t_j's input places p_i, and the number of removed tokens is defined by the number of arcs from p_i to t_j, i.e. the value of $I(p_i, t_j)$. The second step is adding necessary tokens to the transition's output places p_u, and the number of added tokens equals the number of arcs from t_j to the corresponding output places, i.e. $O(p_u, t_j)$. Firing a transition refers to the occurrence of an event or operation. In the preceding example of assembly operation, if the assembly operation occurs (transition fires), a desired number of parts are removed from the part inventories (input places), converted into products, and subsequently added to the product inventory (output place).

The following Petri net concepts are often used:

A node x in Petri net refers to either a place or transition, i.e. $x \in P \cup T$.

a) A directed path is a path formed by a finite sequence of nodes.
b) A directed circuit or loop is a closed directed path, i.e. a path from one node back to itself.
c) A pure directed circuit or an elementary loop is a directed circuit such that each place in the circuit has exactly one input transition and one output transition, and each transition has exactly one input place and one output place.
d) A Petri net is strongly connected if there is a directed path connecting any two pairs of nodes.

Example 3.9 Consider the Petri net in Example 3.8. Define its marking after two firings.

Solution

Given the initial marking M_0, transition t_1 is enabled as $M_0(p_1) \geq I(p_1, t_1)$. Transition t_2 is not enabled since $M_0(p_2) < I(p_2, t_2)$ and $M_0(p_3) < I(p_3, t_2)$. After firing t_1, two tokens are removed from p_1 and one token is placed at p_2, p_3, and p_4, respectively. Hence, after the first firing, we have a new marking:

$$M_1 = \begin{bmatrix} 3 \\ 0 \\ 0 \\ 0 \end{bmatrix} - \begin{bmatrix} 2 \\ 0 \\ 0 \\ 0 \end{bmatrix} + \begin{bmatrix} 0 \\ 1 \\ 1 \\ 1 \end{bmatrix} = \begin{bmatrix} 1 \\ 1 \\ 1 \\ 1 \end{bmatrix}$$

Based on M_1, transition t_2 is enabled since $M_1(p_2) \geq I(p_2, t_2)$ and $M_1(p_3) \geq I(p_3, t_2)$. Transition t_1 is not enabled because $M_1(p_1) < I(p_1, t_1)$. After firing t_2, one token is removed from p_2 and p_3, respectively, and one token is placed at p_4. Hence, after the second firing, we have a new marking:

$$M_2 = \begin{bmatrix} 1 \\ 1 \\ 1 \\ 1 \end{bmatrix} - \begin{bmatrix} 0 \\ 1 \\ 1 \\ 0 \end{bmatrix} + \begin{bmatrix} 0 \\ 0 \\ 0 \\ 1 \end{bmatrix} = \begin{bmatrix} 1 \\ 0 \\ 0 \\ 2 \end{bmatrix}$$

The entire process can be graphically demonstrated as shown in Figure 3.20.

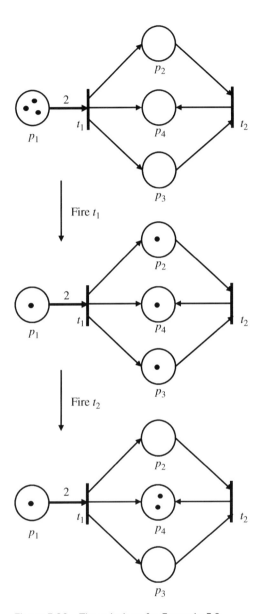

Figure 3.20 The solutions for Example 3.9.

3.2.2 Classical Petri Net

In reference to Example 3.9, marking M_{k+1} can be immediately reached from M_k, i.e. there is no enabling or firing time delay. Such a Petri net is called a classical Petri net. In the preceding section, the structure of a Petri net and the execution rules are discussed. In this section, the general methods to model manufacturing systems using classical Petri net are introduced.

A manufacturing system consists of multiple components, such as material inventories, material handling equipment, and machines. These components and their associated processes need to cooperate to ensure that the manufacturing system can work as expected. It is essential to understand the basic relationships among these processes at the modeling stage. Therefore, some fundamental structures of the Petri net are presented here, followed by two special classes of Petri nets (i.e. state machine Petri nets and marked graphs) and a systematic modeling method for complex system representation using Petri nets.

- Sequential: If the operations occur one after another, then the places and transitions form a sequential structure, as shown in Figure 3.21.
- Concurrent: If the occurrence of an event triggers multiple operations, then the places and transitions form a parallel or concurrent structure. As shown in Figure 3.22, both places p_1 and p_2 are initiated by transition t_1.
- Conflicting: If multiple operations follow the occurrence of an operation, the structure of a place connected by multiple transitions with the corresponding output places is referred to as a conflicting structure, as shown in Figure 3.23.
- Cyclic: If from the last operation in a sequence of operations, it is possible to come back to the initial one, the places and transitions form a cyclic structure, as shown in Figure 3.24.

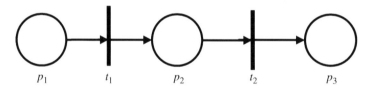

p_1 t_1 p_2 t_2 p_3

Figure 3.21 Sequential structure in Petri net.

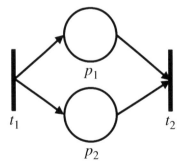

p_1

t_1 t_2

p_2

Figure 3.22 Concurrent structure in Petri net.

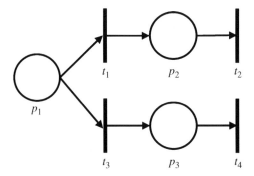

Figure 3.23 Conflicting structure in Petri net.

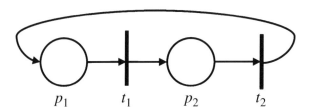

Figure 3.24 Cyclic structure in Petri net.

3.2.2.1 State Machine Petri Net

A state machine, also called a finite-state machine, is regarded as a behavioral model, which can be in exactly one of a finite number of states at any given time. A state machine can be easily modeled through a Petri net by representing each state with a place and introducing a transition to connect two places. Meanwhile, firing a transition results in the change of machine states from one to another. A state machine Petri net, presented by $Z = (P, T, I, O, M)$, has the following properties:

1) For all $p_i \in P$, $t_j \in T$, there is at most one arc connecting a place and a transition:

$$I(p_i, t_j) \leq 1 \text{ and } O(p_i, t_j) \leq 1 \tag{3.82}$$

2) For any $t_j \in T$, it has only one input place and one output place:

$$\sum_{p_i \in P} O(p_i, t_j) = \sum_{p_i \in P} I(p_i, t_j) = 1 \tag{3.83}$$

3) The machine can only be in one state at any given time:

$$\sum_{p_i \in P} M(p_i){}^{\cdot} = 1 \tag{3.84}$$

The procedures for system modeling using a state machine Petri net are summarized as follows:

- List all the possible states and represent them as individual places.
- For each state, consider all possible events that can occur when a system enters this state. Represent these identified events as transitions and insert arcs running from the place to these transitions. Subsequently, determine the succeeding state that the system should be in after the event's occurrence and insert arcs running from the transition to the consequential place. Repeat this procedure until all places are connected.
- Deposit one token into the place that represents the initial state of the system.

According to the state machine Petri net definition, there cannot be concurrent structures, but there can be conflicting structures. An example of a state machine Petri net is demonstrated using the workflow for circuit board soldering. Specifically, the workflow starts by sending the raw materials, i.e. empty circuit board and electronic component, for soldering, followed by an inspection step to verify that the board works as designed. If the circuit board passes the inspection, it is transferred to the final product inventory. Otherwise, the board needs to be returned for reprocessing and inspection until it meets the quality standard. The entire workflow is illustrated in Figure 3.25a.

To model this system using a state machine Petri net, all possible states need to be defined first. Referring to the workflow as shown in Figure 3.25a, four states can be identified, including inside raw material inventory p_1, undergo soldering p_2, under inspection p_3, and inside final product inventory p_4, as shown in Figure 3.25b. The next step is to determine all possible events that can take place in each state. For example, upon completing the board inspection, two possible events can occur, i.e. "transfer it to final product inventory" as a qualified product and "return it for reprocessing" as a defective product. Therefore, one transition should be inserted between p_3 and p_4, and another transition should be inserted between p_3 and p_2. All possible events can be added by repeating the preceding procedures. Finally, a token is deposited at p_1 to represent the initial state of the system. The final state machine Petri net is shown in Figure 3.25c.

3.2.2.2 Marked Graph
A marked graph represents the structure in which every place is associated with only one incoming arc and one outgoing arc. Contradictory to the state machine

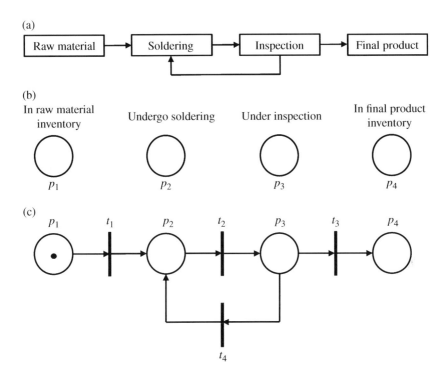

Figure 3.25 (a) The workflow for circuit board soldering, (b) determination of states, and (c) the final state machine Petri net.

Petri net, a marked graph can have concurrent structures but not conflicting structures. Mathematically, a marked graph $Z = (P, T, I, O, M)$ satisfies:

1) For all $p_i \in P$, $t_j \in T$, there is at most one arc connecting a place and a transition:

$$I(p_i, t_j) \leq 1 \text{ and } O(p_i, t_j) \leq 1 \tag{3.85}$$

2) For any $p_i \in P$, it has only one input transition and one output transition:

$$\sum_{t_i \in T} O(p_i, t_j) = \sum_{t_i \in T} I(p_i, t_j) = 1 \tag{3.86}$$

The procedures for system modeling using a marked graph are summarized as follows:

- List all the components in a system.
- For each component, identify a set of sequentially and cyclically executed events. First, treat each event as a transition and arrange the events based on

the order of execution. Then, consider all possible states for this component and define them as places. Finally, connect transitions and places with arcs.

- Merge all common transitions and arcs that represent the same events.
- Deposit tokens in places that represent the initial states of the system.

An example of the marked graph is illustrated by considering a machine composed of three components: loading robotic arm, machine tool, and unloading belt. A workflow of the demonstrative machine is shown in Figure 3.26a. At first,

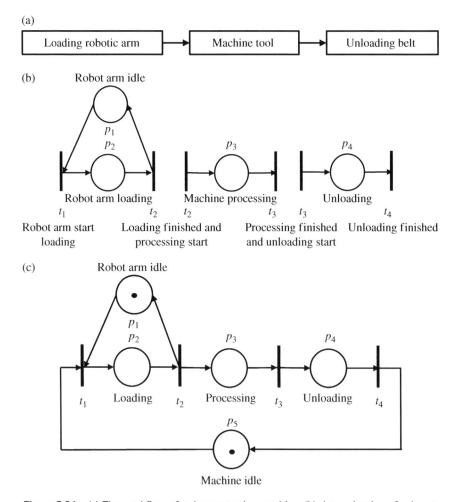

Figure 3.26 (a) The workflow of a demonstrative machine, (b) determination of sub-nets, and (c) the final marked graph.

the robotic arm loads raw material for the machine tool processing. The unloading belt removes the finished product after the machining process. At last, the machine enters the idle state and awaits the next load of raw material. To construct the marked graph, a set of sequentially and cyclically executed events can be firstly identified for each component, as shown in Figure 3.26b. For example, a robotic arm has two events: start loading and finish loading. Next, two states can be defined for the robotic arm, i.e. loading and idle. A cyclic structure (also referred to as a sub-net) can then be formed by connecting the transitions and places. All sub-nuts can be created by repeating the same procedures for all machine components. Subsequently, all common transitions are merged to draw the Petri net, and a token is deposited at p_1 to represent the initial state of the system. The final marked graph is shown in Figure 3.26c.

3.2.2.3 Systematic Modeling Methods

A manufacturing system can be much more complicated in practice, consisting of multiple machines, buffers, material handling equipment, etc. Hence, it is crucial to apply a systematic modeling method for complex manufacturing system. Three representative designing methods for constructing a Petri net with high complexity are demonstrated in the following sections: bottom-up, top-down, and hybrid approaches.

- **Bottom-up**

The bottom-up approach mainly involves two steps: decomposition and composition. In particular, a decomposition step refers to the partition of a complex system into multiple subsystems, which can be modeled as sub-nets. For example, the entire manufacturing system can be divided into several subsystems based on the job type or plant layout. Upon modeling all required subsystems, the separated sub-nets can be combined again to construct a complete Petri net in the composition step. The sub-nets can be connected by sharing places, transitions, arcs, and/or inserting necessary linking components to build a system-level Petri net structure.

For example, consider a manufacturing system consisting of five components: three machines and two buffers, as demonstrated in Figure 3.27. Machine M_1

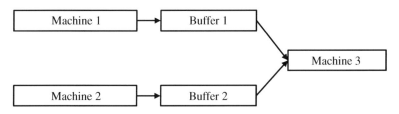

Figure 3.27 The workflow of a system with three machines two buffers.

and M_2 process raw materials and buffer B_1 and B_2 are used to store semifinished parts. Machine M_3 assembles the final products by taking one part from each buffer. Note that it is assumed the two upstream sub-workflows are considered to work in parallel.

In this case, each individual machine and buffer can be treated as a subsystem. Each machine contains a loading robot arm and an unloading belt. The corresponding sub-nets can be designed as shown in Figure 3.28.

More specifically, Figure 3.28a can be applied to represent the sub-net of either M_1 or M_2, since the machine features of M_1 and M_2 are said to be identical. The machine's internal configuration in this example is the same as in Figure 3.26c. Similarly, Figure 3.28b can be used to represent the sub-net structure of either Buffer B_1 or Buffer B_2. Note that Buffer B_1 and B_2 are used to temporally store two types of semifinished parts, denoted as part 1 and part 2, respectively, and the respective buffer capacity is represented by the initial number of tokens in the "Empty space" place. In addition, Machine M_3 is employed for final product assembly, which contains one shared robotic arm to load part 1 and part 2 from Buffers B_1 and B_2. M_3 is initiated to assemble the final product when both parts are available. By combining the aforementioned sub-nets, the complete Petri net for the entire manufacturing system is shown in Figure 3.29.

- **Top-down**

The top-down approach mainly designs a Petri net using a stepwise refinement of places, transitions, and/or substructures. Initially, the structure of the Petri net is relatively simple. Complicated sub-nets subsequently substitute the components in the original Petri net with more details. This procedure needs to be repeated until all desired system components are incorporated in the final Petri net. Some examples can be found in Ref. [7, 9].

- **Hybrid approach**

The hybrid approach is a combination of bottom-up and top-down approaches. The bottom-up approach can ensure the correct structural representations of interactions among different subsystems; meanwhile, the integration with the top-down approach ensures that the Petri net includes enough detailed information for better graphical and computational modeling of the complex system. Some examples can be found in Ref. [6, 9].

3.2.3 Deterministic Timed Petri Net

In the classical Petri net, the enabling and firing transitions are assumed to happen immediately, and thus the classical Petri net does not involve the concept of time. In real manufacturing systems, it takes a finite amount of execution time to complete a

(a)

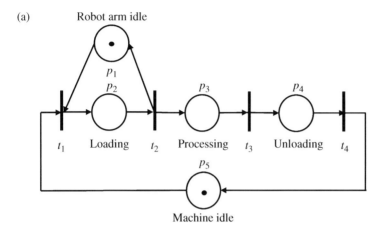

(b)

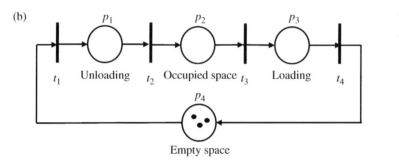

(c)

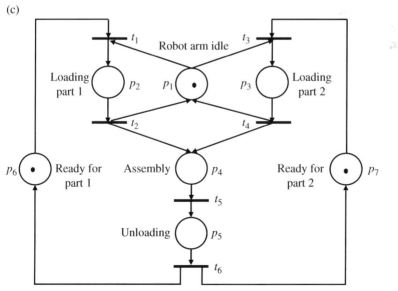

Figure 3.28 Sub-nets for subsystems: (a) Machine M_1 and M_2, (b) buffer B_1 and B_2, and (c) machine M_3.

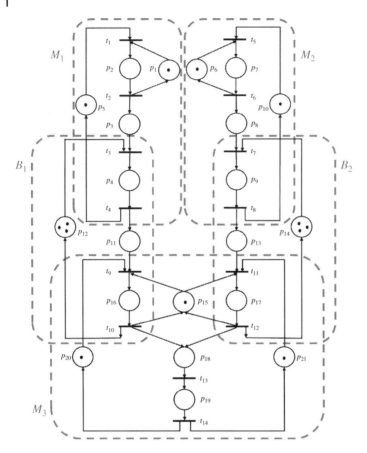

Figure 3.29 The complete Petri net for the entire system.

process or an event. For example, conveying raw material to a machine, returning an empty pallet, and drilling a hole on a part are associated with a process-specific execution time. If the time required to complete an event or process does not change, deterministic timed Petri net can be employed to model this type of manufacturing systems. Their structures must satisfy marked graphs' ones [12, 13].

In a deterministic timed Petri net, transitions, places, and arcs can be assigned with time delays. More specifically,

- If a place is assigned with time τ, then the tokens need to be held for a period of time τ before they can be removed from that place and enable its output transition.
- If an arc running from a place to a transition is assigned with time τ, then it takes time τ to enable the transition after the tokens being removed from the place.

- If a transition is associated with time τ, then the enabled transition needs to wait time τ to fire, i.e. deposit tokens to its output places.
- If an arc running from a transition to a place is associated with time τ, then it takes time τ to deposit new tokens in the output place after the transition's firing.

The above four types of time delay are interconvertible. The following example shows how to convert a place or an arc with time delay into a transition with time delay.

Example 3.10 Consider a Petri net with two places and a transition. Place p_1 connects to transition t_1 and transition t_1 connects to place p_2. Convert a place with time delay and an arc with time delay into a transition with time delay.

Solution

The conversions can be performed in the following situations.

a) Convert a place with time delay into a transition with time delay, as shown in Figure 3.30a.
b) Convert an arc between place and transition with time delay into a transition with time delay, as shown in Figure 3.30b.
c) Convert an arc between transition and place with time delay into a transition with time delay, as shown in Figure 3.30c.

Cycle time is a fundamental performance measure of a deterministic timed Petri net, which is denoted as C. Its calculation is based on two concepts: the total time delay in an elementary circuit or cycle (denoted as T) and the total number of tokens in that elementary (denoted as N). The cycle time of a strongly connected deterministic timed Petri net is defined as

$$C = \max_i \frac{T_i}{N_i} \tag{3.87}$$

where i represents the index of the circuit, and the ratio of T_i to N_i denotes the cycle time of the ith elementary circuit.

For example, consider the marked graph presented in Figure 3.26c. The timing characteristics of the system are shown in Table 3.2. The total time delay and token counts in each circuit are listed in Table 3.3. The system cycle time is 21 time units.

3.2.4 Stochastic Petri Net

In the previous section, the operation time is assumed to be deterministic. In a real manufacturing system, environmental conditions, such as temperature and humidity, and even workers' mood may affect operation duration. The usage of

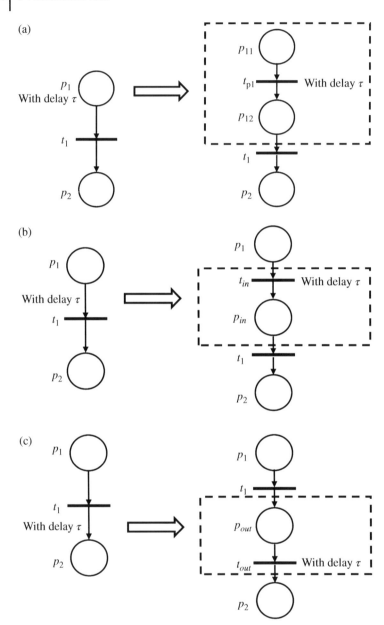

Figure 3.30 Solutions for Example 3.10: (a) convert a place, (b) convert an arc between place and transition, and (c) convert an arc between transition and place.

Table 3.2 The nodes associated time delays.

Nodes	t_1	p_2	t_2	p_3	t_3	p_4	t_4
Time delays	1	3	1	12	1	2	1

Table 3.3 The total time delay and token counts in each circuit.

Circuit	T_i	N_i	T_i/N_i
$t_1p_2t_2p_1t_1$	4	1	4
$t_1p_2t_2p_3t_3p_4t_4p_5t_1$	21	1	21

random variables to represent such duration enables Petri net to handle more general cases than deterministic timed Petri nets. A timed Petri net whose time delay for each transition is assumed to be random and exponentially distributed is referred to as a stochastic Petri net.

A stochastic Petri net can be formally defined as

$$(P, T, I, O, M, \Lambda) \tag{3.88}$$

where delay time in each transition obeys the exponential distribution, and $\Lambda = \{\lambda_1, \lambda_2, ..., \lambda_m\}$ is a set of exponential rates (referred to as firing rates) for transitions $\{t_1, t_2, ..., t_m\}$.

Consider the example in Figure 3.26. Suppose that the machine tool is not 100% reliable and has a probability of breaking down, the new Petri net can be constructed as shown in Figure 3.31, and the interpretation of each node is listed in Table 3.4.

Since this system has a finite number of states (four states), and the time to transition from a state to another obeys a continuous distribution (exponential one), this process is regarded as a continuous Markov chain.

Let states s_i denote the state of p_i being marked with a token for it $\{1, 2, 3, 4\}$, and the state transition rate from the state s_j to s_i is denoted as v_{ij}. The steady-state probabilities can be obtained as follows:

When the system is in state s_1:

$$v_{21} = \lambda_1, \quad v_{31} = 0, \quad v_{41} = 0.$$

Recall that $\sum_i v_{ij} = 0$, hence, $v_{11} = -\lambda_1$

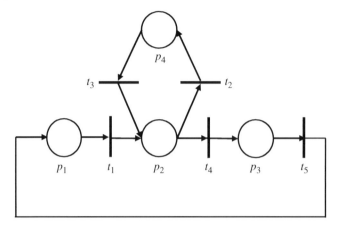

Figure 3.31 Example of stochastic timed Petri net.

Table 3.4 A list of places, transitions, and firing rates.

Place (state)	Interpretation	
p_1	Robotic arm is ready for loading	
p_2	Machine tool is ready for processing	
p_3	Conveyor belt is ready for unloading	
p_4	Machine tool is under maintenance or repair	

Transition (event)	Interpretation	Firing rate
t_1	Loading raw material	$\lambda_1 = 20$
t_2	Machine tool breaks	$\lambda_2 = 0.02$
t_3	Machine tool is repaired	$\lambda_3 = 3$
t_4	Processing	$\lambda_4 = 2$
t_5	Unloading finished product	$\lambda_5 = 20$

When the system is in state s_2:

$$v_{12} = 0, \quad v_{32} = \lambda_4, \quad v_{42} = \lambda_2, \quad v_{42} = -\lambda_2 - \lambda_4.$$

When the system is in state s_3:

$$v_{13} = \lambda_5, \quad v_{23} = 0, \quad v_{43} = 0, \quad v_{33} = -\lambda_5.$$

When the system is in state s_4:

$$v_{14} = 0, \quad v_{24} = \lambda_3, \quad v_{34} = 0, \quad v_{44} = -\lambda_3.$$

Hence, the state transition rate matrix is

$$\nu = \begin{bmatrix} -20 & 0 & 20 & 0 \\ 20 & -2.02 & 0 & 3 \\ 0 & 2 & -20 & 0 \\ 0 & 0.02 & 0 & -3 \end{bmatrix}$$

Assume that the steady-state probabilities are $[\pi_1, \pi_2, \pi_3, \pi_4]^T$, which can be solved by

$$\nu[\pi_1, \pi_2, \pi_3, \pi_4]^T = [\pi_1, \pi_2, \pi_3, \pi_4]^T$$

$$\pi_1 + \pi_2 + \pi_3 + \pi_4 = 1$$

Solving yields

$$\pi_1 = 0.0571, \quad \pi_2 = 0.6298, \quad \pi_3 = 0.0600, \quad \pi_4 = 0.2531$$

3.3 Optimization Methods

In the previous sections, the basic concepts of probability theory are illustrated, followed by the theories and applications of Petri net. The probability theory often plays a supporting role in modeling manufacturing systems due to its ability to positively inform the decision-making processes in industrial applications. Petri net is introduced as a graphical tool for the visualization of manufacturing system behaviors. However, in practice, apart from the appropriate system description, engineers are commonly confronted with the tasks of designing and operating a system with the implementation of the best possible settings while meeting all constraints involved in the system. As a mathematical tool, optimization allows engineers to identify the best design configuration or make the optimal decisions in various operational scenarios considering system constraints and criteria. In the simplest case, optimization seeks the optimal value, i.e. minimum or maximum, of an objective function and determines the associated best set of variables; meanwhile, an optimization process also ensures that the selected variables are within a feasible range or space. Optimization is generally about finding the best available values of some objective functions in the desired domain. The optimization problem can be formulated with a variety of different types of objective functions and constraints.

3.3.1 Fundamentals of Optimization

An objective optimization model involves one objective function and three types of constraints, which can be expressed as

$$\text{Optimize } f(X), \quad X = (x_1, x_2, ..., x_i, ..., x_N) \tag{3.89}$$

Subject to

$$h_j(x) = 0, \quad j = 1, 2, ..., m \tag{3.90}$$

$$g_k(x) \leq 0, \quad k = 1, 2., ..., n \tag{3.91}$$

$$\check{x}_i \leq x_i \leq \hat{x}_i, \quad i = 1, 2, ..., N \tag{3.92}$$

where $f(X)$ is an objective function; X represents a set of decision variables, and N is the total number of decision variables, which is also called the dimension of the optimization problem. $h_j(x)$ denotes an equality constraint; $g_k(x)$ represents an inequality constraint (with the total number of n), $\check{x}_i$ and $\hat{x}_i$ are the lower and upper bounds of decision variable x_i. Additionally, if x_i is unbounded in one or both directions, $\check{x}_i$ and $\hat{x}_i$ could be represented by $-\infty$ and ∞, respectively. There are in total m equality constraints and n inequality ones.

3.3.1.1 Objective Function

The objective function represents the goal of an optimization problem. The desirability of a specific set of variables X as a potential solution is measured by the value of its objective function $f(X)$. The best value of an objective function that one can possibly locate is called the optimal value. "Optimize" in (3.89) can be substituted with either "minimize" or "maximize" depending on specific optimization problems. For example, if the objective function represents the throughput of manufacturing systems, it is regarded as a maximization problem; if it represents the energy consumption in production activities, it is a minimization problem. In practice, the objective functions in most optimization algorithms are set to minimize by default. These algorithms can be employed to solve a maximization problem by transferring the objective function from $f(X)$ to $-f(X)$.

3.3.1.2 Decision Variables

A set of decision variables is also called a solution to an optimization problem, and it determines the value of the objective function. In some scenarios, selecting decision variables to achieve the optimal value of an objective function is even more important than the optimal value itself. In that case, the optimization problem can also be formulated as

$$\underset{X}{\text{argmin}} \, f(X), \quad X = (x_1, x_2, ..., x_i, ..., x_N) \tag{3.93}$$

where argmin denotes the argument of the minimization function. In mathematics, the argument of an objective function is the values of decision variables that must be provided to obtain the optimal value of the objective function.

The decision variables can be categorized into continuous and discrete variables. Examples of continuous variables include selecting the temperature for a chemical reactor or adjusting the spray pressure in a paint shop. On the other hand, discrete decision variables are often applied in problems like determining the number of material handling and associated devices in a manufacturing plant. A binary variable is a special type of discrete variable, which only takes either 0 or 1. Binary variables can be used to represent whether a certain action should be made or not. Indeed, an optimization problem may involve multiple types of decision variables. For example, in a supply chain design problem, whether to build a new warehouse at a specific location can be considered as a binary variable, and its optimal capacity is a continuous variable.

3.3.1.3 Constraints

Unlike an objective function and decision variables, which are indispensable components in an optimization problem, constraints are not necessary for every optimization problem. A problem without constraints is called an unconstrained optimization problem. For the problems with constraints (often referred to as a constrained optimization problem), a solution that satisfies all constraints is called a feasible solution. The set of all feasible solutions is the feasible region or solution space of a problem. The optimal solution to an optimization problem must be a feasible one. In addition, if the solution space of an optimization problem is empty, then this problem is called an infeasible problem. An example of the solution space is shown in Figure 3.32. Consider a two-dimensional optimization problem with

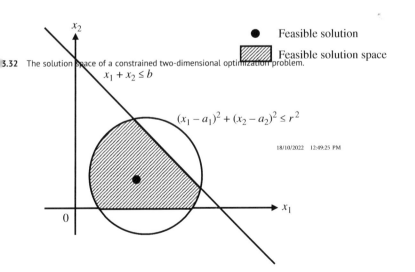

3.32 The solution space of a constrained two-dimensional optimization problem.

Feasible solution

Feasible solution space

$x_1 + x_2 \leq b$

$(x_1 - a_1)^2 + (x_2 - a_2)^2 \leq r^2$

decision variables x_1 and x_2. It has four constraints: $x_1 \geq 0$, $x_2 \geq 0$, $x_1 + x_2 \leq b$, and $(x_1 - a_1)^2 + (x_2 - a_2)^2 \leq r^2$, where a_1, a_2, b, and r are constants. The solution space and an example of a feasible solution are shown as follows.

Many optimization algorithms have been proposed for solving unconstrained problems. To employ these algorithms to a constrained problem, one needs to first convert a constrained optimization problem into an unconstrained one, whose solution ideally converges to the solution of the original constrained problem. The penalty method is a widely used method to convert constrained problems into unconstrained ones by introducing a penalty function to the original objective function. More specifically, the penalty function contains a penalty parameter multiplied by a measure for violating the constraint. The measure is zero when a certain constraint is satisfied; otherwise, a large positive penalty is applied. An example of the penalty method is given as follows.

Consider a constrained problem:

$$\min f(x)$$

subject to

$$g_k(x) \leq 0, \quad k = 1, 2, ..., n$$

The preceding problem can be converted to an unconstrained problem:

$$\min f(x) + \sum_{k=1}^{n} \sigma_k \cdot \max\left(0, g_k(x)\right)^2 \tag{3.94}$$

where σ_k is a large positive number for the kth constraint.

3.3.1.4 Local and Global Optimum

A local optimum is a feasible solution that is optimal in its neighborhood. For example, in an N-dimensional minimization problem with a set of decision variables $X = (x_1, x_2, ..., x_i, ..., x_N)$, a local optimum X^* can be expressed as

$$f(X^*) \leq f(X), \quad x_i^* - \varepsilon_i \leq x_i \leq x_i^* + \varepsilon_i \tag{3.95}$$

where $\varepsilon_i > 0$ is a limited distance that is used to specify the scope of the neighborhood.

If a solution is associated with the best objective function value (either it is maximum or minimum) in a feasible decision space, it is called a global optimum. For example, a global optimum X^* in a minimization problem can be expressed as follows:

$$f(X^*) \leq f(X) \tag{3.96}$$

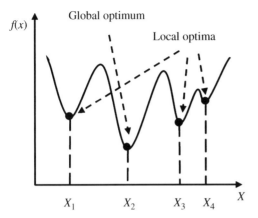

Figure 3.33 Illustration of local and global optima.

Figure 3.33 illustrates an example of a one-dimensional minimization problem. As shown in this figure, X_2 is the global optimum, and X_1, X_3, and X_4 are the local optima. It should be noted that there could be more than one feasible solution that can reach the globally optimal values, and thus an optimization problem could have multiple global optima.

3.3.1.5 Near-optimal Solutions

In practice, it may be challenging to find the exact optimal solution due to the complexity of the optimization problem. Sometimes, a powerful computing device and a long computational time may be required to reach the exact optimal solution. In such cases, engineers need to consider whether the investment in finding an exact optimal solution is worthful, or the near-optimal solutions are also acceptable. The near-optimal solutions are usually very close to the global optimum, but they may require much less computational capacity and time. Figure 3.34 shows an example of near optima in a minimization problem.

3.3.1.6 Single-objective and Multi-objective Optimization

In practice, some optimization problems are formulated with only one objective function, such as minimizing cost or risk and maximizing productivity or profit. This type of optimization problems is called single-objective optimization. On the contrary, in multi-objective optimization, one tries to achieve multiple goals simultaneously, and these goals are often conflicting. For example, engineers may want to reduce machine energy consumption while increasing throughput. The interaction among different objectives leads to a set of solutions, such as Pareto-optimal solutions.

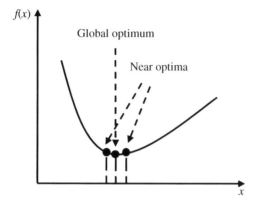

Figure 3.34 Global optimum and near optima.

In a multi-objective optimization problem, the goodness of two solutions can be compared by dominance. If a solution x_1 dominates another solution x_2, x_1 is no worse than x_2 in all objective functions and strictly better than x_2 in at least one objective function. A solution that is not dominated by any other solutions in the feasible solution space is called a Pareto-optimal solution. A set of all Pareto-optimal solutions in the entire feasible solution space is called the Pareto-optimal set. By mapping all solutions of an optimization problem from the feasible solution space to the feasible objective space, the boundary composed of all points in the Pareto-optimal set is called the Pareto-optimal front, as shown in Figure 3.35.

3.3.1.7 Deterministic and Stochastic Optimization

Deterministic optimization, or mathematical programming, is a branch of optimization algorithms aiming to find the optimal solution theoretically. It can be guaranteed that the obtained solution is indeed the global optimum, or the difference between the obtained solution and the global optimum is below a threshold. Usually, deterministic methods involve necessary assumptions and take advantage of the analytical properties of a problem. If an optimization problem is not too complicated, the deterministic methods can find the optimal solution efficiently.

When the complexity of an optimization problem increases, for example, involving complex objective function or constraints, and increasing the dimensions of the problem, it is difficult or even impossible to apply a deterministic method to find the optimal solution. In such cases, stochastic algorithms can be applied to find near-optimal solutions. Stochastic optimization refers to a collection of algorithms that involve random variables. Stochastic algorithms cannot guarantee the convergence to a global optimum, but it usually requires less computational

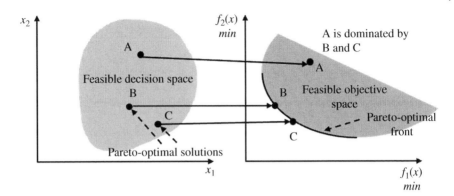

Figure 3.35 The schematic diagram of Pareto-optimal solutions.

capacity and time for complex problems. In particular, metaheuristic algorithms are stochastic optimizers with limited or no assumptions about an optimization problem. Hence, they are widely used in engineering problems, such as complex scheduling problems, and space allocation problems. In the following sections, two widely used metaheuristic algorithms, i.e. genetic algorithms and particle swarm optimization, are introduced.

3.3.2 Genetic Algorithms

A genetic algorithm (GA) was first proposed by Holland in 1975. The basic concept behind GA is Darwin's theory of evolution. The principle of evolution indicates that chromosomal crossover and genetic mutation would occur when parents pass on their genes to the children. Only the fittest group of members with the best genes has the opportunity to survive, reproduce, and pass on their genes under environmental selection pressures. The standard GA begins with a sequence of randomly generated solutions. These solutions undergo evaluations, and a subset of solutions in the current generation is selected to create the next generation of solutions through "crossover" and/or "mutation" operations. This procedure is repeated until a predefined maximal number of iterations is reached. A flow chart of the standard GA is demonstrated in Figure 3.36.

3.3.2.1 Initialization

In GA, a possible solution to an optimization problem is referred to as a chromosome. Each decision variable in the solution is called a gene. For example, in an N-dimensional optimization problem, the chromosome can be expressed as follows:

$$X = (x_1, x_2, ..., x_i, ..., x_N) \tag{3.97}$$

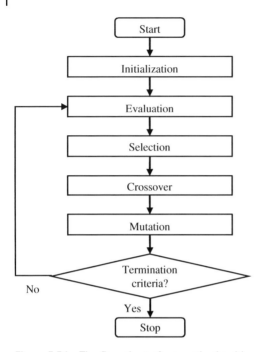

Figure 3.36 The flow chart of a genetic algorithm.

where X denotes a chromosome; x_i is the ith gene, and N is the total number of decision variables or problem dimensions. The initial chromosomes in GA are generated at random, and the total number of chromosomes is called population size. In general, the population of an N-dimensional optimization problem with population size M can be formulated by an $M \times N$ matrix, i.e.

$$
\begin{bmatrix} X_1 \\ X_2 \\ \vdots \\ X_j \\ \vdots \\ X_M \end{bmatrix} = \begin{bmatrix} x_{11} & x_{12} & \cdots & x_{1i} & \cdots & x_{1N} \\ x_{21} & x_{22} & \cdots & x_{2i} & \cdots & x_{2N} \\ \vdots & \vdots & \ddots & \vdots & \ddots & \vdots \\ x_{j1} & x_{j2} & \cdots & x_{ji} & \cdots & x_{jN} \\ \vdots & \vdots & \ddots & \vdots & \ddots & \vdots \\ x_{M1} & x_{M2} & \cdots & x_{Mi} & \cdots & x_{MN} \end{bmatrix}
\tag{3.98}
$$

GA is an iterative process, and the population in each iteration is also referred to as a generation. The population size is a user-defined parameter, and its value also depends on the nature of the problem. It remains constant in all generations. Typically, a population could contain hundreds or thousands of possible solutions.

3.3.2.2 Evaluation

In a general optimization problem, an objective function is used to measure the performance of a specific solution. Similarly, in GA, the performance of each chromosome is evaluated by a fitness function. The chromosome with better performance has a larger chance of surviving and passing on its genes to the offspring. The fitness function is generally designed based upon the objective function. Meanwhile, other factors, such as computational efficiency, chromosome selection methods, and constraints of an optimization problem, should also be jointly considered in the design of fitness functions. For example, the traditional proportional selection method assumes that a better solution has a higher fitness value. Hence, different from an objective function in a minimization problem, the value of the fitness function associated with a better solution should be larger instead of smaller. In addition, the objective function itself cannot judge whether a solution is feasible or not. An infeasible solution, even if it can result in a better value of the objective function, should not be considered as a chromosome with better performance. Therefore, for a constrained problem, the fitness function should be designed with the ability to evaluate the feasibility of an individual chromosome. Generally, fitness function $F(X)$ can be expressed as follows:

$$F(X) = g(f(X)) \tag{3.99}$$

where $f(x)$ is the objective function, and $g(f(x))$ maps the value of an objective function to a nonnegative number.

3.3.2.3 Selection

In each iteration of GA, some chromosomes are selected based on their fitness function values, which are subsequently used for breeding the offspring. Some commonly used parent selection methods include ranking selection, proportionate selection, etc.

In the proportionate selection method, each chromosome has a probability of being selected. The selection probability for the kth chromosome is proportional to its fitness function value, i.e.

$$P_k = \frac{F(X_k)}{\sum\limits_{j=1}^{M} F(X_j)} \tag{3.100}$$

where P_k is the selection probability of the kth solution, and $F(X_k)$ is the corresponding fitness function value. For example, in the case of the population size $M = 3$, and the respective selection probabilities of three chromosomes are $P_1 = 0.6$, $P_2 = 0.3$, and $P_3 = 0.1$. Suppose that a random number $u \in [0, 1]$ is generated. It has following prosperities:

$$P(u \in [0, 0.6)) = 0.6; \quad P(u \in [0.6, 0.9)) = 0.3; \quad P(u \in [0.9, 1]) = 0.1.$$

Therefore, if the generated number u is less than 0.6, the first chromosome is selected; if this number is between 0.6 and 0.9, then the second chromosome is selected; otherwise, the last chromosome is selected.

The rationale of the ranking selection is similar to the proportionate selection. However, the selection probability is proportional to a chromosome's ranking instead of its fitness function. In this case, the selection probability can be computed as follows:

$$P_k = U - (S_k - 1) \cdot Z \tag{3.101}$$

$$U = \frac{Z(M-1)}{2} - \frac{1}{M} \tag{3.102}$$

where P_k is the selection probability of the kth solution; S_k is the ranking of the kth solution, and Z is a user-defined value. Given a population size of M, the best chromosome is assigned with a rank $S_k = 1$, and the worst one as $S_k = M$.

After the selection, the unselected chromosomes are eliminated from the population. Hence, new chromosomes need to be generated to compensate for the removal of unselected chromosomes from the population, and thus the algorithm can progress toward the solution. The selected chromosomes from the previous generation are called parents. After the crossover and mutation operations, new chromosomes, referred to as children, are generated. The next generation is a combination of parents and children, as shown in Figure 3.37.

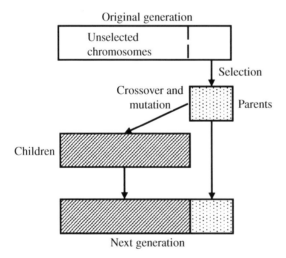

Figure 3.37 The process of constituting the next generation based on the previous generation. *Source:* Adopted from [14].

3.3.2.4 Crossover

The crossover, also referred to as recombination, is a genetic operator used to generate new offspring by swapping parts of two parents' chromosomes. The crossover methods include one-point crossover, two-point one, uniform one, etc.

- One-point crossover: A crossover point on both parents is randomly selected. The sections to the right of the crossover point are exchanged between the two parents' chromosomes, as shown in Figure 3.38a.
- Two-point crossover: Two arbitrary crossover points are selected on both parents. The offspring are generated by exchanging parents' genes between the two points, as shown in Figure 3.38b.
- Uniform crossover: Arbitrary crossover points are selected on both parents. The offspring are generated by exchanging each pair of the selected points along both chromosomes, as shown in Figure 3.38c.

For binary variables, the crossover simply means swapping two values, as shown in the methods mentioned above. Some alternative methods can be applied to

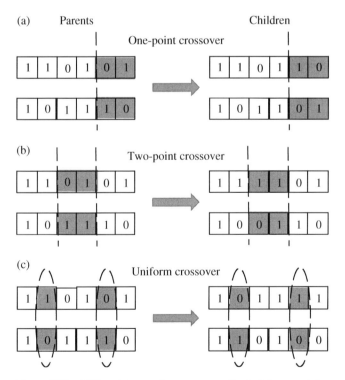

Figure 3.38 Different types of crossover methods: (a) one-point crossover, (b) two-point crossover, and (c) uniform crossover.

create the new genes for real variables (i.e. variables taking real values), such as linear, blend, and simulated binary crossover.

- Linear crossover: given the two parents' genes x_1 and x_2, generate three new genes, and the two best genes are selected as children. The genetic cross can be expressed as

$$
\begin{aligned}
x_1' &= 0.5x_1 + 0.5x_2 \\
x_2' &= 1.5x_1 - 0.5x_2 \\
x_3' &= -0.5x_1 + 1.5x_2
\end{aligned}
\tag{3.103}
$$

- Blend crossover: given the two parents' genes x_1 and x_2, where $x_1 < x_2$, a child is randomly selected in the range $[x_1 - 0.5(x_2 - x_1), \ x_2 + 0.5(x_2 - x_1)]$. Mathematically,

$$
x' = R \text{ and } [x_1 - 0.5(x_2 - x_1), \ x_2 + 0.5(x_2 - x_1)]
\tag{3.104}
$$

- Simulated binary crossover: given the two parents' genes x_1 and x_2, the genetic cross can be expressed as

$$
\begin{aligned}
x_1' &= 0.5[(1 + \beta)x_1 + (1 - \beta)x_2] \\
x_2' &= 0.5[(1 - \beta)x_1 + (1 + \beta)x_2]
\end{aligned}
\tag{3.105}
$$

$$
\beta =
\begin{cases}
(2u)^{\frac{1}{\eta_c + 1}}, & u \le 0.5 \\
\left(\dfrac{1}{2(1-u)}\right)^{\frac{1}{\eta_c + 1}}, & u > 0.5
\end{cases}
\tag{3.106}
$$

where u is a random number, $u \in [0, 1)$, and η_c is the distribution index. A large value of η_c allows the children to be near their parents, and a small value of η_c tends to generate children far from parents.

3.3.2.5 Mutation

Mutations are generally carried out by changing parts of an individual chromosome at random, which increases the population diversity and functions as a possible way to escape from local optima. Figure 3.39 illustrates the concept of the mutation operation.

The preceding example demonstrates a mutation operation on a binary variable. Two mutation methods can be applied to a real variable, i.e. uniform and nonuniform mutation.

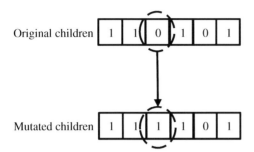

Figure 3.39 An example of the mutation operation.

- Uniform mutation: An existing gene x_i is replaced with a uniform random value drawn between user-specified upper and lower bounds for that gene. Mathematically, the mutated gene x'_i can be expressed as follows:

$$x'_i = \text{R and } (\check{x}_i, \hat{x}_i) \tag{3.107}$$

where $\check{x}_i$ and $\hat{x}_i$ are the lower and upper bounds of decision variable x_i, and Rand $(\cdot, \cdot)$ means a uniform random number generator or function.

- Nonuniform mutation: An existing gene x_i is replaced with a nonuniform random number. Compared with a random choice from a predetermined range, this mutation method increases the probability of generating a new gene close to its successor. This strategy keeps the population from stagnating in the early stages of evolution and allows the algorithm to fine-tune the solution in the later stages. The mutated gene x'_i can be expressed as

$$x'_i = \begin{cases} x_i + \Delta(t, \hat{x}_i - x_i), & u = 0 \\ x_i - \Delta(t, x_i - \check{x}_i), & u = 1 \end{cases} \tag{3.108}$$

$$\Delta(t, y) = y\left(1 - r^{\left(1 - \frac{t}{T}\right)^b}\right) \tag{3.109}$$

where u is a binary random variable; t means the t-th generation; T is the user-specified maximum number of generations; r is a uniform random number drawn from [0, 1], and b is a parameter determining the dependency of mutation on the iteration.

3.3.2.6 Termination Criteria

As a random search method, it is challenging to obtain a specified or ideal convergence criterion for GA. In practice, termination criteria are often used to end

GA iterations. The adequacy of termination conditions significantly influences the proximity of the solution obtained to the optimal solution at the end of the iterations. The termination criteria are generally defined based on factors such as

- The fitness function value is below a predetermined threshold value.
- The number of iterations reaches a given maximum iteration count.
- The fitness function value of the "best-so-far" stabilizes, and there is no significant improvement for many generations.

There are many GA variants that improve the standard GA. Please consult the references [15–21].

3.3.3 Particle Swarm Optimizer (PSO)

The particle swarm optimizer (PSO) was first proposed by Kennedy and Eberhart in 1995. The PSO algorithm has been further evolved with the inspiration of the animal social behaviors such as bird flocking and fish schooling. More specifically, the behaviors of each individual in a swarm include interacting with its surrounding environment and information sharing between each other. In PSO, each feasible solution is designated as a particle or an agent. The algorithm initially generates multiple particles in the solution space, and the location of each particle corresponds to a solution. These particles move around the solution space and cooperate toward the globally optimal solution. PSO is an iterative algorithm, and a fitness function is used to evaluate the performance of each particle in each iteration. The moving direction of each particle is affected by the best location the particle has ever visited (i.e. the personal best position, denoted as *pbest*) and the best location the entire swarm has ever reached (i.e. the global best position, denoted as *gbest*). The basic flow chart of PSO is demonstrated in Figure 3.40.

3.3.3.1 Initialization
In PSO, each feasible solution of an optimization problem can be treated as the position of a particle. Therefore, for an N-dimensional optimization problem, each solution can be represented as a point in the N-dimensional solution space. Each decision variable represents the coordinate of one dimension. PSO starts solving an optimization problem by initializing a group of particles, which is referred to as a swarm. In the case of a swarm consisting of M particles, the positions of all particles can be formulated as a matrix:

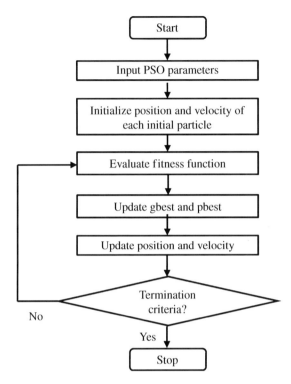

Figure 3.40 The basic flow chart of the PSO algorithm.

$$
\begin{bmatrix} X_1 \\ X_2 \\ \vdots \\ X_j \\ \vdots \\ X_M \end{bmatrix} = \begin{bmatrix} x_{11} & x_{12} & \cdots & x_{1i} & \cdots & x_{1N} \\ x_{21} & x_{22} & \cdots & x_{2i} & \cdots & x_{2N} \\ \vdots & \vdots & \ddots & \vdots & \ddots & \vdots \\ x_{j1} & x_{j2} & \cdots & x_{ji} & \cdots & x_{jN} \\ \vdots & \vdots & \ddots & \vdots & \ddots & \vdots \\ x_{M1} & x_{M2} & \cdots & x_{Mi} & \cdots & x_{MN} \end{bmatrix}
\tag{3.110}
$$

where X_j is the position for the jth particle, and x_{ji} is the coordinate in the ith dimension of the jth particle. The swarm size M is a user-defined parameter, and it varies depending on the complexity of a problem. The initial particles are then randomly distributed across the entire solution space.

In addition to the particle position information, velocity is another important parameter in PSO. Together with its current position, the velocity of a particle can decide where the particle would be in the next iteration. The initial velocity of a particle is usually chosen within a range, and the initial velocity component in the ith dimension of the jth particle can be expressed as

$$v_{ji} = \text{R and}\left(\check{x}_{ji}, \hat{x}_{ji}\right) \tag{3.111}$$

Similar to the position matrix, the velocity information of all particles can be formulated as follows:

$$
\begin{bmatrix} V_1 \\ V_2 \\ \vdots \\ V_j \\ \vdots \\ V_M \end{bmatrix}
=
\begin{bmatrix}
v_{11} & v_{12} & \cdots & v_{1i} & \cdots & v_{1N} \\
v_{21} & v_{22} & \cdots & v_{2i} & \cdots & v_{2N} \\
\vdots & \vdots & \ddots & \vdots & \ddots & \vdots \\
v_{j1} & v_{j2} & \cdots & v_{ji} & \cdots & v_{jN} \\
\vdots & \vdots & \ddots & \vdots & \ddots & \vdots \\
v_{M1} & v_{M2} & \cdots & v_{Mi} & \cdots & v_{MN}
\end{bmatrix}
\tag{3.112}
$$

where V_j is the velocity for the jth particle.

3.3.3.2 Evaluation

Ideally, the particles should only search the feasible solution space toward the optimal solution. However, it is often challenging to completely prevent particles from roaming into the infeasible space, especially when the constraints that define the space boundaries are complicated. Hence, the fitness function $F(X)$ is employed to evaluate the performance of a position, and different positions correspond to different fitness function values. For example, for constrained problems, the penalty method is usually applied to adjust the objective function into a fitness function.

3.3.3.3 Personal and Global Best Positions

After evaluating the position of each particle, the personal best position and global best position should be updated. For a certain particle, the position where it achieved the best fitness value in its historical movement is defined as its personal best position. For the first iteration, the initial positions for all particles are defined as their first personal best positions. A smaller value of fitness function indicates a better solution. Hence, in later iterations, given the jth particle's historical best position pbest$_j$ and its current position X_i, the jth particle's personal best position can be updated by

$$
\text{pbest}_j =
\begin{cases}
\text{pbest}_j, & F\left(\text{pbest}_j\right) \le F\left(X_j\right) \\
X_j, & F\left(X_j\right) < F\left(\text{pbest}_j\right)
\end{cases}
\tag{3.113}
$$

After each particle's personal best position is updated, the next step is to update the global best position. In the first iteration, it is the one with the optimal fitness score. In later iterations, it can be updated by

$$\text{gbest} = \min\left(\text{gbest}, \text{pbest}_1, ..., \text{pbest}_M\right) \tag{3.114}$$

Figure 3.41 shows an example of the personal and global best positions updating in a one-dimensional minimization problem. Let swarm size $M = 3$, and the respective positions of three particles in the first iteration t_1 are denoted as x_1^{t1}, x_2^{t1}, and x_3^{t1}. As shown in Figure 3.41a, x_1^{t1}, x_2^{t1}, and x_3^{t1} are regarded as the personal best positions in the first iteration. Besides, since x_3^{t1} has the smallest fitness score, it is also determined as gbest. In the second iteration, the pbest$_1$ and pbest$_2$ are updated to x_1^{t1} and x_2^{t1}, whereas pbest$_3$ does not change, as shown in Figure 3.41b. In the second iteration, pbest$_2$ is gbest. The same updating strategy is applied in the third iteration, and the results are shown in Figure 3.41c.

3.3.3.4 Updating Velocity and Position

A particle's velocity is updated based on its velocity in the previous iteration, its personal best position, and the global best position. The velocity updating can be performed as follows:

$$v'_{ji} = \omega v_{ji} + c_1 r_1 \left(\text{pbest}_{ji} - x_{ji}\right) + c_2 r_2 \left(\text{gbest}_{ji} - x_{ji}\right) \tag{3.115}$$

where v'_{ji} and v_{ji} are the updated and previous velocities of the jth particle in the ith dimension; x_{ji} is the previous coordinate of the jth particle in the ith dimension; ω is inertia weight; c_1 is a cognitive parameter; c_2 is a social parameter, and r_1 and r_2 are two random numbers between 0 and 1.

- ω is a nonnegative number, which determines the impacts of the previous velocity on particle movement and plays an essential role in the convergence of the population. A large ω can benefit the swarm in exploring and searching a wide range of solution space, but it can also cause difficulties in convergences of particles. On the contrary, a small ω can narrow down the searching space, but it can facilitate particles to converge and fine-tune the solution. To be benefited from both large and small values, a dynamic inertia weight is designed to reach better optimization results. The dynamic inertia weight at iteration t, denoted as ω^t, can be expressed as

$$\omega^t = \omega^0 - \frac{\left(\omega^0 - \omega^T\right)}{T}t \tag{3.116}$$

where ω^0 is the initial inertia weight, and T is the maximal number of iterations.

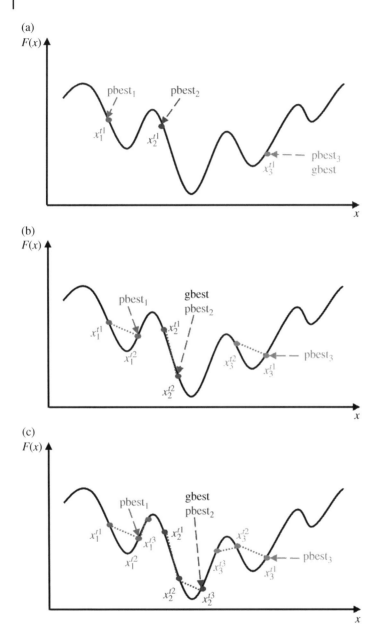

Figure 3.41 An example of pbest and gbest updating in a one-dimensional minimization problem: (a) first iteration, (b) second iteration, and (c) third iteration.

- c_1 and c_2 are two positive coefficients. c_1 represents how much confidence a particle has in its own searching experience, and c_2 represents how much confidence a particle has in other particles. If $c_1 \gg c_2$, all particles only trust their own searching experience; thus, they can be easily trapped and have difficulty in converging to a global/local optimum. If $c_2 \gg c_1$, fast convergence to gbest is guaranteed, such that the particles do not have the opportunity to explore the solution space. In addition to the relative difference between c_1 and c_2, the absolute values of these two parameters can also significantly affect PSO performance. Particularly, lower values of c_1 and c_2 can result in smoother particle movement, which indicates that it takes more time to converge particles and fine-tune the solution. On the other hand, higher values c_1 and c_2 indicate that a particle moves a long distance in every single iteration. Thus, it may directly pass through the region where the optimal solution is located.

After the velocity is updated, the new position can be updated as

$$x'_{ji} = x^t_{ji} + v'_{ji} \tag{3.117}$$

where x'_{ji} and x_{ji} are the updated and previous coordinates of the jth particle in the ith dimension, v'_{ji} is the updated velocity. Figure 3.42 shows an example of the velocity and position updating of the jth particle in a two-dimensional problem.

PSO can also be applied to solve a discrete problem. For example, if x_{ji} is a binary variable, it can be updated as

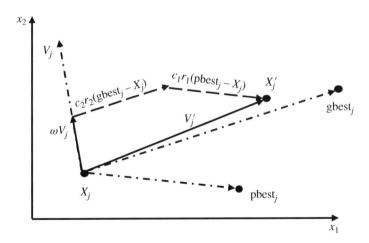

Figure 3.42 Updating the velocity and position of a particle j.

$$S(v_{ji}) = \frac{1}{1 + e^{-v_{ji}}} \tag{3.118}$$

$$x'_{ji} = \begin{cases} 1, & u < S(v_{ji}) \\ 0, & u \geq S(v_{ji}) \end{cases} \tag{3.119}$$

where u is a random variable between $[0, 1]$, and $S(v_{ji})$ is the sigmoid function of velocity v_{ji}. The sigmoid function has the properties that $S(v_{ji}) \to 1$ if $v_{ji} \to \infty$, and $S(v_{ji}) \to 0$ if $v_{ji} \to -\infty$.

3.3.3.5 Termination Criteria

The termination criteria used to stop an iterative search process often depend on the number of iterations, the significance of improvement in solutions between iterations, computational time, etc.

There are many PSO variants that improve the standard PSO. Please consult the references [21–39].

Problems

3.1 The products in a manufacturing plant need to pass three sequential quality inspections before delivery to the warehouse. Suppose that the results from three inspections are independent, and the individual rejection rates of three inspections are 0.02, 0.04, and 0.03. When a product fails an inspection, it needs to be recycled immediately and will not go through the next inspection.
   (a) What is the probability that a product survives the first inspection but is rejected by the second inspection?
   (b) What is the probability that a product cannot be delivered to the warehouse?

3.2 A manufacturing plant has three production lines A, B, and C, which produce the same type of products. The throughputs of three production lines are 30 products per hour, 35 products per hour, and 45 products per hour, respectively, and the associated defective rates are 2%, 3%, and 5%.
   (a) After three production lines running simultaneously for an hour, how many defective products are produced on average?
   (b) If a product is determined to be defective, what is the probability that this product is from production line C?

3.3 The probabilities that a Bernoulli machine is down and up in each time step are 0.1 and 0.9.
   (a) What is the probability that the machine is up for four consecutive steps?
   (b) If the machine is up initially, what is the probability that the first breakdown happens at the fifth step?

3.4 The lifetime of a particular type of electronic devices follows an exponential distribution, and its average life is two years. If 100 devices are installed in a manufacturing plant, and they are independent of each other, what is the probability that at most 20 devices fail during the first year?

3.5 When the machine runs for a long time, what is the percentage of time it is up if
   (a) it is a geometric machine. The transition probability from up to down is 0.2, and the probability from down to up is 0.7.
   (b) it is an exponential machine. The breakdown rate and repair rate are 0.2 and 0.7, respectively.

3.6 A manufacturing system consists of two machines, a robotic arm, and a buffer. The raw material sequentially passes through machine A, buffer, machine B, and is eventually converted to product. The operations of raw material loading and final products unloading are finished by the same robotic arm. The components and the material flow in this system are shown in the following figure. Suppose that the buffer capacity is two. Design a Petri net model for this system.

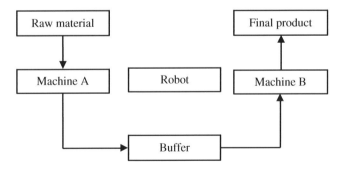

3.7 A stochastic timed Petri net is shown as follows. The firing rate of each transition is given in the following table.

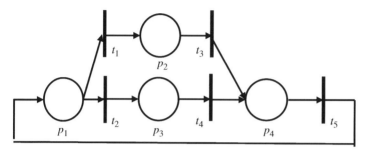

Transition	t_1	t_2	t_3	t_4	t_5
Firing rate	2	3	20	15	5

What is the state transition rate matrix, and what are the steady-state probabilities?

References

1 Siegmund, D.O. (2020). "Probability theory," Encyclopædia Britannica. https://www.britannica.com/science/probability-theory/Probability-distribution (accessed 25 December 2020).

2 Hu, C., Youn, B.D., and Wang, P. (2019). *Engineering Design Under Uncertainty and Health Prognostics*. Springer International Publishing.

3 Ibe, O.C. (2005). *Fundamentals of Applied Probability and Random Processes*, 2e. Elsevier Academic Press.

4 Sericola, B. (2013). *Markov Chains: Theory, Algorithms and Applications*. Wiley.

5 Reisig, W. (2013). *Understanding Petri Nets: Modeling Techniques, Analysis Methods, Case Studies*, vol. 9783642332. Berlin, Heidelberg: Springer-Verlag.

6 Huang, B. and Zhou, M. (2020). *Supervisory Control and Scheduling of Resource Allocation Systems: Reachability Graph Perspective*. Hoboken, NJ: Wiley.

7 Li, Z. and Zhou, M. (2009). *Modeling, Analysis and Deadlock Control of Automated Manufacturing Systems*. Beijing: Science Press.

8 Li, Z. and Zhou, M. (2009). *Deadlock Resolution in Automated Manufacturing Systems: A Novel Petri Net Approach*. New York: Springer Science & Business Media.

9 Zhou, M. and Venkatesh, K. (1999). *Modeling, Simulation, and Control of Flexible Manufacturing Systems: A Petri Net Approach*. Singapore: World Scientific.

10 Zhou, M. (ed.) (1995). *Petri Nets in Flexible and Agile Automation*. London: Kluwer Academic Publishers.

11 Zhou, M. and Dicesare, F. (1993). *Petri Net Synthesis for Discrete Event Control of Manufacturing Systems*. London: Kluwer Academic Publishers.

12 Wu, N. and Zhou, M. (2010). *System Modeling and Control with Resource-Oriented Petri Nets*. New York: CRC Press.

13 Hrúz, B. and Zhou, M. (2007). *Modeling and Control of Discrete Event Dynamic Systems*. London: Springer Science & Business Media.

14 Bozorg-Haddad, O., Solgi, M., and Loáiciga, H.A. (2017). *Meta-Heuristic and Evolutionary Algorithms for Engineering Optimization*. Wiley.

15 Xing, K.Y., Han, L.B., Zhou, M.C., and Wang, F. (2012). Deadlock-free genetic scheduling algorithm for automated manufacturing systems based on deadlock control policy. *IEEE Trans. Syst. Man, Cybern. Part B* 42 (3): 603–615. https://doi. org/10.1109/TSMCB.2011.2170678.

16 Zuo, X., Chen, C., Tan, W., and Zhou, M. (2014). Vehicle scheduling of an urban bus line via an improved multiobjective genetic algorithm. *IEEE Trans. Intell. Transp. Syst.* 1–12. https://doi.org/10.1109/TITS.2014.2352599.

17 Hou, Y., Wu, N., Zhou, M., and Li, Z. (2017). Pareto-optimization for scheduling of crude oil operations in refinery via genetic algorithm. *IEEE Trans. Syst. Man, Cybern. Syst.* 47 (3): 517–530. https://doi.org/10.1109/TSMC.2015.2507161.

18 Mareda, T., Gaudard, L., and Romerio, F. (2017). A parametric genetic algorithm approach to assess complementary options of large scale windsolar coupling. *IEEE/CAA J. Autom. Sin.* 4 (2): 260–272. https://doi.org/10.1109/JAS.2017.7510523.

19 Wang, J., Zhou, M., Guo, X., and Qi, L. (2019). Multiperiod asset allocation considering dynamic loss aversion behavior of investors. *IEEE Trans. Comput. Soc. Syst.* 6 (1): 73–81. https://doi.org/10.1109/TCSS.2018.2883764.

20 Ghahramani, M., Qiao, Y., Zhou, M. et al. (2020). AI-based modeling and data-driven evaluation for smart manufacturing processes. *IEEE/CAA J. Autom. Sin.* 7 (4): 1026–1037. https://doi.org/10.1109/JAS.2020.1003114.

21 Bi, J., Yuan, H., Duanmu, S. et al. (2021). Energy-optimized partial computation offloading in mobile-edge computing with genetic simulated-annealing-based particle swarm optimization. *IEEE Internet Things J.* 8 (5): 3774–3785. https://doi. org/10.1109/JIOT.2020.3024223.

22 Kang, Q., Zhou, M., An, J., and Wu, Q. (2013). Swarm intelligence approaches to optimal power flow problem with distributed generator failures in power networks. *IEEE Trans. Autom. Sci. Eng.* 10 (2): 343–353. https://doi.org/10.1109/TASE.2012.2204980.

23 Gao, K., Cao, Z., Zhang, L. et al. (2019). A review on swarm intelligence and evolutionary algorithms for solving flexible job shop scheduling problems. *IEEE/CAA J. Autom. Sin.* 6 (4): 904–916. https://doi.org/10.1109/JAS.2019.1911540.

24 Wang, J. and Kumbasar, T. (2019). Parameter optimization of interval Type-2 fuzzy neural networks based on PSO and BBBC methods. *IEEE/CAA J. Autom. Sin.* 6 (1): 247–257. https://doi.org/10.1109/JAS.2019.1911348.

25 Zhang, Y., Zhou, P., and Cui, G. (2018). Multi-model based PSO method for burden distribution matrix optimization with expected burden distribution output behaviors. *IEEE/CAA J. Autom. Sin.* 1–7. https://doi.org/10.1109/JAS.2018.7511090.

26 Roy, P., Mahapatra, G.S., and Dey, K.N. (2019). Forecasting of software reliability using neighborhood fuzzy particle swarm optimization based novel neural network. *IEEE/CAA J. Autom. Sin.* 6 (6): 1365–1383. https://doi.org/10.1109/JAS.2019.1911753.

27 Lv, Z., Wang, L., Han, Z. et al. (2019). Surrogate-assisted particle swarm optimization algorithm with Pareto active learning for expensive multi-objective optimization. *IEEE/CAA J. Autom. Sin.* 6 (3): 838–849. https://doi.org/10.1109/JAS.2019.1911450.

28 Pare, S., Kumar, A., Bajaj, V., and Singh, G.K. (2017). A context sensitive multilevel thresholding using swarm based algorithms. *IEEE/CAA J. Autom. Sin.* 6 (6): 1471–1486.

29 Lin, H., Zhao, B., Liu, D., and Alippi, C. (2020). Data-based fault tolerant control for affine nonlinear systems through particle swarm optimized neural networks. *IEEE/CAA J. Autom. Sin.* 7 (4): 954–964. https://doi.org/10.1109/JAS.2020.1003225.

30 Xu, X., Li, J., Zhou, M. et al. (2020). Accelerated two-stage particle swarm optimization for clustering not-well-separated data. *IEEE Trans. Syst. Man, Cybern. Syst.* 50 (11): 4212–4223. https://doi.org/10.1109/TSMC.2018.2839618.

31 Wang, Y. and Zuo, X. (2021). An effective cloud workflow scheduling approach combining PSO and idle time slot-aware rules. *IEEE/CAA J. Autom. Sin.* 8 (5): 1079–1094. https://doi.org/10.1109/JAS.2021.1003982.

32 Liang, X., Li, W., Zhang, Y., and Zhou, M. (2015). An adaptive particle swarm optimization method based on clustering. *Soft Comput.* 19 (2): 431–448. https://doi.org/10.1007/s00500-014-1262-4.

33 Li, J., Zhang, J., Jiang, C., and Zhou, M. (2015). Composite particle swarm optimizer with historical memory for function optimization. *IEEE Trans. Cybern.* 45 (10): 2350–2363. https://doi.org/10.1109/TCYB.2015.2424836.

34 Kang, Q., Liu, S., Zhou, M., and Li, S. (2016). A weight-incorporated similarity-based clustering ensemble method based on swarm intelligence. *Knowledge-Based Syst.* 104: 156–164. https://doi.org/10.1016/j.knosys.2016.04.021.

35 Tian, G., Ren, Y., and Zhou, M. (2016). Dual-objective scheduling of rescue vehicles to distinguish forest fires via differential evolution and particle swarm optimization combined algorithm. *IEEE Trans. Intell. Transp. Syst.* 17 (11): 3009–3021. https://doi.org/10.1109/TITS.2015.2505323.

36 Dong, W. and Zhou, M. (2017). A supervised learning and control method to improve particle swarm optimization algorithms. *IEEE Trans. Syst. Man, Cybern. Syst.* 47 (7): 1135–1148. https://doi.org/10.1109/TSMC.2016.2560128.

37 Kang, Q., Xiong, C., Zhou, M., and Meng, L. (2018). Opposition-based hybrid strategy for particle swarm optimization in noisy environments. *IEEE Access* 6: 21888–21900. https://doi.org/10.1109/ACCESS.2018.2809457.

38 Zhang, J., Zhu, X., Wang, Y., and Zhou, M. (2019). Dual-environmental particle swarm optimizer in noisy and noise-free environments. *IEEE Trans. Cybern.* 49 (6): 2011–2021. https://doi.org/10.1109/TCYB.2018.2817020.

39 Cao, Y., Zhang, H., Li, W. et al. (2019). Comprehensive learning particle swarm optimization algorithm with local search for multimodal functions. *IEEE Trans. Evol. Comput.* 23 (4): 718–731. https://doi.org/10.1109/TEVC.2018.2885075.

4

Mathematical Modeling of Manufacturing Systems

System modeling plays a critical role in the sustainable manufacturing system design and analysis. Appropriate formulation of a manufacturing system model can provide a solid foundation for evaluating system performance, and formulating objective functions and constraints concerning energy efficiency improvement and/or cost savings, which pave the way for determining system performance improvement strategies. In this chapter, the basics of manufacturing system modeling are introduced, followed by detailed discussions on some typical modeling approaches for simple two-machine production lines and complex multi-machine production lines. More specifically, in Section 4.1, the fundamental components in a manufacturing system model are introduced, and the mathematical descriptions of these components and their interactions are presented. Meanwhile, the associated system performance measures are also defined in this section. In Section 4.2, the modeling method used in simple two-machine production lines is illustrated. The closed-form expressions of steady state performance measures are derived. Finally, in Section 4.3, the modeling method used in complex multi-machine serial production lines is demonstrated. Given the difficulties in the derivations of the closed-form expressions for complex systems, an alternative iteration-based approximation method is introduced to analyze their performance. Moreover, an example is provided to demonstrate the use of the iteration-based method in system modeling and the steady-state system performance approximation.

4.1 Basics in Manufacturing System Modeling

4.1.1 Structure of Manufacturing Systems

4.1.1.1 Basic Components

In general, a manufacturing system consists of two primary components: producing unit and material handling equipment. Note that the interpretation of these

Sustainable Manufacturing Systems, First Edition. Lin Li and MengChu Zhou.
© 2023 The Institute of Electrical and Electronics Engineers, Inc.
Published 2023 by John Wiley & Sons, Inc.

components may vary in different types of manufacturing systems. In terms of a producing unit, it may refer to an individual machine or a work cell consisting of multiple machines. Specifically, each machine or work cell conducts a specific manufacturing operation, such as cutting, welding and drilling. On the other hand, the producing unit can also represent a department or workshop in a manufacturing plant, such as a body shop and paint shop. From a system perspective, whatever the producing units' physical implementations may be, they can be modeled by using some common characteristics, such as the time required for a certain operation or average time spent in a specific state. Hence, in this chapter and the following chapters, the producing units are referred to as machines unless otherwise noted.

In a manufacturing system, the material handling equipment can be conveyor belts, forklifts, or automated ground vehicles, which transport parts/products from one machine to another. The parts or semifinished products being transported between different machines are called work in process (WIP). With regard to problems addressed in this book, the essential property of the material handling equipment is its storage capacity. Therefore, the material handling equipment in this book is also referred to as WIP buffer (or just buffer).

4.1.1.2 Structural Modeling

In some manufacturing systems, multiple machines need to cooperate to complete a specific task. Two common structures for representing the machine cooperation are shown as follows [1].

- Parallel machines. Figure 4.1 illustrates the structure of parallel machines, where circles represent machines and rectangles represent buffers. In this case, no single machine has enough processing speed to meet the production requirement, and thus, multiple machines need to operate simultaneously in order to shorten the average processing time.

- Synchronous dependent machines. Figure 4.2 demonstrates the structure of synchronous-dependent machines, where multiple machines perform several synchronous-dependent operations. In this case, the entire operation becomes halted if at least one operation is halted.

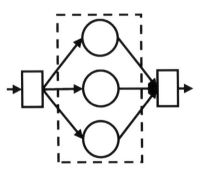

Figure 4.1 The schematic diagram of parallel machines.

In practice, given the structural complexity in the above two structures, one can hardly apply the universal tools for manufacturing system analysis. To address this issue,

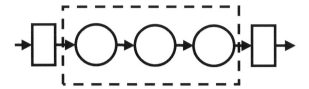

Figure 4.2 The schematic diagram of synchronous-dependent machines.

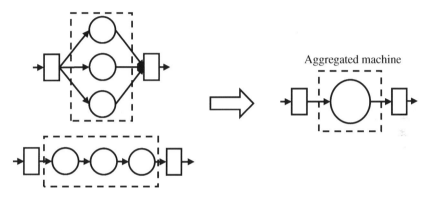

Figure 4.3 Aggregating parallel or synchronous-dependent machines.

the machine structures between two adjacent buffers can be simplified as an aggregated machine, as shown in Figure 4.3, where the machines and buffers are placed alternatively. The simplification process is referred to as structural modeling, and the detailed methods on how to find parameters of the aggregated machine are introduced in Section 4.1.2.

4.1.1.3 Types of Manufacturing Systems

Serial Production Lines. A serial production line is composed of multiple machines and buffers, which are arranged in consecutive order. The schematic diagram of a serial production line with N machines and $N-1$ buffers is shown in Figure 4.4. Similarly, circles represent machines, and rectangles represent buffers.

Assembly Systems. An assembly system is composed of two or more serial production lines, and these serial production lines are then merged at assembly

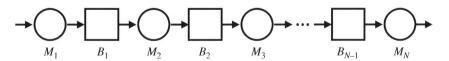

Figure 4.4 The schematic diagram of a serial production line.

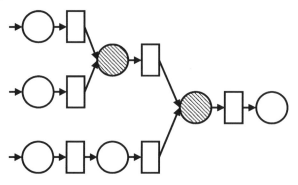

Figure 4.5 The schematic diagram of an assembly system.

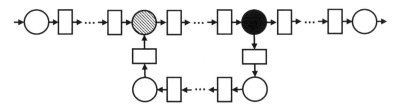

Figure 4.6 The schematic diagram of a production line with rework.

machines, which are represented by shaded circles at the merging points in Figure 4.5. In this case, an assembly machine can be in operation only if all upstream buffers are non-empty.

Rework loop. To ensure the final product quality, inspection machines are often added to some production lines for defective part identification. As shown in Figure 4.6, the black circle represents an inspection machine, and it split parts upon quality inspection. If a part passes the inspection, it is transported to the next machine in the horizontal direction (i.e. a main production line). If the part fails the inspection, it has to undergo repairs through a rework line (i.e. the bypass of the mainline in Figure 4.6) and return to a previous machine for reprocessing. Here, the previous machine refers to the merging point where the rework line merges to the main production line (represented by the shaded circle). The machine at the merging joint can be in operation if at least one upstream buffer is non-empty. A production line with a rework line consists of at least one splitting point and one merging point.

4.1.2 Mathematical Models of Machines and Buffers

The mathematical models of machines and buffers are used to describe the system dynamics, i.e. the system state transition over time. The timing issues and state transition probabilities are discussed next.

4.1.2.1 Timing Issues for Machines

Cycle time (τ) is the time required by a machine to process one part. The cycle time can be a constant or random variable. In mass production systems, such as automobiles and appliances, each machine is designed for a specific process; hence, the cycle time is always a constant or close to a constant. In the following chapters, a constant cycle time is assumed unless otherwise noticed.

Machine capacity (c) represents the number of parts produced by a running machine per unit time. For a constant cycle time τ, the machine capacity can be expressed as

$$c = \frac{1}{\tau} \tag{4.1}$$

Suppose that all machines in a manufacturing system have the same cycle time. In this case, the manufacturing system is called a synchronous system or homogeneous system, and discrete-time modeling is the most suitable method to represent this type of system. In such a system, the time axis can be divided into uniform time steps, and the length of each time step equals the cycle time. All kinds of machine state changes, such as changing from operating to idle or up to down, are assumed to occur either at the beginning or the end of the time step.

If machines in a manufacturing system have their individual and different cycle times, the manufacturing system is referred to as an asynchronous system or inhomogeneous one. To model such a system, one needs to find the greatest common divisor of all cycle times and set it as the unit time step. Otherwise, a continuous-time model should be considered to represent such a system.

4.1.2.2 Machine Reliability Models

Recall that machines commonly have two basic states, up and down. The machine reliability models refer to the probability mass functions (PMFs) or probability density functions (PDFs) of a machine's uptime and downtime in discrete or continuous time.

Bernoulli Reliability Model In the Bernoulli reliability model, the probability distribution of the machine's uptime and downtime follows the Bernoulli distribution. More specifically, the machine is up with a probability of p, and it is down with a probability of $1 - p$. In the Bernoulli reliability model, the machine's state in each time step is independent of its states in all previous time steps. The Bernoulli reliability model of a machine at time step t can be expressed as

$$\begin{aligned} P_{\text{up}}(t) &= p \\ P_{\text{down}}(t) &= 1 - p, \ t = 1, 2, \ldots \end{aligned} \tag{4.2}$$

The Bernoulli reliability model is the most basic reliability model in discrete-time systems. It is simple but still practical, especially when a machine's downtime is very short and comparable with its cycle time.

Exponential Reliability Model   In the exponential reliability model, the probability density function of the machine's uptime and downtime follows the exponential distribution. More specifically, when the machine is up, the machine breaks down in each infinitesimal interval with a constant rate λ. When the machine is down, the machine becomes up in each infinitesimal interval with a constant rate μ. The exponential reliability model can be expressed as

$$f_{\text{up}}(t) = \lambda e^{-\lambda t}$$
$$f_{\text{down}}(t) = \mu e^{-\mu t}$$

(4.3)

The exponential reliability model is a commonly used reliability model in continuous-time systems. It is suitable for systems whose breakdown rate λ and repair rate μ are close to constant.

In practice, it may be difficult to directly measure the parameters for machine reliability models, such as probability p in the Bernoulli model or breakdown rate λ in the exponential model. Indeed, the exact PMF or PDF of machine uptime and downtime is hard to be obtained. Instead, what can be measured are average uptime and downtime, which are generally referred to as the mean time between failures (MTBF) and mean time to repair (MTTR). An example of the time to failure and the time to repair is shown in Figure 4.7.

After recording the machine state changes for a period of time, MTBF and MTTR can be calculated as

$$\text{MTBF} = \frac{\sum(\text{Start of downtime} - \text{Start of uptime})}{\text{Number of failures}}$$

(4.4)

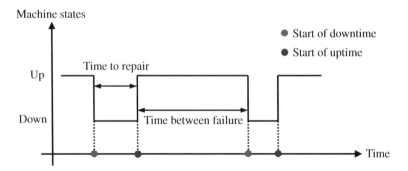

Figure 4.7   The schematic diagram of the time to failure and the time to repair.

$$\text{MTTR} = \frac{\sum(\text{Start of uptime} - \text{Start of downtime})}{\text{Number of repairs}} \tag{4.5}$$

If a machine is assumed to follow the Bernoulli reliability model, parameter p can be estimated as

$$p = 1 - \frac{1}{\text{MTBF}} \quad \text{or} \quad p = \frac{1}{\text{MTTR}} \tag{4.6}$$

If the machine is assumed to follow the exponential reliability model, parameters λ and μ can be estimated as

$$\lambda = \frac{1}{\text{MTBF}} \tag{4.7}$$

$$\mu = \frac{1}{\text{MTTR}} \tag{4.8}$$

4.1.2.3 Parameters of Aggregated Machines

Parallel Machines Before the machine aggregation, the parameters of each original machine should be known. Suppose k parallel machines are identical, i.e. each machine M_i ($i = 1, 2, ..., k$) has the same characteristics, i.e. τ, c, MTBF, and MTTR. Then the corresponding characteristics of the aggregated machine can be calculated as

$$\tau_A^{\|} = \frac{\tau}{k} \tag{4.9}$$

$$c_A^{\|} = kc \tag{4.10}$$

$$\text{MTBF}_A^{\|} = \text{MTBF} \tag{4.11}$$

$$\text{MTTR}_A^{\|} = \text{MTTR} \tag{4.12}$$

In the case where k machines are nonidentical, machine M_i ($i = 1, 2, ...k$) has its own characteristics, i.e. τ_i, c_i, MTBF_i, and MTTR_i. Then the corresponding characteristics of the aggregated machine can be calculated as follows:

$$\tau_A^{\|} = \frac{1}{\sum_{i=1}^{k} \frac{1}{\tau_i}} \tag{4.13}$$

$$c_A^{\|} = \sum_{i=1}^{k} c_i \tag{4.14}$$

$$\text{MTBF}_{\text{A}}^{\parallel} = \frac{\tau_{\text{agg}}^{\text{par}} \sum\limits_{i=1}^{k} \left[\dfrac{1}{\tau_i \text{MTTR}_i} \prod\limits_{j=1, j\neq i}^{k} \left(\dfrac{1}{\text{MTBF}_j} + \dfrac{1}{\text{MTTR}_j} \right) \right]}{\dfrac{1}{k} \sum\limits_{i=1}^{k} \left[\dfrac{1}{\text{MTBF}_i \text{MTTR}_i} \prod\limits_{j=1, j\neq i}^{k} \left(\dfrac{1}{\text{MTBF}_j} + \dfrac{1}{\text{MTTR}_j} \right) \right]} \tag{4.15}$$

$$\text{MTTR}_{\text{A}}^{\parallel} = \frac{\tau_{\text{agg}}^{\text{par}} \sum\limits_{i=1}^{k} \left[\dfrac{1}{\tau_i \text{MTBF}_i} \prod\limits_{j=1, j\neq i}^{k} \left(\dfrac{1}{\text{MTBF}_j} + \dfrac{1}{\text{MTTR}_j} \right) \right]}{\dfrac{1}{k} \sum\limits_{i=1}^{k} \left[\dfrac{1}{\text{MTBF}_i \text{MTTR}_i} \prod\limits_{j=1, j\neq i}^{k} \left(\dfrac{1}{\text{MTBF}_j} + \dfrac{1}{\text{MTTR}_j} \right) \right]} \tag{4.16}$$

Consecutive Dependent Machines Consider the situation where k consecutive dependent machines are identical, each machine M_i ($i = 1, 2, ...k$) is characterized by τ, c, MTBF, and MTTR. Then the corresponding parameters of the aggregated machine can be calculated by

$$\tau_{\text{A}}^{\rightarrow} = \tau \tag{4.17}$$

$$c_{\text{A}}^{\rightarrow} = c \tag{4.18}$$

$$\text{MTBF}_{\text{A}}^{\rightarrow} = \text{MTBF} \left(\frac{\text{MTBF}}{\text{MTBF} + \text{MTTR}} \right)^{k-1} \tag{4.19}$$

$$\text{MTTR}_{\text{A}}^{\rightarrow} = (\text{MTBF} + \text{MTTR}) - \text{MTBF} \left(\frac{\text{MTBF}}{\text{MTBF} + \text{MTTR}} \right)^{k-1} \tag{4.20}$$

In the case where k machines are nonidentical, machine M_i ($i = 1, 2, ...k$) is characterized by τ_i, c_i, MTBF_i, and MTTR_i. Then the corresponding parameters of the aggregated machine can be calculated by

$$\tau_A^{\rightarrow} = \max_i \tau_i \tag{4.21}$$

$$c_A^{\rightarrow} = \min_i c_i \tag{4.22}$$

$$\text{MTBF}_{\text{A}}^{\rightarrow} = \frac{1}{k} \sum_{i=1}^{k} (\text{MTBF}_i + \text{MTTR}_i) \prod_{i=1}^{k} \left(\frac{\text{MTBF}_i}{\text{MTBF}_i + \text{MTTR}_i} \right) \tag{4.23}$$

$$\text{MTTR}_{\text{A}}^{\rightarrow} = \frac{1}{k} \sum_{i=1}^{k} (\text{MTBF}_i + \text{MTTR}_i) \left[1 - \prod_{i=1}^{k} \left(\frac{\text{MTTR}_i}{\text{MTBF}_i + \text{MTTR}_i} \right) \right] \tag{4.24}$$

4.1.2.4 Mathematical Model of Buffers

The state of a buffer represents the number of parts contained in that buffer, and thus, the buffer's state is also referred to as buffer occupancy. The only parameter of a buffer is its capacity, which represents the maximum number of parts that a buffer can store. If the capacity of a buffer is large enough that it can never be fully filled, it is assumed to be an infinite buffer. This assumption can significantly reduce the complexity of system dynamics as one does not need to consider what would happen if buffers are full. However, a large buffer usually requires larger space and more material handling equipment, which may lead to relatively higher operational cost. Hence, in this book, the buffer is considered to have a finite capacity $N < \infty$. The state of such a buffer can be any integer between 0 (empty) and N (full).

Different from the modeling of machines, buffers do not have a timing issue. More specifically, it is assumed that WIP can be placed in a buffer immediately if the buffer is not full. Similarly, it is assumed that WIP stored in a buffer is immediately available to be processed by a machine. Under these assumptions, the timing of state change completely depends on the machine cycle time and machine reliability. Although, in practice, it takes some time for WIP to be loaded into and unloaded from the buffer, the loading and unloading time can be considered as part of the machine cycle time. Therefore, these assumptions can simplify the buffer model without affecting model accuracy.

4.1.2.5 Interaction Between Machines and Buffers
In this book, the following conventions are adopted to standardize the mathematical description of a discrete-time production system.

- The states of machines are determined at the beginning of each time step.
- Given the machine states, the states of buffers are determined at the end of that time step.

In manufacturing systems, the states of machines and buffers can influence each other, i.e. the machine's operation states can affect the buffer state transition. On the other hand, the buffer's occupancy can also affect a machine's operation states due to blockage and starvation.

4.1.2.6 Buffer State Transition
The state of a buffer at time step t is determined by its previous state at time $t - 1$ and its adjacent machines' states at time t. For example, considering a segment of a production line as shown in Figure 4.8, the ith machine is denoted by M_i, and the ith buffer is denoted by B_i. The duration of each time step equals the machine cycle

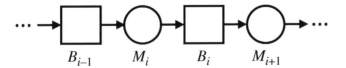

Figure 4.8 The schematic diagram of a segment of a production line.

time, i.e. a machine can process one part in a time step if it is up. Suppose that buffer B_i is neither full nor empty, then the state of buffer B_i at time t can be calculated as

$$c_i(t) = c_i(t-1) + s_i^t - s_{i+1}^t \qquad (4.25)$$

In Eq. (4.25), $c_i(t)$ is the state of buffer B_i at the end of time step t, and s_i^t is the state of machine M_i at the beginning of time step t. If the machine can process operations on parts during this time step, $s_i^t = 1$; otherwise, $s_i^t = 0$.

4.1.2.7 Blockage and Starvation

In the preceding example, a critical assumption is that the buffer is neither full nor empty. Under this assumption, machines M_i and M_{i+1} can always place the part in or remove it from buffer B_i as needed. Hence, the state of buffer B_i does not affect its adjacent machines' operation states in this situation. However, if it is full or empty, blockage or starvation may occur, which may lead to changes in machines' operation states.

Blockage Blockage means that after a machine processes a part, its downstream buffer does not have enough space to store that part. Hence, this part remains in the machine, which prevents the machine from processing the next part. Different situations where the machine is in normal operation or undergoes blockage are illustrated in Figure 4.9.

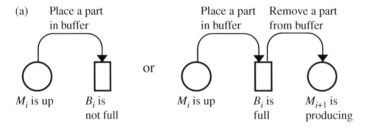

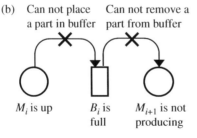

Figure 4.9 Situations where (a) M_i is not blocked, and (b) M_i is blocked.

More specifically, blockage of machine M_i occurs at time t if the following three conditions hold simultaneously: (i) machine M_i is up at the beginning of time t; (ii) machine M_{i+1} does not process any part at the beginning of time t, and (iii) downstream buffer B_i is full at the end of time $t-1$. There are two types of blockage conventions:

- Blocked before service (BBS): when the three blockage conditions are met, the machine is blocked and cannot process a part immediately. In the BBS convention, the space where a machine holds and processes parts is counted as part of its downstream buffer storage room.
- Blocked after service (BAS): when the three blockage conditions are met, the machine can still process one part if its upstream buffer is not empty. The machine becomes blocked if the three conditions persist during the next time step.

It is clear that the capacity of buffers B_i under BBS and BAS conventions can be related by Eq. (4.26).

$$N_i^{\text{BBS}} = N_i^{\text{BAS}} + 1 \tag{4.26}$$

The system modeling procedures under these two conventions are quite similar. The BBS convention is employed in the following chapters as it leads to a more straightforward mathematical description.

Starvation   Starvation refers to the situation where a machine is up, but its upstream buffer is empty, and thus the buffer has no part to process. Different situations where starvation can or cannot occur are illustrated in Figure 4.10.

More specifically, starvation of machine M_i occurs at time t if the following two conditions hold simultaneously: (i) machine M_i is up at the beginning of time t, and (ii) upstream buffer B_{i-1} is empty at the end of time $t-1$.

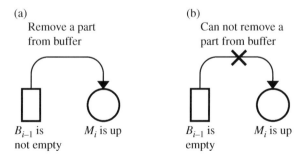

Figure 4.10   Situations where (a) M_i is not starved, and (b) M_i is starved.

4.1.3 Performance Measures

Aided by mathematical models of machines and buffers, the performance measures of a manufacturing system can be calculated and analyzed. The mathematical expressions of the system performance measures are introduced next.

4.1.3.1 Blockage and Starvation

A general relationship among blockage (denoted by $\mathbb{B}$), starvation (denoted by $\mathbb{S}$), and the machine's up and down states are depicted in Figure 4.11. Specifically, when a machine is up, it could be blocked or starved. If the machine is blocked or starved, it cannot process any part, and thus, the machine enters the idle state. Otherwise, if the machine can process during a time step, the machine is busy.

In reference to the example in Figure 4.8, the probabilities that machine M_i is blocked and starved at time t (denoted by $P_i^{\mathbb{B}}(t)$ and $P_i^{\mathbb{S}}(t)$, respectively) can be expressed by

$$P_i^{\mathbb{B}}(t) = P[\{M_i \text{ is up at the beginning of time } t\}$$
$$\cap \{M_{i+1} \text{ is down or idle at the beginning of time } t\}$$
$$\cap \{B_i \text{ is full at the end of time } t-1\}] \qquad (4.27)$$

$$P_i^{\mathbb{S}}(t) = P[\{M_i \text{ is up at the beginning of } t\}$$
$$\cap \{B_{i-1} \text{ is empty at the end of time } t-1\}] \qquad (4.28)$$

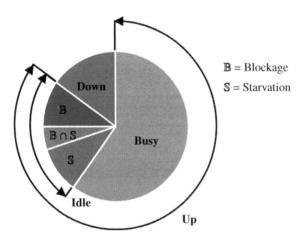

Figure 4.11 The relationship among blockage, starvation, up and down states of a machine.

4.1.3.2 Production Rate and Throughput

The production rate (ρ) of a particular machine is the average number of parts processed by that machine during each cycle time. The production rate of the entire production line is the average number of final products produced in each cycle time. Production rate is an important performance measure in synchronous systems, and the production rate of the entire production line equals the slowest machine's production rate without considering the reliability of machines.

Similarly, the throughput (n) of a machine or a system is the average number of parts processed by the machine or the number of final products produced by the system in a period of time. Throughput can be used to describe the performance of both synchronous and asynchronous systems. In a synchronous system, the relationship between ρ and n can be expressed as

$$n = c \cdot \rho \tag{4.29}$$

where c is the number of cycle time during the production horizon.

In general, due to the interaction between machines and buffers, the production rate or throughput of a manufacturing system is a complex nonlinear function involving all parameters of machines and buffers. Therefore, the closed-form expressions of ρ and n at the steady state are almost impossible to derive for production lines with more than two machines, subject to arbitrary time distributions. For production lines with only two machines, the detailed derivations of the closed-form expression of ρ at steady state are provided in Section 4.2 under some strict assumptions of their time distributions. For more general multi-machine production lines, an approximation method is often used to analyze ρ, which are introduced in Section 4.3.

4.1.3.3 Work-in-process

Work-in-process refers to the semifinished products or parts stored in the buffer. With a slight abuse of terminology, work-in-process of buffer B_i, denoted by WIP_i, represents the average number of parts contained in the buffer.

The total work-in-process of a system (WIP_{SYS}) is the average number of parts contained in all buffers. For example, for a serial production line with N machines and $N-1$ buffers, WIP_{SYS} can be calculated as

$$\text{WIP}_{\text{SYS}} = \sum_{i=1}^{N-1} \text{WIP}_i \tag{4.30}$$

Similar to system production rate and throughput, the closed-form expressions of P_i and WIP_{SYS} are almost impossible to derive. Again, the commonly used techniques for analyzing work-in-process for two-machine production line and complex multi-machine production lines are presented next.

4.2 Two-machine Production Lines

The performance analysis of a manufacturing system is a complex process, and the complexity is determined by the diversity of system states. Along with the increases in the number of machines and buffers, the number of possible system states increases exponentially. Therefore, before addressing the problems for a complex manufacturing system, a simple system consisting of only two machines and one buffer is used to demonstrate the procedures for system modeling and performance measures. Note that in the following discussions, it is assumed that all the machines are Bernoulli machines.

4.2.1 Conventions and Notations

The system layout discussed in this section is illustrated in Figure 4.12, which is a synchronous system with the length of each time step equaling the machine cycle time. The probability that machine M_1 is up in each time step is set as p_1 and that of machine M_2 is p_2. The capacity of buffer B is C.

4.2.1.1 Assumptions

The following assumptions, which are true in most practical production systems, are needed to describe the preceding system:

1) Blocked before service.
2) The first machine is never starved, and the second machine is never blocked.
3) The states of machines are determined at the beginning of each time step, and the state of the buffer is determined at the end of each time step.
4) Whether or not a machine is up or down, its state is independent of the other machine's state.
5) Buffer's state (occupancy) can change at most by one in each time step.

4.2.1.2 Notations

The following notations are adopted to represent the states of machines and buffer [2–4].

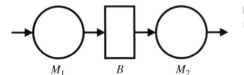

Figure 4.12 The layout of a two-machine production line.

For machines M_i ($i = 1, 2$):

$\left\{M_i^{\mathbb{P}}\right\} \equiv \{M_i \text{ is processing a part}\}$

$\left\{M_i^{\overline{\mathbb{P}}}\right\} \equiv \{M_i \text{ is not processing a part}\}$

$\left\{M_i^{\mathbb{S}}\right\} \equiv \{M_i \text{ is starved}\}$

$\left\{M_i^{\overline{\mathbb{S}}}\right\} \equiv \{M_i \text{ is not starved}\}$

$\left\{M_i^{\mathbb{B}}\right\} \equiv \{M_i \text{ is blocked}\}$ $\quad\quad\quad$ (4.31)

$\left\{M_i^{\overline{\mathbb{B}}}\right\} \equiv \{M_i \text{ is not blocked}\}$

$\left\{M_i^1\right\} \equiv \{M_i \text{ is up}\}$

$\left\{M_i^0\right\} \equiv \{M_i \text{ is down}\}$

According to the second assumption that machine M_1 is never starved and M_2 is never blocked, the relationships among different machine states are illustrated in Figure 4.13. The probabilities of machines being in each state can be mathematically formulated in (4.32).

$$P\left[\left\{M_1^{\mathbb{P}}\right\}\right] \equiv P\left[\left\{M_1^1\right\} \cap \left\{M_1^{\overline{\mathbb{B}}}\right\}\right]$$

$$P\left[\left\{M_1^{\overline{\mathbb{P}}}\right\}\right] \equiv P\left[\left\{M_1^0\right\} \cup \left\{M_1^{\mathbb{B}}\right\}\right]$$

$$P\left[\left\{M_2^{\mathbb{P}}\right\}\right] \equiv P\left[\left\{M_2^1\right\} \cap \left\{M_2^{\overline{\mathbb{S}}}\right\}\right]$$ (4.32)

$$P\left[\left\{M_2^{\overline{\mathbb{P}}}\right\}\right] \equiv P\left[\left\{M_2^0\right\} \cup \left\{M_2^{\mathbb{S}}\right\}\right]$$

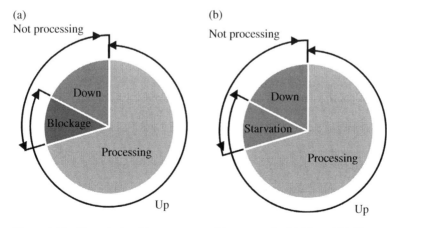

Figure 4.13 The relationships among machine states for (a) M_1 and (b) M_2.

For buffer B:

$$\{c(t) = j\} \equiv \{B \text{ contains } j \text{ parts at the end of time step } t\}$$
$$q_j(t) \equiv P[\{c(t) = j\}] \tag{4.33}$$

4.2.2 State Transition

Given that the buffer state can change at most by one in each time step, the state transition diagram is shown in Figure 4.14.

In Figure 4.14, $Q_{j_2 j_1}(t)$ represents the transition probability from state j_1 at time $t - 1$ to state j_2 at time t. Prior to the deviations of the state transition possibility, the following four lemmas shall be introduced [4].

Lemma 4.1 *Let A, B, and C be three sets. If $C \subseteq B$, then $P(A \cap B \mid C) = P(A \mid C)$*

Proof:

$$P(A \cap B \mid C) = \frac{P(A \cap B \cap C)}{P(C)} = \frac{P(A \cap C)}{P(C)} = P(A \mid C) \tag{4.34}$$

Lemma 4.2 *Let A, B, and C be three sets. If $P(A \mid B \cap C) = 0$, then $P(A \cap B \mid C) = 0$*

Proof:

$$P(A \cap B \mid C) = \frac{P(A \cap B \cap C)}{P(C)} = \frac{P(A \mid B \cap C)P(B \cap C)}{P(C)} = 0 \tag{4.35}$$

Lemma 4.3 *Let A, B, and C be three sets. If $P(A \mid B \cap C) = 1$, then $P(A \cap B \mid C) = P$*
(B | C)

Proof:

$$P(A \cap B \mid C) = \frac{P(A \cap B \cap C)}{P(C)} = \frac{P(A \mid B \cap C)P(B \cap C)}{P(C)} = \frac{P(B \cap C)}{P(C)} = P(B \mid C) \tag{4.36}$$

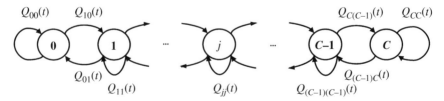

Figure 4.14 State transition diagram of buffer B.

Lemma 4.4 *Let A, B, C and D be four sets. If $CD \subseteq B$, then $P(A \cap B \cap C | D) = P(A \cap C | D)$*

Proof:

$$P(A \cap B \cap C \mid D) = \frac{P(A \cap B \cap C \cap D)}{P(D)} = \frac{P(A \cap C \cap D)}{P(D)} = P(A \cap C \mid D)$$

$$(4.37)$$

4.2.2.1 State Transition Probabilities

Since the buffer state can change at most by one in each time step, the state transition probability $Q_{j_2 j_1}(t) = 0$, if $|j_2 - j_1| > 1$. Otherwise

1) The buffer state transition probability from state 0 to itself is

$$
\begin{aligned}
Q_{00}(t) &= P[\{c(t) = 0\} | \{c(t-1) = 0\}] \\
&= P\left[\left\{M_1^{\overline{\mathbb{P}}}\right\} \cap \left\{M_2^{\overline{\mathbb{P}}}\right\} | \{c(t-1) = 0\}\right] \\
&= P\left[\{M_1^0\} \cup \left\{M_1^{\overline{\mathbb{B}}}\right\} | \{c(t-1) = 0\}\right] \\
&= P\left[\{M_1^0\}\right] \\
&= 1 - p_1
\end{aligned}
$$

$$(4.38)$$

where the elimination of $\left\{M_2^{\overline{\mathbb{P}}}\right\}$ and $\left\{M_1^{\mathbb{B}}\right\}$ is due to

$$
\begin{aligned}
&\{c(t-1) = 0\} \subseteq \left\{M_2^{\overline{\mathbb{P}}}\right\}, (by\ Lemma\ 1) \\
&P\left[\left\{M_1^{\mathbb{B}}\right\} | \{c(t-1) = 0\}\right] = 0
\end{aligned}
$$

$$(4.39)$$

The state transition probability from state 0 to state 1 is

$$
\begin{aligned}
Q_{10}(t) &= P[\{c(t) = 1\} | \{c(t-1) = 0\}] \\
&= P\left[\{M_1^{\mathbb{P}}\} \cap \left\{M_2^{\overline{\mathbb{P}}}\right\} | \{c(t-1) = 0\}\right] \\
&= P\left[\{M_1^1\} \cap \left\{M_1^{\overline{\mathbb{B}}}\right\} | \{c(t-1) = 0\}\right] \\
&= P\left[\{M_1^1\}\right] \\
&= p_1
\end{aligned}
$$

$$(4.40)$$

where the elimination of $\left\{M_2^{\overline{\mathbb{P}}}\right\}$ and $\left\{M_1^{\overline{\mathbb{B}}}\right\}$ is due to

$$
\begin{aligned}
&\{c(t-1) = 0\} \subseteq \left\{M_2^{\overline{\mathbb{P}}}\right\}, (by\ Lemma\ 1) \\
&\{c(t-1) = 0\} \subseteq \left\{M_1^{\overline{\mathbb{B}}}\right\}, (by\ Lemma\ 1)
\end{aligned}
$$

$$(4.41)$$

2) The state transition probability from state j to itself is

$$
\begin{aligned}
Q_{jj}(t) &= P[\{c(t) = j\} | \{c(t-1) = j\}] \\
&= P\Big[\{M_1^{\mathbb{P}}\} \cap \{M_2^{\mathbb{P}}\} \cup \{M_1^{\overline{\mathbb{P}}}\} \cap \{M_2^{\overline{\mathbb{P}}}\} | \{c(t-1) = j\}\Big] \\
&= P[\{M_1^1\} \cap \cancel{\{M_1^{\overline{\mathbb{B}}}\}} \cap \{M_2^1\} \cap \cancel{\{M_2^{\overline{\mathbb{S}}}\}} | \{c(t-1) = j\}] \\
&\quad + P[\{M_1^0\} \cup \cancel{\{M_1^{\mathbb{B}}\}} \cap \{M_2^0\} \cup \cancel{\{M_2^{\mathbb{S}}\}} | \{c(t-1) = j\}] \\
&= P[\{M_1^1\} \cap \{M_2^1\}] + P[\{M_1^0\} \cap \{M_2^0\}] \\
&= p_1 p_2 + (1-p_1)(1-p_2)
\end{aligned}
\tag{4.42}
$$

where the elimination of $\{M_1^{\overline{\mathbb{B}}}\}$, $\{M_2^{\overline{\mathbb{S}}}\}$, $\{M_1^{\mathbb{B}}\}$, and $\{M_2^{\mathbb{S}}\}$ is due to

$$
\begin{aligned}
&\{c(t-1) = j\} \subseteq \{M_1^{\overline{\mathbb{B}}}\}, (by\ Lemma\ 1) \\
&\{c(t-1) = j\} \subseteq \{M_2^{\overline{\mathbb{S}}}\}, (by\ Lemma\ 1) \\
&P[\{M_1^{\mathbb{B}}\} | \{c(t-1) = j\}] = 0 \\
&P[\{M_2^{\mathbb{S}}\} | \{c(t-1) = j\}] = 0
\end{aligned}
\tag{4.43}
$$

The state transition probability from state $j+1$ to state j is

$$
\begin{aligned}
Q_{j(j+1)}(t) &= P[\{c(t) = j\} | \{c(t-1) = j+1\}] \\
&= P\Big[\{M_1^{\overline{\mathbb{P}}}\} \cap \{M_2^{\mathbb{P}}\} | \{c(t-1) = j+1\}\Big] \\
&= P[\{M_1^0\} \cup \cancel{\{M_1^{\mathbb{B}}\}} \cap \{M_2^1\} \cap \cancel{\{M_2^{\overline{\mathbb{S}}}\}} | \{c(t-1) = j+1\}] \\
&= P[\{M_1^0\} \cap \{M_2^1\}] \\
&= (1-p_1)p_2
\end{aligned}
\tag{4.44}
$$

where the elimination of $\{M_1^{\mathbb{B}}\}$ and $\{M_2^{\overline{\mathbb{S}}}\}$ is due to

$$
\begin{aligned}
&P[\{M_1^{\mathbb{B}}\} | \{M_2^{\mathbb{P}}\} \cap \{c(t-1) = j+1\}] = 0, (by\ Lemma\ 2) \\
&\{c(t-1) = j+1\} \subseteq \{M_2^{\overline{\mathbb{S}}}\}, (Lemma\ 1)
\end{aligned}
\tag{4.45}
$$

The state transition probability from state j to state $j+1$ is:

$$
\begin{aligned}
Q_{(j+1)j}(t) &= P[\{c(t) = j+1\} | \{c(t-1) = j\}] \\
&= P\Big[\{M_1^{\mathbb{P}}\} \cap \{M_2^{\overline{\mathbb{P}}}\} | \{c(t-1) = j\}\Big] \\
&= P[\{M_1^1\} \cap \cancel{\{M_1^{\overline{\mathbb{B}}}\}} \cap \{M_2^0\} \cup \cancel{\{M_2^{\mathbb{S}}\}} | \{c(t-1) = j\}] \\
&= P[\{M_1^1\} \cap \{M_2^0\}] \\
&= p_1(1-p_2)
\end{aligned}
\tag{4.46}
$$

where the elimination of $\left\{M_1^{\overline{\mathbb{B}}}\right\}$ and $\left\{M_2^{\mathbb{S}}\right\}$ is due to

$$\{c(t-1) = j\} \subseteq \left\{M_1^{\overline{\mathbb{B}}}\right\},(by\ Lemma\ 1)$$

$$P\left[\{M_2^{\mathbb{S}}\}|\{c(t-1) = j\}\right] = 0$$

(4.47)

3) The state transition probability from state C to itself is

$$
\begin{aligned}
Q_{CC}(t) &= P[\{c(t) = C\}|\{c(t-1) = C\}] \\
&= P\left[\{M_1^{\mathbb{P}}\} \cap \{M_2^{\mathbb{P}}\} \cup \left\{M_1^{\overline{\mathbb{P}}}\right\} \cap \left\{M_2^{\overline{\mathbb{P}}}\right\}|\{c(t-1) = C\}\right] \\
&= P\left[\{M_1^1\} \cap \left\{M_1^{\overline{\mathbb{B}}}\right\} \cap \{M_2^1\} \cap \left\{M_2^{\overline{\mathbb{S}}}\right\}|\{c(t-1) = C\}\right] \\
&\quad + P\left[\{M_2^0\} \cup \left\{M_2^{\mathbb{S}}\right\}|\{c(t-1) = C\}\right] \\
&= P\left[\{M_1^1\} \cap \{M_2^1\}\right] + P\left[\{M_2^0\}\right] \\
&= p_1 p_2 + (1 - p_2)
\end{aligned}
$$

(4.48)

where the elimination of $\left\{M_1^{\overline{\mathbb{P}}}\right\}, \left\{M_1^{\overline{\mathbb{B}}}\right\}, \left\{M_2^{\overline{\mathbb{S}}}\right\}$, and $\left\{M_2^{\mathbb{S}}\right\}$ is due to

$$P\left[\left\{M_1^{\overline{\mathbb{P}}}\right\}|\left\{M_2^{\overline{\mathbb{P}}}\right\} \cap \{c(t-1) = C\}\right] = 1,(by\ Lemma\ 3)$$

$$\{M_2^{\mathbb{P}}\} \cap \{c(t-1) = C\} \subseteq \left\{M_1^{\overline{\mathbb{B}}}\right\},(by\ Lemma\ 4)$$

$$\{c(t-1) = C\} \subseteq \left\{M_2^{\overline{\mathbb{S}}}\right\},(by\ Lemma\ 1)$$

$$P\left[\{M_2^{\mathbb{S}}\}|\{c(t-1) = C\}\right] = 0$$

(4.49)

Note that all the state transition probabilities in the above two-machine production line are static, i.e. the state transition probabilities do not vary with time.

4.2.2.2 System Dynamics

According to the law of total probability, the evolution of the probability that buffer B is in state j at time t can be calculated by solving the following set of equations:

$$q_j(t) = \sum_{k=0}^{C} Q_{jk}\, q_k(t-1), \quad i = 0,1,2,...,c$$

(4.50)

$$\sum_{j=0}^{C} q_j(t) = 1$$

(4.51)

4.2.3 Steady-state Probabilities

Since the transition probabilities are constants, i.e. they are independent of time t, the system presented here can be regarded as a stationary discrete-time Markov chain. Let q_j denote the steady-state probability of buffer in state j. The steady state of the above discrete-time Markov chain can be calculated using the following equations:

$$q_j = \sum_{k=0}^{C} Q_{jk} \, q_k, \quad j = 0,1,2,...,c \tag{4.52}$$

$$\sum_{j=0}^{C} q_j = 1 \tag{4.53}$$

Substitute the transition probabilities discussed in the previous section into (4-52):

$$q_0 = (1-p_1)q_0 + (1-p_1)p_2 q_1$$

$$q_1 = p_1 q_0 + [p_1 p_2 + (1-p_1)(1-p_2)]q_1 + (1-p_1)p_2 q_2$$

$$...$$

$$q_j = p_1(1-p_2)q_{j-1} + [p_1 p_2 + (1-p_1)(1-p_2)]q_j + (1-p_1)p_2 q_{j+1}$$

$$...$$

$$q_C = p_1(1-p_2)q_{C-1} + [p_1 p_2 + (1-p_2)]q_C \tag{4.54}$$

The above equations can be expressed as functions of p_1, p_2, and q_0, i.e.

$$q_1 = \frac{p_1}{(1-p_1)p_2} q_0$$

$$q_2 = \frac{p_1{}^2(1-p_2)}{(1-p_1)^2 p_2{}^2} q_0$$

$$...$$

$$q_j = \frac{p_1{}^j(1-p_2)^{j-1}}{(1-p_1)^j p_2{}^j} q_0 \tag{4.55}$$

$$...$$

$$q_C = \frac{p_1{}^C(1-p_2)^{C-1}}{(1-p_1)^C p_2{}^C} q_0$$

To make the formula more concise, a parameter α is defined as

$$\alpha = \frac{p_1(1-p_2)}{(1-p_1)p_2} \tag{4.56}$$

Substitute α into (4.55), the steady-state probabilities of two-machine production lines can be expressed as

$$q_j = \frac{\alpha^j}{1 - p_2} q_0, \quad j = 1, 2, ..., C \tag{4.57}$$

4.2.3.1 Identical Machines

In the case where two machines are identical, i.e. $p_1 = p_2 = p$, parameter $\alpha = 1$. Substitute (4.57) into (4.53), the PMF of buffer occupancy in two-machine production line with identical Bernoulli machines is

$$q_j = \begin{cases} \dfrac{1-p}{1-p+C}, & j = 0 \\[2mm] \dfrac{1}{1-p+C}, & j = 1, 2, ..., C \end{cases} \tag{4.58}$$

Equation (4.58) implies that in the steady state, the buffer is less likely to be empty, and it has an identical probability for every non-empty state. In addition, when the machine reliability increases, i.e. $p \to 1$, the PMF of the buffer occupancy has the following property:

$$\lim_{p \to 1} q_j = \begin{cases} 0, & j = 0 \\[2mm] \dfrac{1}{C}, & j = 1, 2, ..., C \end{cases} \tag{4.59}$$

Example 4.1 Consider a production line with two identical Bernoulli machines, M_1 and M_2. The probability that the machine is up at each time step is p, and the buffer capacity $C = 5$. Plot the stationary PMF of the buffer occupancy when $p = 0.5$, 0.9, and 0.99.

Solution

According to (4.58) we have

$p = 0.5$:	$p = 0.9$:	$p = 0.99$:
$q_j = \begin{cases} 0.091, & j = 0 \\ 0.182, & j = 1,...,5 \end{cases}$	$q_j = \begin{cases} 0.020, & j = 0 \\ 0.196, & j = 1, ..., 5 \end{cases}$	$q_j = \begin{cases} 0.020, & j = 0 \\ 0.196, & j = 1,...,5 \end{cases}$

As p increases, the chances of empty buffer approach zero, as shown in Figure 4.15, while the chances for buffer values at non-zero values tend be at $1/5 = 0.2$ as predicted by (4.59).

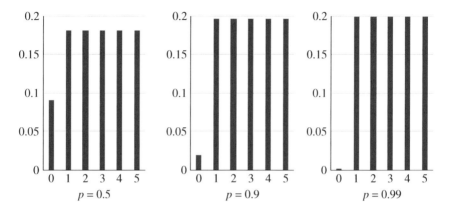

Figure 4.15 PMF of buffer occupancy in two-machine line with identical Bernoulli machines.

4.2.3.2 Nonidentical Machines

When two machines are nonidentical, i.e. $p_1 \neq p_2$ and $\alpha \neq 1$, the probability that the buffer is empty can be calculated as

$$P_0 = \frac{(1-\alpha)(1-p_1)}{1 - \frac{p_1}{p_2}\alpha^C} \tag{4.60}$$

The probabilities that buffer is in other non-empty states can be solved consecutively based on Eq. (4.57). In principle, if the first machine is more reliable than the second machine, i.e. $p_1 > p_2$, the first machine can process more parts than the second machine within the same period. In this case, the buffer tends to be full. On the contrary, if the second machine is more reliable, i.e. $p_1 < p_2$, the buffer tends to be empty. This phenomenon can be further demonstrated using the following example.

Example 4.2 Consider a production line with two Bernoulli machines, M_1 and M_2. The probabilities that two machines are up at each time step are p_1 and p_2, respectively. Suppose that the buffer capacity $C = 10$, plot the stationary PMFs of the buffer state when 1) $p_1 = 0.9$, and $p_2 = 0.8$, and 2) $p_1 = 0.8$, and $p_2 = 0.9$.

Solution

According to (4.57) and (4.60), the PMFs in two cases are shown in Figure 4.16.

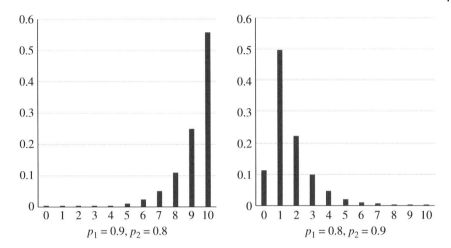

$p_1 = 0.9, p_2 = 0.8$ $p_1 = 0.8, p_2 = 0.9$

Figure 4.16 PMFs of buffer occupancy in a two-machine production line with nonidentical Bernoulli machines.

4.2.4 Performance Measures

4.2.4.1 Blockage and Starvation

For a production line with two Bernoulli machines, it is assumed that the first machine is never starved, and the second machine is never blocked. According to the definitions of blockage and starvation as introduced in Section 4.1, the probabilities that M_1 is blocked and M_2 is starved during time t can be calculated as follows:

$$
\begin{aligned}
P\big[\{M_1^{\mathbb{B}}\}\big] &= P\big[\{M_1^1\} \cap \{M_2^0\} \cap \{c(t-1) = C\}\big] \\
&= p_1(1-p_2)q_C(t-1)
\end{aligned}
\tag{4.61}
$$

$$
\begin{aligned}
P\big[\{M_2^{\mathbb{S}}\}\big] &= P\big[\{M_2^1\} \cap \{c(t-1) = 0\}\big] \\
&= p_2 q_0(t-1)
\end{aligned}
\tag{4.62}
$$

4.2.4.2 Production Rate

The production rate of the first machine during time step t, denoted by $\mathrm{PR}_1(t)$, can be expressed as

$$
\begin{aligned}
\rho_1(t) &= P\big[\{M_1^{\mathbb{P}}\}\big] \\
&= P\Big[\{M_1^1\} \cap \big\{M_1^{\overline{\mathbb{B}}}\big\}\Big] \\
&= P\big[\{M_1^1\}\big]\big(1 - P\big[\{M_2^0\} \cap \{c(t-1) = C\}\big]\big) \\
&= p_1(1 - (1-p_2)q_C(t-1))
\end{aligned}
\tag{4.63}
$$

The production rate of the second machine during time step t, which is also the production rate of the system, can be written as

$$
\begin{aligned}
\rho_{\text{SYS}}(t) = \rho_2(t) &= P\big[\{M_2^{\mathbb{P}}\}\big] \\
&= P\Big[\{M_2^1\} \cap \{M_2^{\bar{S}}\}\Big] \\
&= P\big[\{M_2^1\}\big]\big(1 - P[\{c(t-1) = 0\}]\big) \\
&= p_2(1 - q_0(t-1))
\end{aligned}
\tag{4.64}
$$

The production rate under steady state can be calculated by replacing $q_C(t-1)$ and $q_0(t-1)$ in Eqs. (4.63) and (4.64) with the steady-state probabilities q_C and q_0.

4.2.4.3 Work-in-process

The work-in-process of the buffer B during time step t, denoted by $\text{WIP}(t)$, is

$$
\text{WIP}(t) = \sum_{j=1}^{C} j q_j(t)
\tag{4.65}
$$

The steady-state work-in-process, denoted by WIP, can be calculated by replacing $q_j(t)$ with steady-state probabilities q_j. If two machines are identical, according to Eq. (4.58), WIP can be expressed as

$$
\begin{aligned}
\text{WIP} &= \sum_{j=1}^{C} j \frac{1}{1 - p + C} \\
&= \frac{C(C+1)}{2(1 - p + C)}
\end{aligned}
\tag{4.66}
$$

In the case of two nonidentical machines, according to Eqs. (4.57) and (4.60), WIP can be written as

$$
\text{WIP} = \frac{p_1}{p_2 - p_1 \alpha^C}\left(\frac{1 - \alpha^C}{1 - \alpha} - C\alpha^C\right)
\tag{4.67}
$$

4.3 Multi-machine Production Lines

In the two-machine production line as discussed in the previous section, the state transition probabilities are static, and thus the closed-form expressions of performance measures at the steady state can be derived. However, in a more complex multi-machine production line, the state transition probabilities are dynamic, and the exact analytical formulas for its performance measures are infeasible to be derived. Therefore, different approximate methods, such as aggregation-based method, decomposition-based method, and iteration-based method, are developed to analyze the system performance. In this section, the iteration-based method is introduced since it can be employed in both steady-state and transient analyses. Note that the machines discussed in this section are assumed to be Bernoulli machines.

4.3.1 Assumptions and Notations

The system layout of a typical serial production line with N machines and $N - 1$ buffers is shown in Figure 4.17. The time is slotted, and the length for each time step equals the machine cycle time. Machine M_i ($i = 1, 2, ..., N$) is up in each cycle time with a probability of p_i, and it is down with a probability of $1 - p_i$. The capacity of the buffer B_i is C_i ($i = 1, 2, ..., N - 1$).

4.3.1.1 Assumptions

Similar to the two-machine production lines, the following assumptions are made to describe the system in Figure 4.17 according to [3, 4]:

1) Blocked before service.
2) The first machine is never starved, and the last machine is never blocked.
3) The states of machines are determined at the beginning of each time step, and the states of buffers are determined at the end of each time step.
4) Whether or not a machine is up or down, its state is independent of the other machines' states.
5) Each buffer states (occupancy) can change at most by one in each time step.

4.3.1.2 Notations

In addition to the notations used in two-machine production line, let $\mathbb{S}_i(t)$ and $\mathbb{B}_i(t)$ denote the probabilities of starvation and blockage of machine M_i at time t.

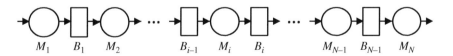

Figure 4.17 The layout of a production line with N machines and $N - 1$ buffers.

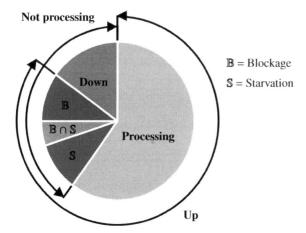

Figure 4.18 The relationships among different machine states.

The relationships among different machine states are depicted in Figure 4.18. The probabilities of machines in each state are mathematically formulated as

$$P[\{M_i^{\mathbb{P}}\}] \equiv P\left[\{M_i^1\} \cap \{M_i^{\overline{\mathbb{B}}}\} \cap \{M_i^{\overline{\mathbb{S}}}\}\right]$$

$$P\left[\{M_i^{\overline{\mathbb{P}}}\}\right] \equiv P[\{M_i^0\} \cup \{M_i^{\mathbb{B}}\} \cup \{M_i^{\mathbb{S}}\}]$$

$$P[\{M_1^{\mathbb{S}}\}] \equiv 0$$

$$P[\{M_N^{\mathbb{B}}\}] \equiv 0$$

$$(4.68)$$

For buffer B_i:

$$\{c_i(t) = j\} \equiv \{B_i \text{ contains } j \text{ parts at the end of time step } t\}$$

$$q_{i,j}(t) \equiv P[\{c_i(t) = j\}] \tag{4.69}$$

4.3.2 State Transition

Let $Q_{i,(j_2|j_1)}(t)$ represent the transition possibility of buffer B_i in state j_2 at time t given that it was in state j_1 at time $t-1$. The state transition diagram for B_i is shown in Figure 4.19. Similar to the two-machine production line, the buffer state can change at most by one in each time step. Therefore, the state transition probability $Q_{i,(j_2|j_1)}(t) = 0$, if $|j_2 - j_1| > 1$. The state transition matrix is shown:

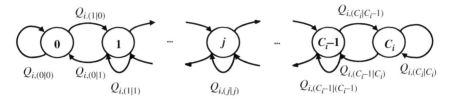

Figure 4.19 State transition diagram of buffer B_i.

$$Q_i(t) = \begin{bmatrix} \ddots & & \ddots & & \ddots & \\ \ddots & Q_{i,(j-1|j-1)}(t) & Q_{i,(j-1|j)}(t) & & 0 & \\ \ddots & Q_{i,(j|j-1)}(t) & Q_{i,(j|j)}(t) & Q_{i,(j|j+1)}(t) & & \ddots \\ & 0 & Q_{i,(j+1|j)}(t) & Q_{i,(j+1|j+1)}(t) & & \ddots \\ & & \ddots & & \ddots & \ddots \end{bmatrix} \qquad (4.70)$$

The state probabilities of buffer B_i at time t can be arranged as a vector $q_i(t)$:

$$q_i(t) = \begin{bmatrix} q_{i,0}(t) \\ q_{i,1}(t) \\ \vdots \\ q_{i,C_i}(t) \end{bmatrix} \qquad (4.71)$$

4.3.2.1 State Transition Probabilities

Lemmas 4.1–4.4, which are previously introduced in Section 4.2, are used to simplify algebraic expressions. Note that the term elimination in expression simplification can be performed as explained previously in Section 4.2.2.

1) The state transition probability from state 0 to itself is

$$\begin{aligned} Q_{i,(0|0)}(t) &= P[\{c_i(t) = 0\}|\{c_i(t-1) = 0\}] \\ &= P\left[\left\{M_i^{\bar{\mathbb{P}}}\right\} \cap \left\{M_{i+1}^{\bar{\mathbb{P}}}\right\}|\{c_i(t-1) = 0\}\right] \\ &= P\left[\{M_i^0\} \cup \{M_i^{\mathbb{B}}\} \cup \{M_i^{\mathbb{S}}\}|\{c_i(t-1) = 0\}\right] \\ &= P\left[\{M_i^0\} \cup \{M_i^{\mathbb{S}}\}\right] \\ &= 1 - p_i + \mathbb{S}_i(t) \end{aligned} \qquad (4.72)$$

The state transition probability from state 0 to state 1 is

$$
\begin{aligned}
Q_{i,(1|0)}(t) &= P[\{c_i(t) = 1\}|\{c_i(t-1) = 0\}] \\
&= P\left[\{M_i^{\mathbb{P}}\} \cap \left\{M_{i+1}^{\overline{\mathbb{P}}}\right\}|\{c_i(t-1) = 0\}\right] \\
&= P\left[\{M_i^1\} \cap \left\{M_i^{\overline{\mathbb{B}}}\right\} \cap \left\{M_i^{\overline{\mathbb{S}}}\right\}|\{c_i(t-1) = 0\}\right] \\
&= P\left[\{M_i^1\} \cap \left\{M_i^{\overline{\mathbb{S}}}\right\}\right] \\
&= p_i - \mathbb{S}_i(t)
\end{aligned}
\tag{4.73}
$$

2) The state transition probability from state j to itself is

$$
\begin{aligned}
Q_{i,(j|j)}(t) &= P[\{c_i(t) = j\}|\{c_i(t-1) = j\}] \\
&= P\left[\{M_i^{\mathbb{P}}\} \cap \{M_{i+1}^{\mathbb{P}}\} \cup \left\{M_i^{\overline{\mathbb{P}}}\right\} \cap \left\{M_{i+1}^{\overline{\mathbb{P}}}\right\}|\{c_i(t-1) = j\}\right] \\
&= P\left[\{M_i^1\} \cap \left\{M_i^{\overline{\mathbb{B}}}\right\} \cap \left\{M_i^{\overline{\mathbb{S}}}\right\} \cap \{M_{i+1}^1\} \cap \left\{M_{i+1}^{\overline{\mathbb{B}}}\right\} \cap \left\{M_{i+1}^{\overline{\mathbb{S}}}\right\}|\{c_i(t-1) = j\}\right] \\
&\quad + P\left[\{M_i^0\} \cup \left\{M_i^{\overline{\mathbb{B}}}\right\} \cup \{M_i^{\mathbb{S}}\} \cap \{M_{i+1}^0\} \cup \{M_{i+1}^{\mathbb{B}}\} \cup \left\{M_{i+1}^{\overline{\mathbb{S}}}\right\}|\{c_i(t-1) = j\}\right] \\
&= P\left[\{M_i^1\} \cap \left\{M_i^{\overline{\mathbb{S}}}\right\} \cap \{M_{i+1}^1\} \cap \left\{M_{i+1}^{\overline{\mathbb{B}}}\right\}\right] \\
&\quad + P\left[\{M_i^0\} \cup \{M_i^{\mathbb{S}}\} \cap \{M_{i+1}^0\} \cup \{M_{i+1}^{\mathbb{B}}\}\right] \\
&= [p_i - \mathbb{S}_i(t)][p_{i+1} - \mathbb{B}_{i+1}(t)] + [1 - p_i + \mathbb{S}_i(t)][1 - p_{i+1} + \mathbb{B}_{i+1}(t)]
\end{aligned}
\tag{4.74}
$$

The state transition probability from state $j+1$ to state j is

$$
\begin{aligned}
Q_{i,(j|j+1)}(t) &= P[\{c_i(t) = j\}|\{c_i(t-1) = j+1\}] \\
&= P\left[\left(\left\{M_i^{\overline{\mathbb{P}}}\right\} \cap \{M_{i+1}^{\mathbb{P}}\}\right)|\{c_i(t-1) = j+1\}\right] \\
&= P\left[\{M_i^0\} \cup \left\{M_i^{\overline{\mathbb{B}}}\right\} \cup \{M_i^{\mathbb{S}}\} \cap \{M_{i+1}^1\} \cap \left\{M_{i+1}^{\overline{\mathbb{B}}}\right\} \cap \left\{M_{i+1}^{\overline{\mathbb{S}}}\right\}|\{c_i(t-1) = j+1\}\right] \\
&= P\left[\{M_i^0\} \cup \{M_i^{\mathbb{S}}\} \cap \{M_{i+1}^1\} \cap \left\{M_{i+1}^{\overline{\mathbb{B}}}\right\}\right] \\
&= [1 - p_i + \mathbb{S}_i(t)][p_{i+1} - \mathbb{B}_{i+1}(t)]
\end{aligned}
\tag{4.75}
$$

The state transition probability from state j to state $j+1$ is

$$
\begin{aligned}
Q_{i,(j+1|j)}(t) &= P[\{c_i(t) = j+1\}|\{c_i(t-1) = j\}] \\
&= P\left[\{M_i^{\mathbb{P}}\} \cap \left\{M_{i+1}^{\overline{\mathbb{P}}}\right\}|\{c_i(t-1) = j\}\right] \\
&= P\left[\{M_i^1\} \cap \left\{M_i^{\overline{\mathbb{B}}}\right\} \cap \left\{M_i^{\overline{\mathbb{S}}}\right\} \cap \{M_{i+1}^0\} \cup \{M_{i+1}^{\mathbb{B}}\} \cup \left\{M_{i+1}^{\overline{\mathbb{S}}}\right\}|\{c_i(t-1) = j\}\right] \\
&= P\left[\{M_i^1\} \cap \left\{M_i^{\overline{\mathbb{S}}}\right\} \cap \{M_{i+1}^0\} \cup \{M_{i+1}^{\mathbb{B}}\}\right] \\
&= [p_i - \mathbb{S}_i(t)][1 - p_{i+1} + \mathbb{B}_{i+1}(t)]
\end{aligned}
\tag{4.76}
$$

3) The state transition probability from state C_i to itself is

$$Q_{i,(C_i|C_i)}(t) = P[\{c_i(t) = C_i\}|\{c_i(t-1) = C_i\}]$$

$$= P\left[\{M_i^{\mathbb{P}}\} \cap \{M_2^{\mathbb{P}}\} \cup \left\{M_1^{\overline{\mathbb{P}}}\right\} \cap \left\{M_2^{\overline{\mathbb{P}}}\right\}|\{c_i(t-1) = C_i\}\right]$$

$$= P\left[\{M_i^1\} \cap \left\{M_i^{\overline{\mathbb{B}}}\right\} \cap \left\{M_i^{\overline{\mathbb{S}}}\right\} \cap \{M_{i+1}^1\} \cap \left\{M_{i+1}^{\overline{\mathbb{B}}}\right\} \cap \left\{M_{i+1}^{\overline{\mathbb{S}}}\right\}|\{c_i(t-1) = C_i\}\right]$$

$$+ P[\{M_{i+1}^0\} \cup \{M_{i+1}^{\mathbb{B}}\} \cup \{M_{i+1}^{\overline{\mathbb{S}}}\}|\{c_i(t-1) = C_i\}]$$

$$= P\left[\{M_i^1\} \cap \{M_i^{\mathbb{S}}\} \cap \{M_{i+1}^1\} \cap \left\{M_{i+1}^{\overline{\mathbb{B}}}\right\}\right] + P[\{M_{i+1}^0\} \cup \{M_i^{\mathbb{B}}\}]$$

$$= [p_i - \mathbb{S}_i(t)][p_{i+1} - \mathbb{B}_{i+1}(t)] + [1 - p_{i+1} + \mathbb{B}_{i+1}(t)]$$

$$(4.77)$$

4.3.2.2 System Dynamics

Similar to the two-machine production line, the evolution of the probability that buffer B_i is in state j at time t can be written as

$$q_i(t) = Q_i(t)q_i(t-1), \quad \forall i = 1,2,...,N-1 \tag{4.78}$$

$$\sum_{j=0}^{C_i} q_{i,j}(t) = 1, \quad \forall i = 1,2,...,N-1 \tag{4.79}$$

Note that in multi-machine production lines, the state transition probabilities are dynamic, i.e. the state transition probabilities are functions of time. The system discussed here is a nonstationary Markov chain, and it is infeasible to derive closed-form expressions of steady state probabilities for such complex systems.

4.3.3 Performance Measures

4.3.3.1 Blockage and Starvation

In a serial production line with multiple Bernoulli machines, the last machine is never blocked, i.e.,

$$\mathbb{B}_N(t) = P\left[\{M_N^{\mathbb{B}}\}\right] = 0 \tag{4.80}$$

The probability of machine M_i $(i = 1, ..., N - 1)$ being blocked can be calculated based on the definition of blockage:

$$
\begin{aligned}
\mathbb{B}_i(t) &= P\big[\{M_i^{\mathbb{B}}\}\big] \\
&= P\Big[\{M_i^1\} \cap \big\{M_i^{\mathbb{P}}\big\} \cap \{c_i(t-1) = C_i\}\Big] \\
&= P\big[\{M_i^1\} \cap \{M_{i+1}^0\} \cup \{M_{i+1}^{\mathbb{B}}\} \cup \{M_{i+1}^{\mathbb{S}}\} \cap \{c_i(t-1) = C_i\}\big] \\
&= p_i(1 - p_{i+1} + \mathbb{B}_{i+1}(t))q_{i,C_i}(t-1)
\end{aligned}
$$

$$(4.81)$$

According to the second assumption, the first machine is never starved, i.e.

$$
\mathbb{S}_1(t) = P\big[\{M_1^{\mathbb{S}}\}\big] = 0 \tag{4.82}
$$

The probability of machine M_i $(i = 2, ..., N)$ being starved during time t is

$$
\begin{aligned}
\mathbb{S}_i(t) &= P\big[\{M_i^{\mathbb{S}}\}\big] \\
&= P\big[\{M_i^1\} \cap \{c_{i-1}(t-1) = 0\}\big] \\
&= p_i q_{i-1,0}(t-1)
\end{aligned}
\tag{4.83}
$$

4.3.3.2 Production Rate

The production rate of the first machine M_1 during time t is

$$
\rho_1(t) = P\big[\{M_1^{\mathbb{P}}\}\big] = P\Big[\{M_1^{\mathbb{P}}\} \cap \big\{M_1^{\bar{\mathbb{B}}}\big\}\Big] = p_1 - \mathbb{B}_1(t) \tag{4.84}
$$

The production rate of the last machine during time step t, which is also the production rate of the system, can be expressed as

$$
\rho_{\text{SYS}}(t) = \rho_N(t) = P\big[\{M_N^{\mathbb{P}}\}\big] = P\Big[\{M_N^1\} \cap \big\{M_N^{\bar{\mathbb{S}}}\big\}\Big] = p_N - \mathbb{S}_N(t) \tag{4.85}
$$

For the other machine M_i $(i = 2, ..., N - 1)$, the respective production rate can be calculated as

$$
\begin{aligned}
\rho_i(t) &= P\big[\{M_i^{\mathbb{P}}\}\big] \\
&= P\Big[\{M_i^1\} \cap \big\{M_i^{\bar{\mathbb{B}}}\big\} \cap \big\{M_i^{\bar{\mathbb{S}}}\big\}\Big] \\
&= p_i - \mathbb{B}_i(t) - \mathbb{S}_i(t) + P\big[\{M_i^{\mathbb{S}}\} \cap \{M_i^{\mathbb{B}}\}\big]
\end{aligned}
\tag{4.86}
$$

At the steady state, according to the conservation of material flow, production rates for all machines should be the same, that is

$$
\rho_1 = \rho_i = \rho_N, \quad i = 2,3,...,N-1 \tag{4.87}
$$

where PR_i is the production rate for machine M_i under the steady state. Hence, at the steady state, the probability of the intersection between starvation and blockage can be calculated as

$$
\begin{aligned}
P\left[\left\{M_i^{\mathbb{S}}\right\} \cap \left\{M_i^{\mathbb{B}}\right\}\right] &= \rho_1 - p_i + \mathbb{B}_i + \mathbb{S}_i \\
&= \rho_N - p_i + \mathbb{B}_i + \mathbb{S}_i
\end{aligned}
\tag{4.88}
$$

4.3.3.3 Work-in-process

The work-in-process of buffer B_i at the end of time step t, denoted by $\mathrm{WIP}_i(t)$, can be expressed as

$$
\mathrm{WIP}_i(t) = \sum_{j=1}^{C_i} j q_{i,j}(t)
\tag{4.89}
$$

The total work-in-process of the system at the end of a time step t, denoted by $\mathrm{WIP}_{\mathrm{SYS}}(t)$, is

$$
\mathrm{WIP}_{\mathrm{SYS}}(t) = \sum_{i=1}^{N-1} \mathrm{WIP}_i(t)
\tag{4.90}
$$

4.3.4 System Modeling with Iteration-based Method

The iteration-based method is capable of analyzing the system performance in both transient and steady states. The system state probabilities and system measures are updated in each time step. A flowchart of the iteration-based method is shown in Figure 4.20.

- Initialization. The first step is the determination of the system layout, the maximum capacity C_i of each buffer, the reliability p_i of each machine, and the initial buffer occupancy.
- Update $\mathbb{S}(t)$ and $\mathbb{B}(t)$. The probabilities of starvation and blockage should be updated first in each iteration. Note that according to Eq. (4.83), $\mathbb{B}_i(t)$ is a function of $\mathbb{B}_{i+1}(t)$. Hence, the probability of blockage should be updated from the last machine to the first machine.
- Update state transition matrix and state probabilities. Since the state transition probabilities at time t involve $\mathbb{S}(t)$ and $\mathbb{B}(t)$, they can be updated when the probabilities of starvation and blockage are known.
- Update performance measures. The system performance can be updated in each time step. However, this updating process is not necessary and can be skipped if one only inquires about the steady state performance.
- Termination criteria. When the number of iterations exceeds a predefined maximum number, the iteration is terminated. If one inquires about the steady state system performance, the termination criteria could be that the difference between two adjacent system states is smaller than a pre-set threshold value.

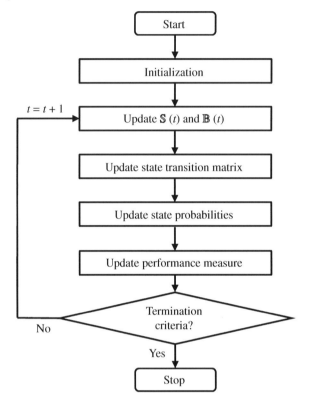

Figure 4.20   A flowchart of an iteration-based method.

The following example is designed to demonstrate the use of an iterative-based method in system modeling.

Example 4.3   Consider a serial production line with four Bernoulli machines. The machine reliabilities are [0.9, 0.8, 0.8, 0.9]. The buffer capacities are [3, 2, 3]. The initial number of parts in each buffer is [1, 0, 3].
Find the system state in step t_1 manually:

a) Plot the evolution of state probabilities and WIP in the first 50 steps.
b) Approximate the steady state probabilities for each buffer and the system production rate.

Solution

a) Formulate three state vectors as follows:

$$q_1(t) = \begin{bmatrix} q_{1,0}(t) \\ q_{1,1}(t) \\ q_{1,2}(t) \\ q_{1,3}(t) \end{bmatrix}, \quad q_2(t) = \begin{bmatrix} q_{2,0}(t) \\ q_{2,1}(t) \\ q_{2,2}(t) \end{bmatrix}, \quad q_3(t) = \begin{bmatrix} q_{3,0}(t) \\ q_{3,1}(t) \\ q_{3,2}(t) \\ q_{3,3}(t) \end{bmatrix}$$

According to the initial condition:

$$q_1(t_0) = \begin{bmatrix} 0 \\ 1 \\ 0 \\ 0 \end{bmatrix}, \quad q_2(t_0) = \begin{bmatrix} 1 \\ 0 \\ 0 \end{bmatrix}, \quad q_3(t_0) = \begin{bmatrix} 0 \\ 0 \\ 0 \\ 1 \end{bmatrix}$$

Update the probability of starvation:

$$\mathbb{S}_1(t_1) = 0$$
$$\mathbb{S}_2(t_1) = p_2 q_{1,0}(t_0) = 0.8 * 0 = 0$$
$$\mathbb{S}_3(t_1) = p_3 q_{2,0}(t_0) = 0.8 * 1 = 0.8$$
$$\mathbb{S}_4(t_1) = p_4 q_{3,0}(t_0) = 0.9 * 0 = 0$$

Update the probability of blockage from the last machine to the first machine:

$$\mathbb{B}_4(t_1) = 0$$
$$\mathbb{B}_3(t_1) = p_3(1 - p_4 + \mathbb{B}_4(t_1))q_{3,3}(t_0) = 0.8 * (1 - 0.9 + 0) * 1 = 0.08$$
$$\mathbb{B}_2(t_1) = p_2(1 - p_3 + \mathbb{B}_3(t_1))q_{2,2}(t_0) = 0.8 * (1 - 0.8 + 0.08) * 0 = 0$$
$$\mathbb{B}_1(t_1) = p_1(1 - p_2 + \mathbb{B}_2(t_1))q_{1,3}(t_0) = 0.9 * (1 - 0.8 + 0) * 0 = 0$$

Update the transition probabilities for buffer B_1:

$$Q_{1,(0|0)}(t_1) = 1 - p_1 + \mathbb{S}_1(t_1) = 0.1$$
$$Q_{1,(1|0)}(t_1) = p_1 - \mathbb{S}_1(t_1) = 0.9$$
$$Q_{1,(0|1)}(t_1) = Q_{1,(1|2)}(t_1) = Q_{1,(2|3)}(t_1)$$
$$= [1 - p_1 + \mathbb{S}_1(t_1)][p_2 - \mathbb{B}_2(t_1)] = 0.08$$
$$Q_{1,(1|1)}(t_1) = Q_{1,(2|2)}(t_1)$$
$$= [p_1 - \mathbb{S}_1(t_1)][p_2 - BL_2(t_1)] + [1 - p_1 + \mathbb{S}_1(t_1)][1 - p_2 + \mathbb{B}_2(t_1)] = 0.74$$
$$Q_{1,(2|1)}(t_1) = Q_{1,(3|2)}(t_1)$$
$$= [p_1 - \mathbb{S}_1(t_1)][1 - p_2 + \mathbb{B}_2(t_1)] = 0.18$$
$$Q_{1,(3|3)}(t_1) = [p_1 - \mathbb{S}_1(t_1)][p_2 - \mathbb{B}_2(t_1)] + [1 - p_2 + \mathbb{B}_2(t_1)] = 0.92$$

Hence, the state transition matrix for B_1 can be written as:

$$Q_1(t_1) = \begin{bmatrix} 0.1 & 0.08 & 0 & 0 \\ 0.9 & 0.74 & 0.08 & 0 \\ 0 & 0.18 & 0.74 & 0.08 \\ 0 & 0 & 0.18 & 0.92 \end{bmatrix}$$

Similarly, $Q_2(t_1)$ and $Q_3(t_1)$ can be calculated and shown as follows:

$$Q_2(t_1) = \begin{bmatrix} 0.2 & 0.144 & 0 \\ 0.8 & 0.632 & 0.144 \\ 0 & 0.224 & 0.856 \end{bmatrix} \quad Q_3(t_1) = \begin{bmatrix} 1 & 0.9 & 0 & 0 \\ 0 & 0.1 & 0.9 & 0 \\ 0 & 0 & 0.1 & 0.9 \\ 0 & 0 & 0 & 0.1 \end{bmatrix},$$

The system states in time step t_1 can be calculated by:

$$q_1(t_1) = Q_1(t_1)q_1(t_0) = \begin{bmatrix} 0.1 & 0.08 & 0 & 0 \\ 0.9 & 0.74 & 0.08 & 0 \\ 0 & 0.18 & 0.74 & 0.08 \\ 0 & 0 & 0.18 & 0.92 \end{bmatrix} \cdot \begin{bmatrix} 0 \\ 1 \\ 0 \\ 0 \end{bmatrix} = \begin{bmatrix} 0.08 \\ 0.74 \\ 0.18 \\ 0 \end{bmatrix}$$

$$q_2(t_1) = Q_2(t_1)q_2(t_0) = \begin{bmatrix} 0.2 & 0.144 & 0 \\ 0.8 & 0.632 & 0.144 \\ 0 & 0.224 & 0.856 \end{bmatrix} \cdot \begin{bmatrix} 1 \\ 0 \\ 0 \end{bmatrix} = \begin{bmatrix} 0.2 \\ 0.8 \\ 0 \end{bmatrix}$$

$$q_3(t_1) = Q_3(t_1)q_3(t_0) = \begin{bmatrix} 1 & 0.9 & 0 & 0 \\ 0 & 0.1 & 0.9 & 0 \\ 0 & 0 & 0.1 & 0.9 \\ 0 & 0 & 0 & 0.1 \end{bmatrix} \cdot \begin{bmatrix} 0 \\ 0 \\ 0 \\ 1 \end{bmatrix} = \begin{bmatrix} 0 \\ 0 \\ 0.9 \\ 0.1 \end{bmatrix}$$

b) By repeating the above process 50 times using Microsoft's Excel or other programming software, the evolution of state probabilities and WIP are shown in Figure 4.21.

c) Let the termination criterion be that the difference between two adjacent system states is smaller than 1%. According to the results obtained in part (b), the steady state probabilities are

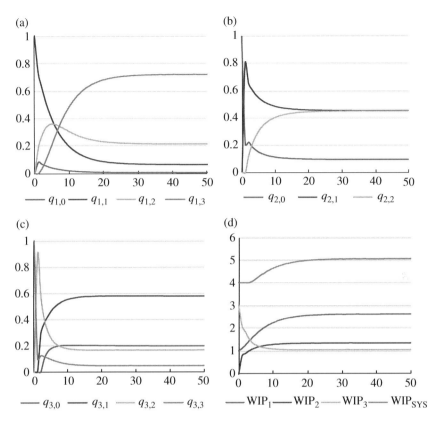

Figure 4.21 Evolution of state probabilities for (a) buffer B_1, (b) buffer B_2, (c) buffer B_3, and system measure (d) WIP.

$$q_1 = \begin{bmatrix} 0.005 \\ 0.063 \\ 0.212 \\ 0.720 \end{bmatrix}, \quad q_2 = \begin{bmatrix} 0.092 \\ 0.454 \\ 0.454 \end{bmatrix}, \quad q_3 = \begin{bmatrix} 0.198 \\ 0.581 \\ 0.171 \\ 0.050 \end{bmatrix}$$

The system production rate at steady state can be calculated by

$$\rho_{SYS} = p_4 - \mathbb{S}_4 = p_4 - p_4 q_{3,0} = 0.9 - 0.9 * 0.198 = 0.722$$

Note that the steady state probabilities and system performance measures are independent of the initial system states. For example, the work-in-process of the system converges to the same value with different initial conditions: [1, 0, 3], [0, 0, 0], [3, 2, 3], as shown in Figure 4.22.

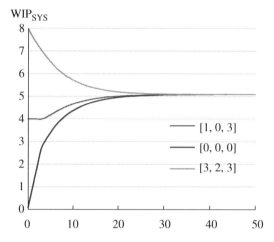

WIP$_{SYS}$

Figure 4.22 Evolution of WIP$_{SYS}$ with different initial conditions.

legend:
— [1, 0, 3]
— [0, 0, 0]
⋯ [3, 2, 3]

4.4 Production Lines Coupled with Material Handling Systems

In the previous two sections, two fundamental assumptions are made in the system modeling methods: (i) the first machine is never starved, and (ii) the last machine is never blocked. The first assumption is true if the material handling system (MHS) can always guarantee the timely delivery of raw materials or other resources to the first machine. The second assumption is true if the MHS can always remove the finished products from the last machine on time. These assumptions are practical only if MHS is perfect or nearly perfect. If it is not, its impact on the production system has to be considered in system modeling and analysis. In this section, the above two crucial assumptions are released and a new modeling method for the production system coupled with a material handling system is introduced. The integrated modeling method derived here is established with the iteration-based method, which also demonstrates the advantages of this method over other system modeling methods (e.g. aggregation-based method and decomposition-based method) that cannot release these two assumptions. Note that the machines discussed in this section are assumed to be Bernoulli machines.

4.4.1 Assumptions and Notations

The system layout of a typical serial production line coupled with MHS is shown in Figure 4.23. Considering the scenarios where the first machine can be starved and the last machine can be blocked, two virtual buffers are created to model these conditions. In Figure 4.23, the gray rectangles represent the virtual buffers. The buffer before the first machine is denoted by B_0, and the buffer after the last

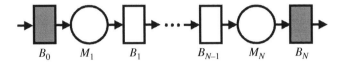

Figure 4.23 System layout of a production system coupled with MHS.

machine is denoted by B_N. The empty probability of B_0 is used to model the situation where the incoming raw materials are not available. The full probability of buffer B_N is used to model the situation where the finished parts cannot be delivered on time. The virtual buffer B_0 or B_N is empty or full at each time step t with the probability of $p_0(t)$ or $p_N(t)$.

Similar to the production line as discussed in Section 4.3, all machines have unique cycle time, and the length for each time step equals the machine cycle time. Machine M_i ($i = 1, 2, ..., N$) is up in each cycle time with a probability of p_i, and it is down with a probability of $1 - p_i$. The capacity of the buffer B_i is C_i ($i = 1, 2, ..., N - 1$).

4.4.1.1 Assumptions
Except for the two assumptions discussed above, all the others can be adopted to describe the system in Figure 4.23 according to [5]:

1) Blocked before service.
2) The states of machines are determined at the beginning of each time step, and the states of buffers are determined at the end of each time step.
3) Whether or not a machine is up or down, its state is independent of the other machines' state.
4) Buffers' states (occupancy) can change at most by one in each time step.

4.4.1.2 Notations
All the notations are the same as those in Section 4.3.

4.4.2 State Transition and Performance

The expressions of state transition probabilities and system dynamics are the same as those in the multi-machine production line model in Section 4.3. The differences induced by the integration of MHS are demonstrated through the following performance measures.

4.4.2.1 Blockage and Starvation
In a serial production line coupled with MHS, the blockage can occur in the last machine. In reference to the definition of blockage in (4.27), the last machine M_N is

blocked at time t if M_N is up at the beginning of time t and virtual buffer B_N is full at the end of time $t - 1$, i.e.

$$\mathbb{B}_N(t) = P[\{M_N^{\mathbb{B}}\}] = p_N p_N(t) \tag{4.91}$$

The blockage probability for other machine M_i ($i = 1, ..., N - 1$) is the same as shown in Eq. (4.81).

In accordance with the definition of starvation in Eq. (4.28), the first machine M_1 is starved at time t if M_1 is up at the beginning of time t while the virtual buffer B_0 is empty at the end of time $t - 1$, i.e.

$$\mathrm{ST}_1(t) = P[\{M_1 \text{ st}\}] = p_1 p_0(t) \tag{4.92}$$

The starvation probability for other machine M_i ($i = 2, ..., N$) is the same as shown in (4.83).

4.4.2.2 Production Rate

The estimate of system production rate can be expressed as

$$
\begin{aligned}
\rho_{\mathrm{SYS}}(t) = \rho_N(t) &= P[\{M_N^{\mathbb{P}}\}] = P\left[\{M_N^1\} \cap \{M_N^{\bar{\mathbb{B}}}\} \cap \{M_N^{\bar{\mathbb{S}}}\}\right] \\
&= p_N\left(1 - \frac{\mathbb{S}_N(t)}{p_N}\right)\left(1 - \frac{\mathbb{B}_N(t)}{p_N}\right) = (p_N - \mathbb{S}_N(t))(1 - p_N(t))
\end{aligned} \tag{4.93}
$$

The expressions of work-in-process are the same as shown in (4.89) and (4.90).

Example 4.4   Consider a serial production line with four Bernoulli machines as discussed in Example 4.3. Suppose MHS is not perfect and the virtual buffer B_0 is empty at each time step with a constant probability $p_0 = 0.1$; the virtual buffer B_N is full at each time step with a constant probability $p_N = 0.1$.

Find the system state in step t_1 manually:

a) Plot the evolution of state probabilities and WIP in the first 50 steps.
b) Approximate the steady state probabilities for each buffer and the system production rate.

Solution

a) The initial three state vectors can be written as

$$
q_1(t_0) = \begin{bmatrix} 0 \\ 1 \\ 0 \\ 0 \end{bmatrix}, \qquad q_2(t_0) = \begin{bmatrix} 1 \\ 0 \\ 0 \\ 0 \end{bmatrix}, \qquad q_3(t_0) = \begin{bmatrix} 0 \\ 0 \\ 0 \\ 1 \end{bmatrix}
$$

Update the probability of starvation:

$$\mathbb{S}_1(t_1) = p_1 p_0 = 0.9 * 0.1 = 0.09$$
$$\mathbb{S}_2(t_1) = p_2 q_{1,0}(t_0) = 0.8 * 0 = 0$$
$$\mathbb{S}_3(t_1) = p_3 q_{2,0}(t_0) = 0.8 * 1 = 0.8$$
$$\mathbb{S}_4(t_1) = p_4 q_{3,0}(t_0) = 0.9 * 0 = 0$$

Update the probability of blockage from the last machine to the first machine:

$$\mathbb{B}_4(t_1) = p_4 p_L = 0.9 * 0.1 = 0.09$$
$$\mathbb{B}_3(t_1) = p_3(1 - p_4 + \mathbb{B}_4(t_1))q_{3,3}(t_0) = 0.8 * (1 - 0.9 + 0.09) * 1 = 0.152$$
$$\mathbb{B}_2(t_1) = p_2(1 - p_3 + \mathbb{B}_3(t_1))q_{2,2}(t_0) = 0.8 * (1 - 0.8 + 0.152) * 0 = 0$$
$$\mathbb{B}_1(t_1) = p_1(1 - p_2 + \mathbb{B}_2(t_1))q_{1,3}(t_0) = 0.9 * (1 - 0.8 + 0) * 0 = 0$$

Update the transition probabilities for buffer B_1:

$$Q_{1,(0|0)}(t_1) = 1 - p_1 + \mathbb{S}_1(t_1) = 0.19$$
$$Q_{1,(1|0)}(t_1) = p_1 - \mathbb{S}_1(t_1) = 0.81$$
$$Q_{1,(0|1)}(t_1) = Q_{1,(1|2)}(t_1) = Q_{1,(2|3)}(t_1)$$
$$= [1 - p_1 + \mathbb{S}_1(t_1)][p_2 - \mathbb{B}_2(t_1)] = 0.152$$
$$Q_{1,(1|1)}(t_1) = Q_{1,(2|2)}(t_1)$$
$$= [p_1 - \mathbb{S}_1(t_1)][p_2 - \mathbb{B}_2(t_1)] + [1 - p_1 + \mathbb{S}_1(t_1)][1 - p_2 + \mathbb{B}_2(t_1)] = 0.686$$
$$Q_{1,(2|1)}(t_1) = Q_{1,(3|2)}(t_1)$$
$$= [p_1 - \mathbb{S}_1(t_1)][1 - p_2 + \mathbb{B}_2(t_1)] = 0.162$$
$$Q_{1,(3|3)}(t_1) = [p_1 - \mathbb{S}_1(t_1)][p_2 - \mathbb{B}_2(t_1)] + [1 - p_2 + \mathbb{B}_2(t_1)] = 0.848$$

Hence, the state transition matrix for B_1 can be written as

$$Q_1(t_1) = \begin{bmatrix} 0.19 & 0.152 & 0 & 0 \\ 0.81 & 0.686 & 0.152 & 0 \\ 0 & 0.162 & 0.686 & 0.152 \\ 0 & 0 & 0.162 & 0.848 \end{bmatrix}$$

Similarly, $Q_2(t_1)$ and $Q_3(t_1)$ can be calculated as

$$Q_2(t_1) = \begin{bmatrix} 0.2 & 0.1296 & 0 \\ 0.8 & 0.5888 & 0.1296 \\ 0 & 0.2816 & 0.8704 \end{bmatrix}, \quad Q_3(t_1) = \begin{bmatrix} 1 & 0.81 & 0 & 0 \\ 0 & 0.19 & 0.81 & 0 \\ 0 & 0 & 0.19 & 0.81 \\ 0 & 0 & 0 & 0.19 \end{bmatrix}$$

The system states in time step t_1 can be calculated as

$$q_1(t_1) = Q_1(t_1)q_1(t_0) = \begin{bmatrix} 0.152 \\ 0.686 \\ 0.162 \\ 0 \end{bmatrix}$$

$$q_2(t_1) = Q_2(t_1)q_2(t_0) = \begin{bmatrix} 0.2 \\ 0.8 \\ 0 \end{bmatrix}$$

$$q_3(t_1) = Q_3(t_1)q_3(t_0) = \begin{bmatrix} 0 \\ 0 \\ 0.81 \\ 0.19 \end{bmatrix}$$

b) By repeating the above process 50 times using Excel or other programming software, the evolution of state probabilities and WIP are given in Figure 4.24.

c) Let the termination criteria be that the difference between two adjacent system states is smaller than 1%. According to the results obtained in part (b), the steady state probabilities are

$$q_1 = \begin{bmatrix} 0.030 \\ 0.180 \\ 0.297 \\ 0.493 \end{bmatrix}, \quad q_2 = \begin{bmatrix} 0.101 \\ 0.450 \\ 0.449 \end{bmatrix}, \quad q_3 = \begin{bmatrix} 0.139 \\ 0.439 \\ 0.264 \\ 0.158 \end{bmatrix}$$

Compared with the steady state probabilities obtained in Example 4.3, when considering the imperfect MHS, the first buffer tends to be empty, and the last buffer tends to be full.

The system production rate at steady state can be calculated by

$$PR_{SYS} = (p_4 - ST_4) * (1 - p_L) = 0.697$$

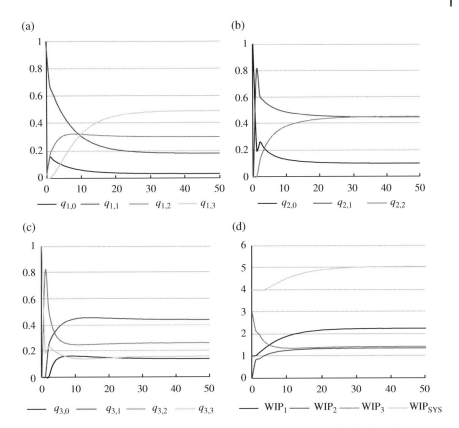

Figure 4.24 Evolution of state probabilities for (a) buffer B_1, (b) buffer B_2, (c) buffer B_3, and system measure (d) WIP.

Problems

4.1 Consider a production line with two Bernoulli machines. Suppose that the buffer capacity $C = 3$, $p_1 = 0.9$, and $p_2 = 0.8$.

(a) Draw the state transition diagram of the Markov chain that describes this system, and then determine its transition probabilities.

(b) Calculate the stationary probability of each state and the steady state production rate of this system.

(c) Increase the buffer capacity from 3 to 10, calculate the production rate and WIP. Do you think increasing the buffer capacity can significantly improve the system performance?

4.2   Consider a serial production line with four Bernoulli machines. Suppose the buffer capacities are [2, 3, 3], and machine reliabilities are [0.8, 0.9, 0.8, 0.9].

(a) Approximate the production rate of this system at the steady state.

(b) Calculate the average buffer occupancy of each buffer.

(c) Suppose that apart from the current buffer allocation plan, engineers come up with two new plans for buffer capacities: [2,4,2] and [3,2,3]. Among these three buffer allocation plans, which one can help achieve the highest steady state production rate?

References

1 Jingshan, L. and Meerkov, S.M. (2008). *Product Systems Engineering*. Springer Science & Business Media.

2 Hou, Y., Li, L., Ge, Y. et al. (2018). A new modeling method for both transient and steady-state analyses of inhomogeneous assembly systems. *J. Manuf. Syst.* 49: 46–60.

3 Ge, Y., Li, L., and Wang, Y. (2017). Modeling of Bernoulli production line with the rework loop for transient and steady-state analysis. *J. Manuf. Syst.* 44: 22–41. https://doi.org/10.1016/j.jmsy.2017.04.002.

4 Wang, Y. and Li, L. (2015). A novel modeling method for both steady-state and transient analyses of serial bernoulli production systems. *IEEE Trans. Syst. Man, Cybern. Syst.* 45: 97–108. https://doi.org/10.1109/TSMC.2014.2327561.

5 Zhou, Z., Li, L., and Lin, J. (2014). Integrated analytical modeling for production systems coupled with material handling systems. *IEEE Trans. Syst. Man, Cybern. Syst.* 44: 1067–1076. https://doi.org/10.1109/TSMC.2013.2292569.

5

Energy Efficiency Characterization in Manufacturing Systems

The production rate and energy consumption of a manufacturing system have drawn great attention from both researchers and practitioners. In the Chapter 4, the modeling and analysis of production rate of typical manufacturing systems have been discussed. In this chapter, the modeling and characterization of the energy consumption in manufacturing systems are discussed, followed by diverse types of modeling approaches in energy cost estimation. More specifically, in Section 5.1, based on the inter-process dependency or the machines' operation schemes, the energy consumption modeling approaches are classified into three major categories: operation-based, component-based, and system-based energy modeling. In reference to the demand-side management as introduced in Chapter 2, multiple electricity pricing options are available in practice, and the energy cost may vary under different electricity tariffs. In Section 5.2, three types of electricity tariffs are introduced, i.e. flat rate, time-of-use, and critical peak price. The corresponding energy cost models for manufacturing systems are also demonstrated with illustrative examples.

5.1 Energy Consumption Modeling

The energy models are used to characterize the energy consumption of manufacturing systems, which are useful tools to help engineers identify the major energy consumers and determine synthetic strategies toward improved system energy efficiency. The energy consumption of a manufacturing system can be complicated, which requires studies at various temporal and spatial levels. Figure 5.1 illustrates the major methods for energy consumption modeling in manufacturing systems. In particular, if the energy consumption behaviors of different machines are independent, i.e. the energy-related parameters of one machine do not affect the other machines' energy consumption patterns, the energy consumed in each machine can be calculated individually. Then, the energy consumption of the entire system can be modeled by summing up the energy consumption of each machine.

Sustainable Manufacturing Systems, First Edition. Lin Li and MengChu Zhou.
© 2023 The Institute of Electrical and Electronics Engineers, Inc.
Published 2023 by John Wiley & Sons, Inc.

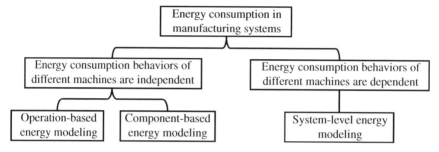

Figure 5.1 Classifications of energy consumption modeling methods.

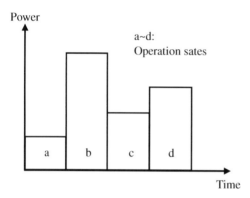

Figure 5.2 Illustration of the operation-based energy model.

In the manufacturing systems with independent machines, two types of energy modeling approaches can be applied: operation-based energy modeling and component-based energy modeling. The operation-based modeling represents the relationships between energy use and machine operation states, which can be employed to find the energy-efficient production schedule. The component-based energy modeling decomposes the machine energy consumption into component energy consumption, which can help machine designers and engineers identify the most energy-consuming component and adjust process parameters toward improved energy efficiency. On the other hand, if the energy consumption behaviors of different machines are dependent, i.e. there exist some parameters that can affect the energy consumption rates of multiple machines, a set of parameters that can optimize the energy consumption of a single machine does not necessarily optimize the energy consumption of the entire system. In this case, the system-level energy modeling methods realizing the interrelationships among multiple machines should be applied.

5.1.1 Operation-based Energy Modeling

An operation-based energy modeling method focuses on describing the influence of manufacturing operation states on machine energy consumption. As shown in Figure 5.2, the machine energy consumption during a period of time equals the summation of energy consumed in each operation state [1].

In the simplest case, assume that the power required for machine M_i to process a part is a constant P_i. The total energy consumed during T time steps, denoted by E_i, can be calculated by

$$E_i = TP_i\tau \qquad (5.1)$$

where τ is the duration of each time step.

It should be noted that although the assumption of constant power requirement for a machine is rarely true in prac-

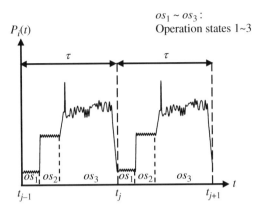

Figure 5.3 Illustration of the machine energy profile.

tice, the idea for treating energy consumption in each time step as a constant is still a valid method to simplify the energy modeling. As illustrated in Chapter 2, a machine may go through multiple operation states to process a part, such as idle, ready for processing, and processing. The power requirement for machine M_i in each operation state may be different and can be formulated as a function of time, denoted by $P_i(t)$, as shown in Figure 5.3.

As shown in Figure 5.3, the machine has three operation states in a processing cycle, i.e. os_1, os_2, and os_3. Since machine M_i performs the same operations in every cycle, the energy profile of the same operation states in different cycles can be similar. Hence, machine M_i's energy consumption in each cycle, denoted by e_i, can be treated as a constant and can be calculated using

$$e_i \equiv \int_{t_{j-1}}^{t_j} P_i(t)\mathrm{d}t = \int_{t_j}^{t_{j+1}} P_i(t)\mathrm{d}t \qquad (5.2)$$

The total energy consumption of M_i during T time steps can be calculated as

$$E_i = Te_i \qquad (5.3)$$

The operation-based energy modeling reveals the relationship between machine energy consumption and machine operation states, and thus it can be used in energy-efficient operation scheduling problems. The first step to formulate the scheduling problem is to define a control variable. In reference to the manufacturing system models as discussed in Chapter 4, a binary control variable $s_i(t)$ is defined as

$$s_i(t) = \begin{cases} 1, & \text{if machine } M_i \text{ is turned on at time step } t \\ 0, & \text{if machine } M_i \text{ is turned off at time step } t \end{cases} \qquad (5.4)$$

If machine M_i is turned off at time step t, M_i does not consume energy. Otherwise, if M_i is turned on, its states are described by the machine reliability model. More specifically, when the machine is up at time t, no matter it is producing, starved, or blocked, the associated energy consumption is e_i; when M_i is down, it does not consume energy. The relationships between the control variable and machine states are depicted in Figure 5.4.

Suppose that machine M_i follows the Bernoulli distribution with the reliability of p_i. Define a new parameter $p_i(t)$ in terms of reliability p_i and control variable $s_i(t)$, as shown in Eq. (5.5).

$$p_i(t) = p_i s_i(t) \tag{5.5}$$

In particular, if the machine is turned on, the machine is in an "up" state with the probability of p_i; if the machine is turned off, then it cannot be in an "up" state, and the corresponding probability $p_i(t) = 0$. To involve the control variable in a manufacturing system model, one can simply replace the original p_i with $p_i(t)$. Then, the energy consumption of M_i during T time steps can be calculated as

$$E_i = \sum_{t=1}^{T} e_i p_i(t) \tag{5.6}$$

Considering a serial production line with N Bernoulli machines, the production scheduling problem can be formulated as an optimization problem to minimize the total energy consumption with the objective function shown as follows:

$$\min_{s_i(t)} \sum_{i=1}^{N} E_i \tag{5.7}$$

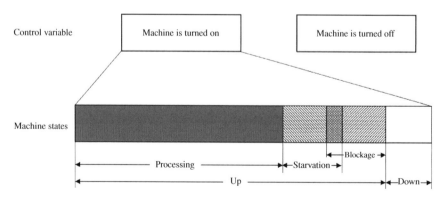

Figure 5.4 The relationships between the control variable and machine states.

Clearly, turning all machines off would result in zero energy consumption, but it also leads to zero production yield. Therefore, the optimization problem should obey certain constraints, such as the production yield constraint:

$$\sum_{t=1}^{T} \rho_{\text{SYS}}(t) \geq Y \tag{5.8}$$

where $\rho_{\text{SYS}}(t)$ is the system production rate at time t (the mathematical expression of $\rho_{\text{SYS}}(t)$ is previously discussed in Chapter 4), and Y is a user-defined target production yield.

5.1.2 Component-based Energy Modeling

A component-based energy modeling method characterizes the energy consumption of a machine at its component level. As shown in Figure 5.5, the machine energy consumption during a period of time equals the summation of energy consumed by different components.

To formulate a component-based energy model, the first step is to identify all energy-consuming components in a machine, as shown in Figure 5.6. Suppose that a machine consists of n components. The energy consumption of this machine, denoted by E, during period $[t_0, t_1]$ can be calculated as

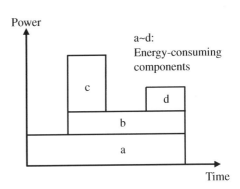

$$E_i = \int_{t_0}^{t_1} P_i(t)dt \tag{5.9}$$

$$E = \sum_{i=1}^{n} E_i \tag{5.10}$$

where $P_i(t)$ is the dynamic energy consumption rate of the ith component, and E_i is the energy consumption of the ith component.

Figure 5.5 Illustration of a component-based energy model.

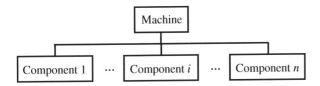

Figure 5.6 Schematic diagram of a machine consisting of n components.

The component-based energy model can help machine designers and engineers identify inefficient energy-consuming components. Since the energy model is based on the energy consumption of each component, it allows the inclusion of relationships between process parameters and component energy consumption in the model. Therefore, this type of modeling method can be used to improve machine energy efficiency through process parameter optimization.

Suppose that a machine has m adjustable process parameters. Let $(s_1, s_2, ..., s_m)$ denote these parameters and treat them as control variables in the optimization problem. The energy consumption of each component can be formulated as a function of these control variables, i.e.

$$E_i \equiv f_i(s_1, s_2, ..., s_m) \tag{5.11}$$

$$\min_{s_1,...,s_m} \sum_{i=1}^{n} E_i \tag{5.12}$$

Equation (5.12) gives the objective function of the optimization problem. The solution to the optimization problem should obey some essential constraints. The first type of constraints is induced by the performance limitations of a machine. Hence, the possible value of the control variable s_j, $(j = 1, ..., n)$ should be limited by certain boundaries $\left[\check{S}_j, \hat{S}_j\right]$, i.e.

$$\check{S}_j \leq s_j \leq \hat{S}_j, \quad j = 1,...,n \tag{5.13}$$

The second type of constraint comes from the production- or quality-related requirements. Mathematically, they can be formulated as K non-equality and L equality constraints, i.e.

$$g_k(s_1, ..., s_m) \leq 0, \quad k = 1, ..., K \tag{5.14}$$

$$h_l(s_1, ..., s_m) = 0, \quad l = 1, ..., L \tag{5.15}$$

The following example demonstrates the formulation of a component-based energy model and an optimization problem for a 3D printing machine.

Example 5.1   Consider a stereolithography-based 3D printing machine, as shown in Figure 5.7. In general, the AM machine fabricates a part in a layer-by-layer fashion. To start part fabrication, a control software is used to generate a layer-wise building file that compiles all *build* commands based on the part geometry design and printing parameter settings. The generated file is then sent to a 3D printer to direct the operations in the machine working chamber. The machine working chamber consists of three major energy-consuming components: galvanometer, ultraviolet (UV) light, and build platform motor. Based on the *build* commands, the UV laser beam is deflected by the galvanometer onto the surface of the liquid

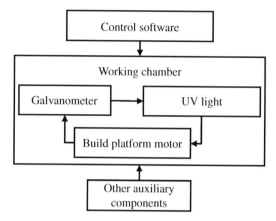

Figure 5.7 Schematic diagram of the major components in a stereolithography-based 3D printing machine.

printing material and solidifies the material upon exposure to UV laser. After the first layer is finished, the *build* platform ascends by the distance of a layer thickness for the next layer printing. These steps are repeated until the entire object is fabricated. In addition, some other auxiliary components (e.g. machine cooling and material tank tilting) are also required in the printing process [2].

For the purpose of the component-based modeling method illustration, in this example, only two control variables are discussed in detail, i.e. the layer thickness d and irradiance (intensity) of the UV laser beam P_{UV}.

The energy consumed by the galvanometer, denoted by E_{galvo}, can be calculated by

$$E_{galvo} = \sum_{i=1}^{K} \frac{L_i C_1}{\nu_{UV}}$$

$$\nu_{UV} = C_2 \times \frac{P_{UV}}{e_{max}}$$

$$e_{max} = C_3 \times \exp(d)$$

where K is the total number of *build* layers; ν_{UV} is the printing speed (which is a function of P_{UV} and d); L_i is the total scanning length in the ith layer; e_{max} is the laser energy density per unit area (which is a function of d); C_1, C_2, and C_3 are constant coefficients.

The energy consumed by the UV light, denoted by E_{UV}, can be calculated by

$$E_{UV} = C_4 \times \sum_{i=1}^{K} S_i \times e_{max}$$

where S_i is the *build* area in the ith layer (which is determined by the product geometry and is independent of the control variables), and C_4 is a constant coefficient.

The energy consumed by the *build* platform motor, denoted by E_{platform}, can be calculated as

$$E_{\text{platform}} = \frac{C_5}{d} + C_6$$

where C_5 and C_6 are constant coefficients.

To minimize the energy consumption of a 3D printing machine, the objective function of the optimization problem can be formulated as

$$\min_{P_{\text{UV}},\, d} \left(E_{\text{software}} + E_{\text{UV}} + E_{\text{galvo}} + E_{\text{platform}} + E_{\text{others}} \right)$$

where E_{software} and E_{others} represent the energy consumed by the software and other auxiliary components. They are not related to control variables and can be treated as constants in this example.

Considering the machine specifications, the layer thickness and irradiance of the UV laser beam should be within a certain range [3]:

$$d^L \leq d \leq d^U$$
$$P_{\text{UV}}^L \leq P_{\text{UV}} \leq P_{\text{UV}}^U$$

In addition, the laser energy density per unit area needs to surpass a threshold, denoted by e_{critical}, to ensure that the liquid printing material can be solidified. Furthermore, the printing speed cannot be too slow or too fast. If the printing speed is too slow, it may induce material over-solidification, which would affect the part quality and dimension accuracy. If the printing speed is too fast, it can result in a semisolidified part, affecting the part quality, surface finish, and mechanical performance. Let $\check{\nu}_{\text{UV}}$ and $\hat{\nu}_{\text{UV}}$ denote the minimum and maximum acceptable printing speeds. The following three constraints should be satisfied in this optimization problem.

$$e_{\text{critical}} - e_{\text{max}} \leq 0$$
$$\check{\nu}_{\text{UV}} - \nu_{\text{UV}} \leq 0$$
$$\nu_{\text{UV}} - \hat{\nu}_{\text{UV}} \leq 0$$

5.1.3 System-level Energy Modeling

In the energy modeling methods introduced in the previous two sections, the energy usage patterns in different machines are assumed to be independent. The energy consumption of the entire system is calculated through the summation of the energy consumed in each machine. This assumption is appropriate for some manufacturing systems. For example, if the first machine drills a hole on a part and the second one cuts a chamfer, the energy consumed by these two machines are not related. However, in some manufacturing systems, such as the manufacturing systems for biofuel production that involve multistep chemical and/or bio-reactions, some internal factors may affect the energy consumptions of multiple production units. In this case, it is necessary to employ a system-level energy modeling method to investigate the system energy consumption behaviors and develop strategies to improve the system energy efficiency. Since the manufacturing system discussed here may involve both machines (such as the cutting or grinding machines) and reactors (where the chemical- and/or bioreactions take place), in this section, a more general term "process" instead of "machine" is used to represent a production unit.

In system-level energy modeling of the aforementioned systems, the first step is to divide the factors (also referred to as variables) into two categories: intra-process variables and inter-process variables. The former is only related to one process and can affect the energy consumption of that process. The latter is involved in multiple processes and can have direct impacts on the energy consumptions in these processes. For example, assume that a system consists of N processes. Let α denote a set of inter-process variables and let β_i, ($i = 1, ..., N$) denote the intra-process variable set for the i-th process. The relationships among inter-process variables, intra-process variables, and processes are illustrated in Figure 5.8.

Based on the above example, let E_i denote the energy consumed in the ith process, then E_i can be formulated as a function of the set of inter-process variables α and the corresponding intra-process variable set β_i:

$$E_i \equiv f_i(\alpha, \beta_i) \tag{5.16}$$

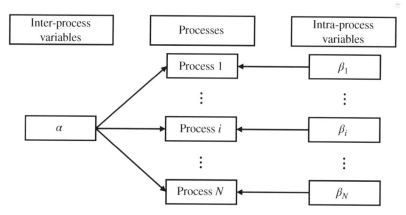

Figure 5.8 The relationship among inter-process variables, intra-process variables, and processes.

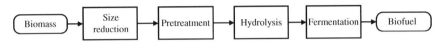

Figure 5.9 Block diagram of the biofuel production from biomass.

To improve the system energy efficiency, the objective function of the optimization problem can be formulated as

$$\min_{\alpha, \beta_1, \dots, \beta_n} \sum_{i=1}^{N} E_i \tag{5.17}$$

The solution to this optimization problem should meet the production or quality constraints, which are formulated based on the nature of the manufacturing system. The following example illustrates the formulation of a system-level energy model and an optimization problem of a biofuel manufacturing system.

Example 5.2 Consider a biofuel manufacturing system as shown in Figure 5.9. This system consists of four processes: size reduction, pretreatment, hydrolysis, and fermentation. The biomass feedstocks, such as corn stover and switchgrass, are first cut or ground into fine particles through a size reduction process. The fine particles undergo subsequent three treatment processes and are converted into the final biofuel. The entire system structure, in practice, involves many subprocesses and variables. To simplify the modeling procedure, in this example, only two inter-process variables are discussed in detail: the feedstock particle size d and the initial input water volume ω.

The energy consumption in the size reduction process, denoted by $E_{reduction}$, can be calculated based on the size of feedstocks before and after grinding, which can be expressed as

$$E_{reduction} = C_1 \ln\left(\frac{d_0}{d}\right)$$

where C_1 is a constant coefficient and d_0 is the feedstock size before grinding.

In a pretreatment process, the mixture of corn stover particles, water, and other reactants needs to be heated to a required temperature. The heating energy, denoted by $E_{heating}$, can be calculated as

$$E_{heating} = g_1 \omega + g_2$$

where g_1 and g_2 are two functions that are related to variables other than ω and d.

In a hydrolysis and fermentation process, depending on whether the reactions are endothermic or exothermic, extra energy may be required to supply or absorb the heat to keep the mixture at a certain temperate. The energy needed for the reaction, denoted by E_{reaction}, is related to the mass of reactant before and after the reaction, which are denoted by R_0 and R_1, respectively. E_{reaction} can be formulated as

$$E_{\text{reaction}} = C_2 \frac{R_0 - R_1}{R_0}$$

where C_2 is a constant coefficient. To find R_1 based on R_0, the dynamic of the reaction can be represented using the following differential equation:

$$\frac{dR}{dt} = f(C(t))$$

$$C(t) = 1 - \left(\frac{6a(1+a)}{9 + 9a + C_3a} \right) \exp \left(-\frac{C_4 t}{(d/2)^2} \right)$$

$$a = \frac{3C_5 \omega}{4\pi(d/2)^3}$$

The differential equation is a function of $C(t)$, where $C(t)$ is a function of inter-process variables ω and d, C_3, C_4, and C_5 are constant coefficients. Therefore,

$$R_1 = R_0 + \int \frac{dR}{dt} dt$$

In this example, it can be observed that variable ω is related to E_{heating} and E_{reaction}, whereas variable d is related to $E_{\text{size reduction}}$ and E_{reaction}.

Let β and E_{other} denoted the set of variables and energy consumptions that does not directly relate to ω and d. The objective function of the optimization problem of the system energy efficiency improvement can be formulated by

$$\min_{\omega, d, \beta} \left(E_{\text{size reduction}} + E_{\text{heating}} + E_{\text{reaction}} + E_{\text{other}} \right)$$

5.2 Energy Cost Modeling

In practice, the utility companies usually provide various electricity tariffs to meet the needs of different customers. The relationships between energy consumption and energy cost may vary under different electricity tariffs. Depending on the electricity tariff, the optimal production strategy that can minimize the energy consumption does not necessarily reduce the energy cost. Therefore, energy cost modeling is crucial in energy and production optimization problems.

In general, the electricity cost consists of three parts:

$$C = C_{\text{E}} + C_{\text{D}} + C_{\text{F}} \tag{5.18}$$

where C is the total energy cost of the system; C_E is the energy consumption cost; C_D is the demand cost, and C_F is the fixed cost.

In the following sections, three different electricity tariffs are introduced, i.e. flat rate, time-of-use (TOU) rate, and critical peak price (CPP) rate. The details of energy cost calculation are demonstrated as follows.

5.2.1 Energy Cost Under Flat Rate

In flat rate electricity tariffs, the energy consumption rate ($/kWh) and demand rate ($/kW) remain unchanged during the contract period. The energy consumption cost and demand cost can be calculated as follows.

5.2.1.1 Energy Consumption Cost

The calculation of energy consumption cost under flat rate electricity tariffs is straightforward. In a manufacturing system with N machines, let c_E denote the constant energy consumption rate and $e_i(t)$ denote the energy consumption of machine M_i at time t. The energy consumption cost of this manufacturing system, denoted by C_E, can be calculated as follows:

$$C_E = \sum_{t \in T} \sum_{i \in I} e_i(t) c_E \tag{5.19}$$

where I is the set of machine index, i.e. $I = [1, 2, ..., N]$, and T is the set of system production time. For example, suppose a manufacturing system operates from 8:00 to 12:00 a.m. with two shifts, and the machine cycle time is 0.25 hour; thus, the total operation time of 16 hours can be slotted into 64 uniform time steps and $T = [1, 2, ..., 64]$.

5.2.1.2 Demand Cost

The calculation of demand cost is related to the value of the peak demand, which is the highest average power (in kW) measured in any time interval t_D (usually 15 minutes) during the billing period. Let d_T denote the peak demand during production time T, and it can be determined using the following "window sliding" methods.

Assume the machine cycle time is τ, and Figure 5.10 illustrates three window sliding strategies where $1 < t_D/\tau < 2$. In Figure 5.10a, a window with the length t_D slides from left to right with an interval τ. With this sliding strategy, the left boundary of the window is always aligned with the beginning of a cycle time. In Figure 5.10b, the window slides from right to left with an interval τ, and the right boundary of the window is always aligned with the end of a cycle time. In Figure 5.10c, the window slides continuously from left to right or from right to left.

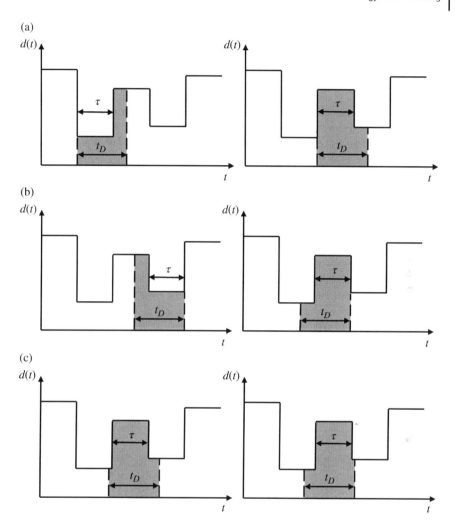

Figure 5.10 Illustrations of three window sliding strategies. (a) Sliding from left to right, (b) sliding from right to left, and (c) sliding continuously from left to right or from right to left. *Source:* Adapted from [4].

At any moment, the average demand can be calculated by dividing the area under the time window, i.e. the gray area in Figure 5.10, with the length of t_D.

The actual peak demand d_T is the highest average demand obtained by the third strategy. The highest average demand obtained using the first two methods are denoted by $d_{T, a}$ and $d_{T, b}$, respectively. It is clear that the actual d_T is always equal to the larger value of $d_{T, a}$ and $d_{T, b}$. Similarly, this rule can also be applied in the cases where $0 < t_D/\tau \leq 1$ and $t_D/\tau \geq 2$.

Let l be the ceiling integer number of the time steps in any interval of length t_D, that is

$$l = \left\lceil \frac{t_D}{\tau} \right\rceil \tag{5.20}$$

In a manufacturing system with N machines, the peak demand d_T can be calculated as follows:

$$d_T = \max\left(d_{T,a}, d_{T,b}\right)$$

$$d_{T,a} = \max_{t \in T} \frac{\sum_{t_1 = t}^{t+l-1} \sum_{i=1}^{N} e_i(t_1) - \left(l - \frac{t_D}{\tau}\right) \sum_{i=1}^{N} e_i(t+l-1)}{t_D} \tag{5.21}$$

$$d_{T,b} = \max_{t \in T} \frac{\sum_{t_2 = t-l+1}^{t} \sum_{i=1}^{N} e_i(t_2) - \left(l - \frac{t_D}{\tau}\right) \sum_{i=1}^{N} e_i(t-l+1)}{t_D}$$

Usually, the demand cost is calculated based on the monthly peak demand. Let c_D denote the constant demand rate, and the demand cost of the preceding system in a month, denoted by C_D, can be calculated as

$$C_D = d_T c_D \tag{5.22}$$

Example 5.3 A manufacturing system has one machine only. The system parameters are set as production time $T = \{1, 2, 3\}$, cycle time $\tau = 0.25$ hour, window length $t_D = 1.2\tau$, the respective energy consumption at $t = 1, 2, 3$ is $e(1) = 10$ kWh, $e(2) = 20$ kWh, and $e(3) = 15$ kWh, as shown in Figure 5.11. In addition, the energy consumption in each time step outside the production time is 0 kWh. Find peak demand d_T of this system.

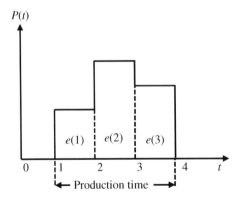

Figure 5.11 Energy profile of the system in Example 5.3.

Solution

First, we compute

$$l = \left\lceil \frac{t_D}{\tau} \right\rceil = 2$$

Then, we have

$$d_{T,a} = \max_{t \in T} \frac{\sum_{t_1 = t}^{t+l-1} \sum_{i=1}^{N} e_i(t_1) - \left(l - \frac{t_D}{\tau}\right) \sum_{i=1}^{N} e_i(t+l-1)}{t_D}$$

$$= \max_{t \in T} \frac{\sum_{t_1 = t}^{t+1} e(t_1) - 0.8 e(t+1)}{t_D}$$

$$= \max \left\{ \frac{\sum_{t_1=1}^{2} e(t_1) - 0.8 e(2)}{t_D}, \frac{\sum_{t_1=2}^{3} e(t_1) - 0.8 e(3)}{t_D}, \frac{\sum_{t_1=3}^{4} e(t_1) - 0.8 e(4)}{t_D} \right\}$$

$$= \max \left\{ \frac{e(1) + 0.2 e(2)}{0.25}, \frac{e(2) + 0.2 e(3)}{0.25}, \frac{e(3) + 0.2 e(4)}{0.25} \right\}$$

$$= \max\{56, 92, 60\}$$

$$= 92$$

$$d_{T,b} = \max_{t \in T} \frac{\sum_{t_2 = t-l+1}^{t} \sum_{i=1}^{N} e_i(t_2) - \left(l - \frac{t_D}{\tau}\right) \sum_{i=1}^{N} e_i(t-l+1)}{t_D}$$

$$= \max_{t \in T} \frac{\sum_{t_2 = t-1}^{t} e(t_2) - 0.8 e(t-1)}{t_D}$$

$$= \max \left\{ \frac{\sum_{t_2=0}^{1} e(t_2) - 0.8 e(0)}{t_D}, \frac{\sum_{t_2=1}^{2} e(t_2) - 0.8 e(1)}{t_D}, \frac{\sum_{t_2=2}^{3} e(t_2) - 0.8 e(2)}{t_D} \right\}$$

$$= \max \left\{ \frac{e(1) + 0.2 e(0)}{0.25}, \frac{e(2) + 0.2 e(1)}{0.25}, \frac{e(3) + 0.2 e(2)}{0.25} \right\}$$

$$= \max\{40, 88, 76\}$$

$$= 88$$

Hence, we have the system's peak demand:

$$d_T = \max(d_{T,a}, d_{T,b}) = 92$$

5.2.2 Energy Cost Under Time-of-use Rate

In time-of-use (TOU) rate electricity tariffs, the energy consumption and demand rates are different at different time of a day. Besides, the utility companies may also provide different TOU rates depending on the days of a week and a season. The energy consumption cost and demand cost can be calculated as follows.

5.2.2.1 Energy Consumption Cost

In general, the time of a day under a TOU rate electricity tariff can be divided into three periods: on, mid, and off-peak periods. Let T_{on}, T_{mid}, and T_{off} denote the sets of production time slots during each period and let T denote the set of all production time slots. Then we have

$$T = T_{on} \cup T_{mid} \cup T_{off} \tag{5.23}$$

Each utility company may have its own definition of time periods in the TOU tariff. For example, the Pacific Gas and Electric Company® (PGE) [5] defines the time periods as shown in Table 5.1.

Example 5.4 A manufacturing system operates with two shifts (from 8:00 to 12:00 a.m.), and the machine cycle time is 0.25 hour. Suppose that this plant joins the TOU tariff as listed in Table 5.1.

Find out T_{on}, T_{mid}, T_{off}, and T for

a) Tuesday, 5 August, and
b) Monday, 20 February.

Table 5.1 Definition of the time periods under one of PGE's TOU tariffs.

Summer (Service from 1 May through 31 October)		
On-peak	12:00 p.m. to 6:00 pm	Monday through Friday
Mid-peak	8:30 a.m. to 12:00 p.m. and 6:00 p.m. to 9:30 p.m.	Monday through Friday
Off-peak	9:30 p.m. to 8:30 a.m.	Monday through Sunday
Winter (Service from 1 November through 30 April)		
Mid-peak	8:30 a.m. to 9:30 p.m.	Monday through Friday
Off-peak	9:30 p.m. to 8:30 a.m.	Monday through Sunday

Solution:

a) Under the provided operation schedule, the operation time can be slotted into 64 uniform time steps and thus $T = \{1, 2, ..., 64\}$. The different time periods in a summer weekday can be demonstrated in Figure 5.12a. Therefore,

$$T_{on} = \{17, 18, ..., 40\}$$

$$T_{mid} = \{3, 4, .., 16, 41, 42, ..., 54\}$$

$$T_{off} = \{1, 2, 55, 56, ..., 64\}$$

b) Similar to the solution for part a), $T = \{1, 2, ..., 64\}$. The different time periods in a winter weekday can be illustrated in Figure 5.12b. Therefore,

$$T_{on} = \emptyset$$

$$T_{mid} = \{3, 4, ..., 54\}$$

$$T_{off} = \{1, 2, 55, 56, ..., 64\}$$

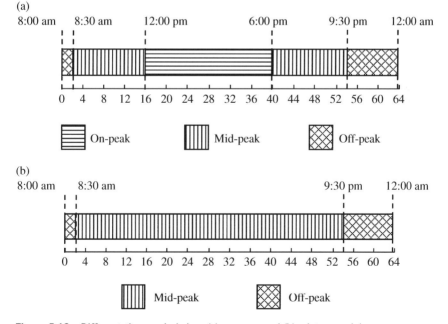

Figure 5.12 Different time periods in a (a) summer and (b) winter weekday.

Under the TOU rate electricity tariffs, let $c_{E,on}$, $c_{E,mid}$, and $c_{E,off}$ denote the energy consumption rate during the on, mid, and off-peak period, respectively. The energy consumption rate $c_E(t)$ can be expressed as

$$c_E(t) = \begin{cases} c_{E,on}, & \text{if } t \in T_{on} \\ c_{E,mid}, & \text{if } t \in T_{mid} \\ c_{E,off}, & \text{if } t \in T_{off} \end{cases} \tag{5.24}$$

In a manufacturing system with N machines, let $e_i(t)$ denote the energy consumption of machine M_i at time t; let I denote the set of machine index ($I = \{1, 2, ..., N\}$). The energy consumption cost of this manufacturing system, denoted by C_E, can be calculated as

$$C_E = \sum_{t \in T} \sum_{i \in I} e_i(t) c_E(t) \tag{5.25}$$

5.2.2.2 Demand Cost

In the TOU rate electricity tariffs, the demand rate $c_D(t)$ is also different in each period. Let $c_{D,on}$, $c_{D,mid}$, and $c_{D,off}$ denote the demand rate during the on, mid, and off-peak period, respectively. The demand rate $c_D(t)$ can be expressed as

$$c_D(t) = \begin{cases} c_{D,on}, & \text{if } t \in T_{on} \\ c_{D,mid}, & \text{if } t \in T_{mid} \\ c_{D,off}, & \text{if } t \in T_{off} \end{cases} \tag{5.26}$$

In a manufacturing system with N machines, the demand cost C_D can be calculated as

$$C_D = \max(C_{D,a}, C_{D,b})$$

$$C_{D,a} = \max_{t \in T} \frac{\sum_{t_1=t}^{t+l-1} \sum_{i=1}^{N} e_i(t_1) c_D(t_1) - \left(l - \frac{t_D}{\tau}\right) \sum_{i=1}^{N} e_i(t+l-1) c_D(t+l-1)}{t_D}$$

$$C_{D,b} = \max_{t \in T} \frac{\sum_{t_2=t-l+1}^{t} \sum_{i=1}^{N} e_i(t_2) c_D(t_2) - \left(l - \frac{t_D}{\tau}\right) \sum_{i=1}^{N} e_i(t-l+1) c_D(t-l+1)}{t_D}$$

$$\tag{5.27}$$

where $C_{D,a}$ and $C_{D,b}$ are the demand costs obtained based on the first and second "window sliding" strategies demonstrated in Section 5.2.1.

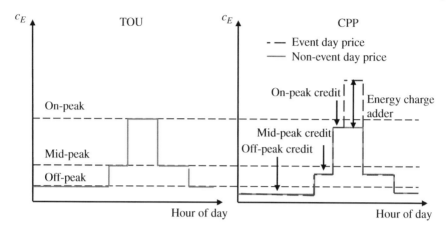

Figure 5.13 Comparisons between TOU and CPP. *Source:* Adapted from [6].

5.2.3 Energy Cost Under Critical Peak Price (CPP)

The critical peak price (CPP) rate is an overlay on the TOU pricing, as shown in Figure 5.13. In CPP rate electricity tariffs, the energy consumption rate is affected by the CPP events. The CPP events are defined by the utility company, and they are most likely to occur when extremely hot weather prompts the peak air-conditioning usage and strains power supplies, or a natural disaster cuts access to electricity resources [7]. The customers who join the CPP tariff need to pay a much higher energy consumption rate during the critical peak period on an event day. The maximum number of event days per year and the start and end time of the critical peak period on an event day are determined before consumers sign the contracts. However, the specific date that such an event occurs cannot be predetermined, and customers are notified one day to one week before the event. Although the energy consumption rate is much higher during the critical peak hours on an event day, the customers are offered a discounted price on other non-event days. Therefore, if the customers are able to adjust their production plan based on the CPP events, compared to TOU tariffs, CPP tariffs can lead to a further reduction in their electricity bill [6].

5.2.3.1 Energy Consumption Cost

In CPP tariffs, the definitions of the on, mid, and off-peak periods are the same as in TOU tariffs. In addition, let T_{cri} denote the set of production time slots in the critical peak period. On a nonevent day, a critical peak rate is not applied and thus

T_{cri} is an empty set. On an event day, the total production time T consists of T_{cri} and the other time sets. Therefore, in general, the following relationships among different time sets should be satisfied.

$$T = T_{on} \cup T_{mid} \cup T_{off} \cup T_{cri} \tag{5.28}$$

To calculate the energy consumption cost in CPP tariffs, first, let $c_{E,on}$, $c_{E,mid}$, and $c_{E,off}$ denote the original energy consumption rates during on, mid, and off-peak periods, respectively. Then, let $c_{E,on}^-$, $c_{E,mid}^-$, and $c_{E,off}^-$ denote the energy consumption charge credits during these three periods. If a CPP event occurs, an additional energy consumption rate needs to be applied, and it is also referred to as the energy charge adder (as shown in Figure 5.13). Let $c_{E,cri}^+$ denote the energy charge adder. The energy consumption rate at time t, denoted by $c_E(t)$, can be expressed as

$$c_E(t) = \begin{cases} c_{E,on} - c_{E,on}^-, & \text{if } t \in T_{on} \\ c_{E,mid} - c_{E,mid}^-, & \text{if } t \in T_{mid} \\ c_{E,off} - c_{E,off}^-, & \text{if } t \in T_{off} \\ c_{E,on} + c_{E,cri}^+, & \text{if } t \in T_{cri} \end{cases} \tag{5.29}$$

Similar to the energy consumption cost calculation in TOU tariffs, in a manufacturing system with N machines, let $e_i(t)$ denote the energy consumption of machine M_i at time t, and I represent the set of machine indices. The energy consumption cost of this manufacturing system, denoted by C_E, can be calculated as

$$C_E = \sum_{t \in T} \sum_{i \in I} e_i(t) c_E(t) \tag{5.30}$$

5.2.3.2 Demand Cost

In CPP rate electricity tariffs, customers can have the demand charge credits in each period. Let $c_{D,on}$, $c_{D,mid}$, and $c_{D,off}$ denote the original demand rate during on, mid, and off-peak periods, respectively. In addition, let $c_{D,on}^-$, $c_{D,mid}^-$, and $c_{D,off}^-$ denote the demand charge credits during three periods. Different utility companies have their own CPP tariffs. Many companies, such as PGE [5], Southern California Edison (SCE) [8], and San Diego Gas & Electric (SDGE) [9] do not have the demand charge adder in their CPP tariffs. Therefore, the demand rate $c_D(t)$ can be expressed as

$$c_D(t) = \begin{cases} c_{D,on} - c_{D,on}^-, & t \in T_{on} \cup T_{cri} \\ c_{D,mid} - c_{D,mid}^-, & t \in T_{mid} \\ c_{D,off} - c_{D,off}^-, & t \in T_{off} \end{cases} \tag{5.31}$$

The discounted TOU-based demand charge, denoted by $C_{\mathrm{D}}^{q_0}$, can be expressed as

$$C_{\mathrm{D}} = \max(C_{\mathrm{D,a}}, C_{\mathrm{D,b}})$$

$$C_{\mathrm{D,a}} = \max_{t \in T} \frac{\sum_{t_1=t}^{t+l-1} \sum_{i=1}^{N} e_i(t_1) c_{\mathrm{D}}(t_1) - \left(l - \frac{t_{\mathrm{D}}}{\tau}\right) \sum_{i=1}^{N} e_i(t+l-1) c_{\mathrm{D}}(t+l-1)}{t_{\mathrm{D}}}$$

$$C_{\mathrm{D,b}} = \max_{t \in T} \frac{\sum_{t_2=t-l+1}^{t} \sum_{i=1}^{N} e_i(t_2) c_{\mathrm{D}}(t_2) - \left(l - \frac{t_{\mathrm{D}}}{\tau}\right) \sum_{i=1}^{N} e_i(t-l+1) c_{\mathrm{D}}(t-l+1)}{t_{\mathrm{D}}}$$

$$(5.32)$$

where $C_{\mathrm{D,a}}^{q_0}$ and $C_{\mathrm{D,b}}^{q_0}$ are the discounted demand cost obtained based on the first and second "window sliding" strategies.

Example 5.5 A manufacturing plant operates with two shifts (from 8:00 to 12:00 a.m.), the machine cycle time is 0.25 hour, and the plant operates 21 days per month. The manager can choose either TOU or CPP tariff (the detailed rates are listed in Table 5.2). The fixed cost is \$630/month in two tariffs. Suppose that during operation time the system requires a constant power demand of 50 kW, which tariff should be selected if

a) one CPP event is expected to occur during a month (21 workdays), and
b) two event days are expected to occur.

Solution

a) Energy cost under TOU tariff:
   The energy consumed in one hour is 50 kW × 1 hour = 50 kWh
   In each workday, the system operates six hours during the on-peak period, seven hours during the mid-peak period, and three hours during the off-peak period. The energy consumption cost in this month is

   $$C_{\mathrm{E}} = 21 \times 50 \times ((6 \times 0.16) + (7 \times 0.11) + (3 \times 0.07)) = 2037$$

   Since the system operates with constant power demand, the peak demand is 50 kW. The demand cost in this month is

   $$C_{\mathrm{D}} = 50 \times (18.37 + 4.25 + 0) = 1131$$

   The total energy cost under the TOU tariff is

   $$C = 2037 + 1131 + 630 = 3798$$

Table 5.2 TOU or CPP tariff rates during on, mid, and off-peak periods.

	On-peak	Mid-peak	Off-peak	Critical-peak
Time	12:00 pm–6:00 pm	8:30 am–12:00 pm 6:00 pm–9:30 pm	9:30 pm–8:30 am	2:00 pm–6:00 pm
TOU energy consumption rates ($/kWh)	0.16	0.11	0.07	–
CPP energy consumption credits ($/kWh)	–	–	–	–
TOU demand rates ($/kW)	18.37	4.25	–	–
CPP demand credits ($/kW)	6.37	1.38	–	–
Adder during CPP event ($/kWh)	–	–	–	1.2

Energy cost under CPP tariff:

There are 20 nonevent days in this month, and there is no energy consumption credit. On the event day, the system operates two hours during the on-peak period, seven hours during the mid-peak period, three hours during the off-peak period, and four hours during the critical peak period. The energy consumption cost in this month is

$$C_E = 50 \times ((20 \times 6 + 2) \times 0.16 + (21 \times 7) \times 0.11 + (21 \times 3) \times 0.07 + 4 \times (1.2 + 0.16)) = 2277$$

The demand cost in this month is

$$C_D = 50 \times ((18.37 - 6.37) + (4.25 - 1.38) + 0) = 743.5$$

The total energy cost under CPP tariff is

$$C = 2277 + 743.5 + 630 = 3650.5$$

Therefore, the CPP tariff should be selected.

b) Energy cost under TOU tariff is still $3798, and the demand cost under CPP is unchanged. The energy consumption cost under CPP is

$$C_E = 50 \times ((19 \times 6 + 2 \times 2) \times 0.16 + (21 \times 7) \times 0.11 + (21 \times 3) \times 0.07 + (4 \times 2) \times (1.2 + 0.16)) = 2517$$

The total energy cost under the CPP tariff is

$$C = 2517 + 743.5 + 630 = 3890.5$$

Therefore, if two events are expected to occur in a month, the TOU tariff should be selected.

Problems

5.1   What is the difference among operation-based energy modeling, component-based modeling, and system-level energy modeling methods?

5.2   In Example 5.3, find out T_{on}, T_{mid}, T_{off}, and T for a summer weekend day.

5.3   In Example 5.4, if the manufacturing system operates with one shift (from 8:00 to 4:00 p.m.), which tariff should be selected if one CPP event is expected to occur during a month.

References

1 Yun, L., Li, L., and Ma, S. (2022). Demand response for manufacturing systems considering the implications of fast-charging battery powered material handling equipment. *Appl. Energy* 310: 118550.

2 Han, M., Yang, Y., and Li, L. (2020). Energy consumption modeling of 4D printing thermal-responsive polymers with integrated compositional design for material. *Addit. Manuf.* 34: 101223.

3 Han, M., Yang, Y., and Li, L. (2021). Techno-economic modeling of 4D printing with thermo-responsive materials towards desired shape memory performance. *IISE Trans.* 1–13.

4 Wang, Y. and Li, L. (2014). Time-of-use based electricity cost of manufacturing systems: modeling and monotonicity analysis. *Int. J. Prod. Econ.* 156: 246–259.

5 "Tariffs," *Pacific Gas & Electric*. https://www.pge.com/tariffs/index.page. (accessed 11 February 2021).

6 Wang, Y. and Li, L. (2016). Critical peak electricity pricing for sustainable manufacturing: modeling and case studies. *Appl. Energy* 175: 40–53.

7 SDG&E (2012). Critical peak pricing. https://www.sdge.com/businesses/savings-center/energy-management-programs/demand-response/critical-peak-pricing (accessed 11 February 2021).

8 "Rates & Pricing Choices," *Southern California Edison*. https://www.sce.com/regulatory/tariff-books/rates-pricing-choices. (accessed 11 February 2021).

9 "Current and Effective Tariffs," *San Diego Gas & Electric*. https://www.sdge.com/rates-and-regulations/current-and-effective-tariffs. (accessed 11 February 2021).

Part III

Energy Management in Typical Manufacturing Systems

6

Electricity Demand Response for Manufacturing Systems

In reference to Chapter 2, electricity is a form of energy that cannot be effectively stored in bulk, and therefore, it must be generated, distributed, and consumed immediately. In order to meet the demand during peak periods, a huge array of expensive equipment, including generators, transformers, wires, and substations, has to be kept on constant standby; otherwise, the grid may become unstable, and blackouts may occur. Therefore, in addition to the electricity consumption cost, the utility providers also charge the industrial customers for their electricity demand to recover the costs of these investments [1]. These charges sometimes are even higher than the electricity consumption cost [2, 3]. Proper management of electricity demand can lower the customers' demand cost and reduce the demand load in the grid. In this chapter, the management strategies of electricity demand from a customer side are introduced. More specifically, in Section 6.1, the time-of-use (TOU) pricing for manufacturing systems is studied, and a survey of 43 TOU programs is presented. Case studies are conducted to compare the energy cost for industry consumers under the flat rate tariffs and TOU ones. The results indicate that if industry consumers can actively adjust their production schedules, they are more likely to save energy cost by participating in TOU programs. In Section 6.2, a detailed production rescheduling method for a TOU program is demonstrated. This method aims to minimize the energy cost under TOU tariffs while maintaining the desired production yield. Finally, Section 6.3 provides the status quo of critical peak pricing (CPP) for industry consumers. Case studies are presented to compare the energy costs under TOU and CPP tariffs. The results indicate that based on the existing CPP programs, customers who have production flexibility in shifting their major electricity use are more likely to reduce their energy cost in CPP programs.

6.1 Time-of-use Pricing for Manufacturing Systems

To date, TOU pricing is one of the most prevalent demand response programs. Other than the traditional passive flat rates with continuous electricity supply, TOU pricing is particularly suitable for entry-level customers who have little knowledge of demand response programs [4]. Despite that TOU pricing is the easiest implementation of demand response, customers must be able to access the electricity tariffs and understand their terms to fully realize the benefits of implementing TOU pricing. Therefore, in this section, a detailed survey of 43 TOU programs offered by US utilities is presented. Besides, case studies are conducted under various industrial scenarios, and the changes in electricity cost when customers switch from flat rate tariffs to TOU ones are discussed. The survey results and case studies can guide existing customers to get the best from their TOU programs and provide necessary information to help potential customers decide whether to participate in such programs. Furthermore, the results can also provide a reference of exemplary tariffs to utility companies that are interested in designing new TOU programs.

6.1.1 Introduction to TOU

According to the FERC's (Federal Energy Regulatory Commission) National Action Plan [5] and Implementation Proposal [6] on demand response, customer education and pricing information transparency are critical to the success of demand response programs. The National Action Plan and Implementation Proposal suggest establishing a web-based clearinghouse to serve as a centralized location for collecting available information on demand response programs. The information should include regulatory documents, program tariffs, and other general information. As a supporting effort, the National Institute of Standards and Technology (NIST) of the U.S. Department of Commerce has taken actions to develop standards for the provision of energy price information [7]. Such information would be helpful for customers to predict electricity costs under various demand response programs [8–11].

The importance of customer education and pricing information transparency is also evidenced by the experiences obtained with existing demand response programs. For example, Ipsos MORI (a market research company) has conducted a nationwide survey in the United Kingdom to better understand customer experiences with TOU programs [12]. Among all the surveyed TOU users, 15% of them expressed their upset to the TOU tariffs; 33% of users did not understand their metering and bills very well or not at all; 35% of users believed that their tariffs were not suitable for their household needs; 38% of users were likely to be spending more on electricity than they needed to, and in general, over 50% of users believed additional information or advice would help them make better use of the tariffs. This survey concludes that it would be most helpful for the customers to obtain information about when the electricity price is lower in the TOU programs. Another example is related to a survey of the US utilities' experiences with real-time pricing, which is conducted by the Lawrence Berkeley National Laboratory [7, 8]. One of the significant findings of the survey is that even if customers need help understanding and managing price risks, such technical assistance from utilities is limited. More specifically, only one-third of the utilities offer technical support for customers to identify the best production strategies under demand response programs. The Berkeley Lab's survey implies that in order to extend demand response program participation, sufficient resources should be devoted to developing a customer education program and making the pricing information more transparent.

To better understand the TOU tariffs provided by utility suppliers, a survey of TOU programs offered by the 43 largest utilities in the United States is presented in this chapter. These programs include a wide range of TOU tariff designs. The detailed rate schedule information from each utility is identified and tabulated for future reference. In addition, these TOU tariffs are compared with the traditional flat rate tariffs such that one can estimate the benefits of switching from the flat rate to the TOU rate.

6.1.2 Survey of TOU Pricing in US Utilities

The surveyed utilities are chosen from 43 states in the United States, and one largest utility in terms of the total number of customers enrolled in its TOU program is selected from each state. The complete survey results of the TOU programs are provided in the Appendix 3.A. Particularly, the utility names, websites, names of their TOU tariffs, and counterpart flat rate tariffs are provided in Table 3.A.1. Detailed pricing information of the TOU tariffs and the counterpart flat rate tariffs is summarized in Tables 3.A.2 and 3.A.3, respectively. In summary, the survey results can be interpreted as follows:

- Compared with the flat rate, the TOU rate structure is generally more complicated. In TOU rate, the entire day is usually divided into on- and off-peak periods, and in some seasons, there may be a mid-peak period. The on-peak period mainly consists of hours during the daytime. The off-peak period primarily consists of hours during the nighttime, weekends, and holidays.
- The energy consumed in different periods is charged at different prices. The energy consumption rate ($/kWh) and demand rate ($/kW) during the on-peak period are generally higher than those during mid-peak and off-peak periods.
- If the maximum power demand occurs during the on-peak period, the customer is charged based on the on-peak demand rate. If the maximum power demand occurs during mid-peak or off-peak periods, the mid- or off-peak demand rates are applied, respectively.
- Customer fees, i.e. fixed monthly charge, for TOU tariffs are higher than or equal to the counterpart flat rate tariffs. The differences in customer fees can be

attributed to technical upgrades for metering as well as the usage and billing information communication.

- The demand metering intervals, i.e. time intervals used to calculate the average demand, in TOU tariffs and the counterpart flat rate tariffs may not always be the same. The commonly used demand metering interval is 15 minutes, and the least used interval is 60 minutes.

6.1.3 Comparison of Energy Cost Between Flat Rate and TOU Rates

In this section, the benefits of switching from flat rate tariffs to the corresponding TOU tariffs are evaluated. Chapter 5 can be referred to for details of the energy cost modeling methods adopted in this section (i.e. Section 5.2).

In the comparative case studies, three manufacturing systems with different sizes are considered, as shown in Table 6.1. The base demand represents the power used to support lighting and HVAC systems to maintain a suitable working environment in the facility. The demand for a single machine and the total number of machines can vary depending on the system size. The total peak demand of the system is the sum of the base demand and the demand for individual machines. The total peak demand of the small, medium, and large systems is 25 kW, 450 kW, and 1200 kW, respectively. Therefore, at least one system would fall into the demand range of any utility surveyed in Table 3.A.1. Only one size of the system is selected for the case study of each state's tariffs. The selection of system size is listed in Table 6.2.

In the case studies, three different scenarios are considered, as summarized in Table 6.3. Three different types of production scheduling during a workday are considered under each scenario: one work shift (8 hours), two work shifts (16 hours), and three work shifts (24 hours). In particular, Scenario 0 is the baseline case where the flat rate tariffs are adopted, and the work shift starts at 8:00 a.m. In Scenario 1, the work shift begins at 8:00 a.m., and the TOU tariffs are employed. In Scenario 2, the customer not only adopts the TOU rate but also

Table 6.1 Industrial systems used in case studies.

System size	Base demand (kW)	Demand for a single machine (kW)	Number of machines	Total peak demand (kW)
Small (S)	5	5	4	25
Medium (M)	50	50	8	450
Large (L)	150	70	15	1200

Table 6.2 Selection of system size in case studies (S – small, M – medium, L – large).

State	AL	AR	AZ	CA	CO	CT	DE	FL	GA	HI	IA
System size	*M*	*M*	*M*	*S*	*S*	*S*	*S*	*M*	*M*	*M*	*M*
State	IL	IN	KS	KY	LA	MA	ME	MI	MN	MO	MS
System size	*M*	*M*	*L*	*M*	*L*	*S*	*S*	*M*	*M*	*M*	*L*
State	NC	ND	NE	NH	NJ	NM	NV	NY	OH	OK	PA
System size	*M*	*M*	*M*	*S*	*L*	*M*	*S*	*M*	*M*	*M*	*S*
State	SC	SD	TN	TX	VA	VT	WA	WI	WV	WY	
System size	*M*	*M*	*S*	*M*	*M*	*M*	*L*	*S*	*M*	*L*	

Table 6.3 Three different scenarios.

Scenario	Tariff	Work shift starting time
Scenario 0	Flat rate in Table 3.A.2	8:00 a.m.
Scenario 1	TOU rate in Table 3.A.3	8:00 a.m.
Scenario 2	TOU rate in Table 3.A.3	Optimal starting time

actively seeks the optimal shift starting time to take advantage of the low-electricity price during the off-peak period.

Figure 6.1a shows the energy cost savings by switching from Scenario 0 to Scenario 1 if the one-shift production is adopted. As shown in Figure 6.1a, the cost savings are negative in most cases and the average cost saving is −10.9%, which indicates that costs are actually increased due to the switch. The reason is that the eight-hour shift from 8 a.m. to 4 p.m. falls into the on-peak period of many TOU tariffs. Figure 6.1b shows the electricity cost savings by switching from Scenario 0 to Scenario 2 if one-shift production is adopted. It can be observed that in most cases, the electricity costs are significantly reduced, and the average cost saving is 37.1%. This is because the eight-hour work shift is relatively short, and the eight-hour work shift can entirely fall into the off-peak period by adjusting the work schedule.

Figure 6.2 shows the energy cost savings by switching from Scenario 0 to Scenario 1 and Scenario 2 if two-shift production is adopted. The two-shift cases show similar results in comparison to the one-shift cases, but the absolute value of the changes in energy cost saving becomes smaller. Especially for the cases of switching from Scenario 0 to Scenario 2, the average cost saving drops from 37.1 to 13.6%. The substantial decrease is due to the increased working hours from 8 to 16 hours.

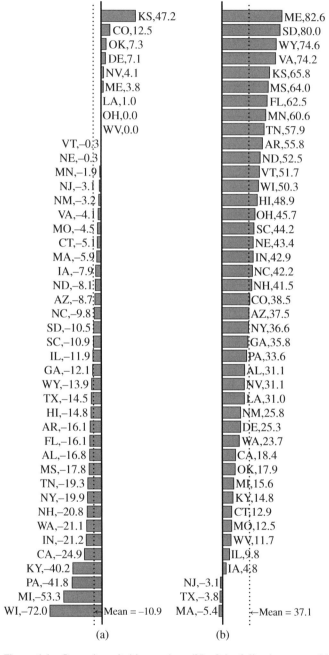

Figure 6.1 Scenario switching savings (%) of the following cases: (a) One-shift, Scenario 0→1; and (b) One-shift, Scenario 0→2. *Source:* [15]/With permission of Elsevier.

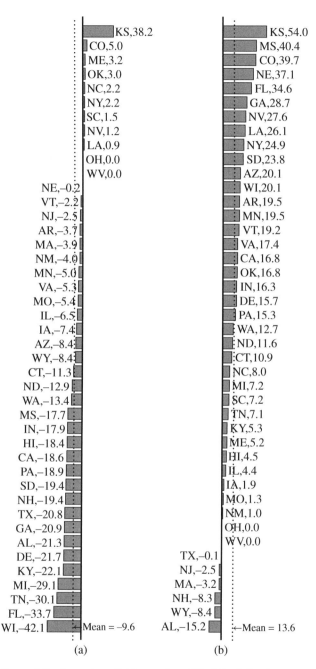

Figure 6.2 Scenario switching savings (%) of the following cases: (a) Two-shift, Scenario 0→1; and (b) two-shift, Scenario 0→2. *Source:* [15]/With permission of Elsevier.

In the 43 TOU tariffs, the average duration of off-peak periods is 15 hours, and many off-peak periods are divided into different segments during a day, which indicates that a significant portion of working hours falls into the on-peak period under many TOU tariffs. As a result, the benefit of the low prices during the off-peak periods cannot be fully utilized.

Figure 6.3 shows the energy cost savings by switching from Scenario 0 to Scenario 1 and Scenario 2 if the three-shift production is adopted. It can be observed that the cost differences between Scenario 0 and Scenario 1 or 2, i.e. the magnitudes of the positive or negative values in Figure 6.3, tend to be smaller. In addition, in the case of the three-shift production, the savings associated with the scenario switching demonstrated in Figure 6.3a and b are the same. The reason is that the three-shift production lasts for a whole day (24 hours), and the energy consumption and power demand are the same regardless of the starting time of the work shift. Therefore, the benefits of adopting TOU tariffs are compromised. The average electricity cost saving of the 43 cases is −3.0%, indicating no significant difference between TOU and flat rate tariffs. This result is consistent with the design principles behind the TOU and flat rate tariffs of many utility companies, i.e. both types of tariffs should reflect the average cost of electricity generation and transmission.

In summary, the above case studies imply that the cost savings can vary greatly, ranging from −72.0% to +82.6%, depending on the specific tariffs and the switching strategy involved in the industrial systems. The customers who switch from Scenario 0 to Scenario 1 likely face the risk of increased electricity costs. The possible explanation is that the customers simply switch from flat rate tariffs to TOU ones without adjusting the production schedules to further take advantage of the low electricity charges during the off-peak period. The customers who switch from Scenario 0 to Scenario 2 would get the most benefit if they adopt one- or two-shift production schedules as these customers actively adjust their production schedules to benefit from TOU tariffs. The TOU tariffs may lose their advantages over flat rate tariffs for customers who require a 24-hour production. For these customers, switching from flat rate tariffs to TOU tariffs may even induce a cost increase.

For utilities currently offering or considering TOU tariffs toward effective demand response, increasing customer participation is critical to the success of such programs. The case studies discussed above indicate that the customers can save more money by joining the TOU programs with a relatively lower off-peak price and shorter on-peak periods. Therefore, the TOU programs with these properties are likely to become popular among customers. In addition, the marketing campaigns of the demand response programs should pay more attention to the customers who have production flexibility in shifting their primary electricity use (e.g. those who run the production on a one or two-shift basis).

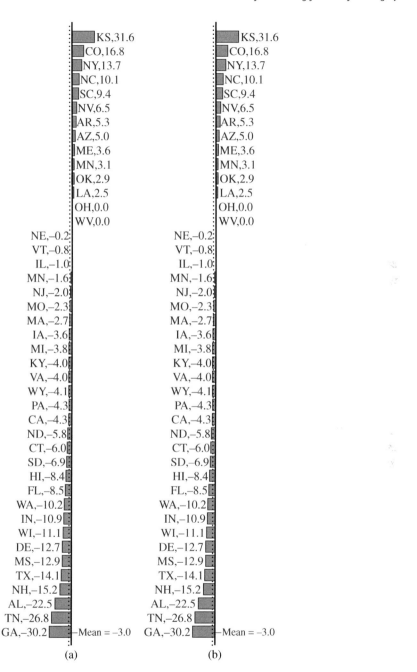

Figure 6.3 Scenario switching savings (%) of the following cases: (a) three-shift, Scenario 0→1; and (b) three-shift, Scenario 0→2. *Source:* [15]/With permission of Elsevier.

6.2 TOU-based Production Scheduling for Manufacturing Systems

In the previous section, the case studies indicate that the customers who can actively adjust their production schedules are more likely to benefit from the TOU tariffs. In those case studies, the production rescheduling strategies simply mean changing the work shift period. In this section, a more detailed production scheduling approach under the TOU tariffs is demonstrated, aiming to control each machine's operation states to achieve the minimum energy cost while maintaining the desired production yield. Note that the production scheduling problem discussed in this section is built upon the manufacturing system modeling methods introduced in Chapter 4, as well as the energy consumption and cost modeling methods presented in Chapter 5.

6.2.1 Manufacturing Systems Modeling

In this section, a serial production line with Bernoulli machines is employed to demonstrate the production scheduling problem under the TOU tariff. The layout of a serial production line with N machines and $N-1$ buffers is shown in Figure 6.4, where the squares represent buffers, the circles stand for machines, and the direction of arrows indicates the moving directions of the processed parts or production tasks.

In reference to the modeling of manufacturing systems demonstrated in Chapter 4, the transition matrix for buffer B_i at time step t, denoted by $Q_i(t)$, can be written as:

$$Q_i(t) = \begin{bmatrix} \ddots & \ddots & & \ddots & & \\ \ddots & Q_{i,(j-1|j-1)}(t) & Q_{i,(j-1|j)}(t) & 0 & & \\ \ddots & Q_{i,(j|j-1)}(t) & Q_{i,(j|j)}(t) & Q_{i,(j|j+1)}(t) & \ddots & \\ & 0 & Q_{i,(j+1|j)}(t) & Q_{i,(j+1|j+1)}(t) & \ddots & \\ & & \ddots & & \ddots & \ddots \end{bmatrix} \quad (6.1)$$

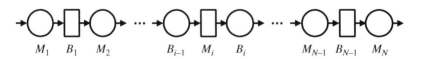

Figure 6.4 A typical serial production system with N machines and $N-1$ buffers.

where $Q_{i,(j_2|j_1)}(t)$ represents the transition possibility of buffer B_i in state j_2 at time t given that it was in state j_1 at time $t - 1$. The state probabilities of buffer B_i at time t can be arranged as a vector $q_i(t)$:

$$q_i(t) = \begin{bmatrix} q_{i,0}(t) \\ q_{i,k}(t) \\ \vdots \\ q_{i,C_i}(t) \end{bmatrix} \tag{6.2}$$

where $q_{i,k}(t)$ ($k = 0, 1, 2, \ldots, C_i$) represents the probability that buffer B_i contains k parts at time t; and C_i is the maximal capacity of B_i.

Let $s_i(t)$ denote the control signal of machine i at time step t, which can be formulated as

$$s_i(t) = \begin{cases} 1, & \text{if machine } M_i \text{ is turned on at time step } t \\ 0, & \text{if machine } M_i \text{ is turned off at time step } t \end{cases} \tag{6.3}$$

Supposing that the reliability of Bernoulli machine M_i is p_i, by jointly considering machine reliability and control signals, machine M_i is up at time step t with the probability of $p_i(t) = s_i(t)p_i$, and down (or turned off) with the probability of $1 - p_i(t) = 1 - s_i(t)p_i$. Let $p_i^S(t)$ and $p_i^B(t)$ denote machine M_i's starvation probability and blockage probability, respectively. The transition probability of buffer B_i during time step t can be written as

$$Q_{i,(0|0)}(t) = 1 - p_i(t) + p_i^S(t)$$
$$Q_{j,(1|0)}(t) = p_i(t) - p_i^S(t)$$
$$Q_{i,(j|j)}(t) = \left[1 - p_i(t) + p_i^S(t)\right]\left[1 - p_{i+1}(t) + p_{i+1}^B(t)\right]$$
$$\qquad + \left[p_i(t) - p_i^S(t)\right]\left[p_{i+1}(t) - p_{i+1}^B(t)\right]$$
$$Q_{i,(j|j+1)} = \left[1 - p_i(t) + p_i^S(t)\right]\left[p_{i+1}(t) - p_{i+1}^B(t)\right]$$
$$Q_{i,(j+1|j)}(t) = \left[p_i(t) - p_i^S(t)\right]\left[1 - p_{i+1}(t) + p_{i+1}^B(t)\right]$$
$$Q_{i,(C_i|C_i)}(t) = \left[1 - p_{i+1}(t) + p_{i+1}^B(t)\right] + \left[p_i(t) - p_i^S(t)\right]\left[p_{i+1}(t) - p_{i+1}^B(t)\right]$$
$$\tag{6.4}$$

Since the sum of all probability should be 1, the evolution of the probability that buffer B_i is in state j at time t can be calculated by solving

$$q_i(t) = Q_i(t)q_i(t-1) \tag{6.5}$$

$$\sum_{j=0}^{C_i} q_{i,j}(t) = 1 \tag{6.6}$$

The probability of machine M_i being starved or blocked can be calculated as

$$p_i^s(t) = p_i(t)q_{i-1,0}(t-1) \tag{6.7}$$

$$p_i^B(t) = p_i(t)q_{i,C_i}(t-1)\left[1 - p_{i+1}(t) + p_{i+1}^B(t)\right] \tag{6.8}$$

Note that the first machine is never starved, and the last machine is never blocked. The production rate of the system during time step t is equal to the production rate of the last machine. It can be calculated as

$$\eta_{\text{SYS}}(t) = \eta_N(t) = p_N(t) - p_N^S(t) \tag{6.9}$$

The average cumulative production of the system during the planning horizon of T time steps, denoted by $\bar{\eta}$, can be calculated by

$$\bar{\eta} = \sum_{t=1}^{T} \eta_{\text{SYS}}(t) \tag{6.10}$$

6.2.2 Energy Consumption and Energy Cost Modeling

The electricity consumption of the serial production line with Bernoulli machines (as shown in Figure 6.4) at time step t, denoted by $E(t)$, can be calculated as

$$E(t) = \sum_{i=1}^{N} e_i p_i(t) \tag{6.11}$$

where e_i denotes machine M_i's energy consumption in each time step.

The total electricity consumption of the system during the planning horizon of T time steps, denoted by E, can be calculated as

$$E = \sum_{t=1}^{T} E(t) \tag{6.12}$$

Assuming that the machine cycle time is τ, the power demand of the system at time step t, denoted by $d_T(t)$, is

$$d_T(t) = \frac{E(t)}{\tau} \tag{6.13}$$

Let l be the ceiling integer number of the time steps in any interval with length t_D, i.e.

$$l = \left\lceil \frac{t_D}{\tau} \right\rceil \tag{6.14}$$

The billable power demand of the system, denoted by d_T, is the highest average power (with the unit of kW) measured in any on-peak interval t_D during the planning horizon, which can be expressed as

$$d_T = \max(d_{T,a}, d_{T,b})$$

$$d_{T,a} = \max_{t \in \mathbb{T}_{on}} \frac{\sum\limits_{t_1 = t}^{t+l-1} E(t_1) - \left(l - \dfrac{t_D}{\tau}\right) E(t+l-1)}{t_D} \tag{6.15}$$

$$d_{T,b} = \max_{t \in \mathbb{T}_{on}} \frac{\sum\limits_{t_2 = t-l+1}^{t} E(t_2) - \left(l - \dfrac{t_D}{\tau}\right) E(t-l+1)}{t_D}$$

where $d_{T,a}$ and $d_{T,b}$ are the highest average demand obtained by using two "window sliding" methods introduced in Section 5.2, and $\mathbb{T}_{on}$ is the set of time steps during the on-peak period.

The total electricity cost is the summation of electricity consumption cost C_E, demand cost C_D, and fixed cost C_F.

$$C = C_E + C_D + C_F \tag{6.16}$$

For the manufacturing system as shown in Figure 6.4, the total electricity consumption cost of the system over the planning horizon is

$$C_E = \sum_{t \in T} E(t) c_E(t) \tag{6.17}$$

where $c_E(t)$ is the TOU consumption rate ($/kWh) at time step t.

The cost of the power demand of the system during the planning horizon is

$$C_D = \max\left(C_{D,a}, C_{D,b}\right)$$

$$C_{D,a} = \max_{t \in T} \frac{\sum\limits_{t_1 = t}^{t+l-1} E(t_1) c_D(t_1) - \left(l - \dfrac{t_D}{\tau}\right) E(t+l-1) c_D(t+l-1)}{t_D} \tag{6.18}$$

$$C_{D,b} = \max_{t \in T} \frac{\sum\limits_{t_2 = t-l+1}^{t} E(t_2) c_D(t_2) - \left(l - \dfrac{t_D}{\tau}\right) E(t-l+1) c_D(t-l+1)}{t_D}$$

where $C_{D,a}$ and $C_{D,b}$ are the demand costs obtained by using two "window sliding" methods introduced in Section 5.2, $c_D(t)$ is the TOU demand rate ($/kW) at time step t.

6.2.3 Production Scheduling for TOU-based Demand Response

6.2.3.1 Production Scheduling Problem Formulation

For the manufacturing system as shown in Figure 6.4, a production schedule should be created to determine when the machine(s) can be temporarily shut down to reduce energy consumption or energy cost. Meanwhile, the production schedule should also ensure that the average cumulative production target, denoted by $\bar{\eta}_0$, can be achieved during the planning horizon. The production scheduling problem is an optimization problem where the control signal $s_i(t)$ of each machine i is a decision variable. The following optimization problems with different objectives can be formulated.

- Formulation 1: Minimizing the total energy consumption while maintaining a certain amount of average cumulative production that is not lower than the required level, i.e.

$$\min_{s_i(t)} E, \quad \text{subject to} : \bar{\eta} \geq \bar{\eta}_0 \tag{6.19}$$

- Formulation 2: Minimizing the total energy cost while maintaining a certain amount of average cumulative production that is not lower than the required level, i.e.

$$\min_{s_i(t)} C, \quad \text{subject to} : \bar{\eta} \geq \bar{\eta}_0 \tag{6.20}$$

By solving these two problems, optimal schedules of control signals can be obtained by setting each machine in the "on" or "off" status to achieve their corresponding objectives.

6.2.3.2 PSO Algorithm for Near-optimal Solutions

The calculations of the energy consumption E and the energy cost C are straightforward. However, the calculation of the average cumulative production CP is highly nonlinear as it relies on the iteration-based method to untangle the complicated interactions among all the machines and buffers. In addition, the decision variable $s_i(t)$ is binary, not continuous. The involvement of the highly nonlinear $\bar{\eta}$ and the binary decision variable increases the complexity of the optimization problem and makes it almost impossible to find the exact optimal solution. Therefore, a metaheuristic algorithm, i.e. particle swarm optimization (PSO), can be used to find the near-optimal solutions for this complicated optimization problem.

Recall PSO as introduced in Section 3.3. Its first step is the initialization of a position matrix and a velocity matrix. For a manufacturing system with N machines (i.e. N decision variables) and planning horizon of T time steps, the dimension of the solution space is NT. Suppose that the swarm contains M particles, the size of position and velocity matrices are $M \times NT$. Since the decision variable $s_i(t)$ is binary, each element in the position matrix is either 0 or 1. During each search iteration, position x and velocity v of each particle are updated as follows:

$$v'_{ji} = \omega v_{ji} + c_1 r_1 \left(p_j - x_{ji} \right) + c_2 r_2 \left(g - x_{ji} \right) \tag{6.21}$$

$$x'_{ji} = \begin{cases} 1, & u < \dfrac{1}{1 + e^{-v_{ji}}} \\ 0, & \text{otherwise} \end{cases} \qquad (6.22)$$

where x_{ji} and v_{ji} are the position and velocity of the jth particle in the ith dimension before the update; x'_{ji} and v'_{ji} are position and velocity after the update; p_j is the jth particle's personal best position; g is the global best position; ω is the inertia weight; c_1 is the cognitive parameter, and c_2 is the social parameter; r_1, r_2, and u are three random numbers between 0 and 1. PSO terminates if it reaches the maximal iteration count N_T.

6.2.3.3 Case Study Setup

In this section, a serial production line with three machines and two buffers is applied to demonstrate a production scheduling problem. It is assumed that the manufacturing system operates 16 hours per day (from 8:00 to 12:00 a.m.). Suppose the cycle time $\tau = 0.25$ hour for each machine, and the operation horizon can be slotted into $T = 64$ uniform time steps with the time interval $t_D = 0.25$ hour. The following parameters of machines and buffers are adopted: when a machine is up at each time step, the respective energy consumption is $e_1 = e_2 = e_3 = 6.25$ KWh; machine reliabilities are $p_1 = p_2 = p_3 = 0.95$, and buffer capacities are set as $C_1 = C_2 = 5$. The same production schedule is repeated every working day, with a goal of an average cumulative production of $\bar{\eta}_0 = 45$ under a fresh start (i.e. all the buffers are empty at the beginning of a working day).

The TOU pricing profile is obtained from the Orange and Rockland Utilities [16], and the data are provided in Table 6.4. Normally, the demand cost and fixed cost are calculated on a monthly basis. Hence, to obtain the daily equivalents, the TOU demand rates and the fixed cost listed in Table 6.4 should be divided by 21, which is the average number of working days in a month.

Table 6.4 Representative TOU pricing profiles.

Season	Type	Time of day	Consumption rate ($/kWh)	Demand rate ($/kW)	Fixed cost ($)
Summer (Jun–Sep)	Off-peak	7 p.m. to 1 p.m.	0.08274	0	51.42
	On-peak	1 p.m. to 7 p.m.	0.16790	18.80	
Winter (Oct–May)	Off-peak	9 p.m. to 10 a.m.	0.08274	0	
	On-peak	10 a.m. to 9 p.m.	0.11224	8.12	

The PSO parameters are set as swarm size $M = 1000$; maximal number of iterations $N_T = 2000$; inertia weight $\omega = 1$; cognitive parameter $c_1 = 2$ and social parameter $c_2 = 2$.

6.2.3.4 Optimal Production Schedules

The best production schedules in summer and winter for Formulation 1 (i.e. energy consumption minimization) and Formulation 2 (i.e. energy cost minimization), are demonstrated in this section.

The best solutions for Formulation 1 are illustrated in Figure 6.5. The objective and constraint values corresponding to the production schedules for Formulation 1 are listed in Table 6.5. The detailed results are explained as follows.

The control signals for the three machines are represented by s_1, s_2, and s_3 in Figure 6.5. These signals are changing over time, which leads to changes in the expected values of performance measures W_{SYS}, η_{SYS}, E, and d_T.

- At the beginning of production, three machines are sequentially turned on. The first machine is turned on at 8:00 a.m., while the second and the third machines are turned on with gradually increasing delays. More specifically, at the beginning of production, the first and second buffers are empty; thus, even if the second and third machines were turned on immediately, they would be starved and cannot produce any products as it takes time for buffers to build up. Hence, this schedule can reduce energy consumption without affecting the average cumulative production.

- At the end of production, the three machines are turned off sequentially. The first machine shuts down first, while the second and third machines are turned off with gradually increasing delays. This schedule can reduce the energy consumption on the first and the second machines as it takes time to deplete the built-up inventories in the first and second buffers.

- Since the objective of Formulation 1 is to minimize energy consumption, different electricity rates during on-peak and off-peak periods do not affect the production schedule. More specifically, other than the beginning and end periods of the production, the machines are controlled such that the respective energy consumption is reduced intermittently and randomly throughout the entire planning horizon (as shown by the plots of E in Figures 6.5 and 6.6), no matter it is in the on-peak period or off-peak one.

- Although demand $d_T(t)$ can be as low as 0 kW (i.e. all three machines are turned off at the same time step), peak demand $\hat{d}_T$ of the system remains at 75 kW. This is because the peak demand is the highest average power demand in any 15-minute interval during the on-peak period.

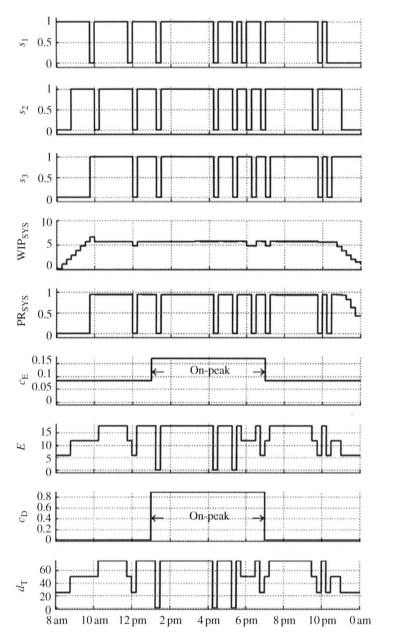

Figure 6.5 Best schedules for Formulation 1 based on the TOU rates in summer.
Source: [17]/With permission of Elsevier.

Table 6.5 The objective and constraint values corresponding to the best schedules for Formulation 1.

Formulation	E (kWh)	d_T (kW)	C_E ($)	C_D ($)	C($)	CP
Formulation 1 (summer)	872.813	75	103.993	67.143	171.135	45.217
Formulation 1 (winter)	872.813	75	94.458	29.000	123.458	45.217

- The demand charge can be a significant part of the utility bill. In Table 6.5, demand cost C_D accounts for 39.23% of total energy cost C in summer. The reason is that Formulation 1 does not consider demand, and, therefore, the production schedule has no effect of reducing this part of the charge.

The best solutions for Formulation 2 are demonstrated in Figures 6.7 and 6.8. The objective and constraint values corresponding to the production schedules for Formulation 2 are listed in Table 6.6. The detailed result explanations are illustrated as follows.

- At the beginning and end of production, the control signals of machines are similar to those in Formulation 1.
- Since the objective of Formulation 2 is to minimize the energy cost, other than the beginning and the end periods of production, the machines are shut down only during the time when the consumption and demand rates are higher (i.e. during the on-peak period), as shown by the plots of E and d_T in Figures 6.7 and 6.8.
- As shown in Table 6.6, the summer power demand d_T is 50 kW. In comparison with the summer power demand in Formulation 1, it is significantly reduced by avoiding simultaneous operation of all three machines during the on-peak period.
- The winter electricity demand d_T remains at 75 kW, which is the same as that of Formulation 1. This is because the on-peak period in winter is much longer than in summer. In order to meet the cumulative production target, it is impossible to avoid the simultaneous operation of all three machines during the long on-peak periods in winter.
- In summer, although the total energy consumption E increases slightly, the energy consumption cost C_E is reduced by switching the machine operation time from the on-peak period to the off-peak period. In addition, the production schedule can significantly reduce the demand charge C_D. In conclusion, Formulation 2 jointly responds to both energy consumption and power demand, while Formulation 1 only responds to energy consumption.

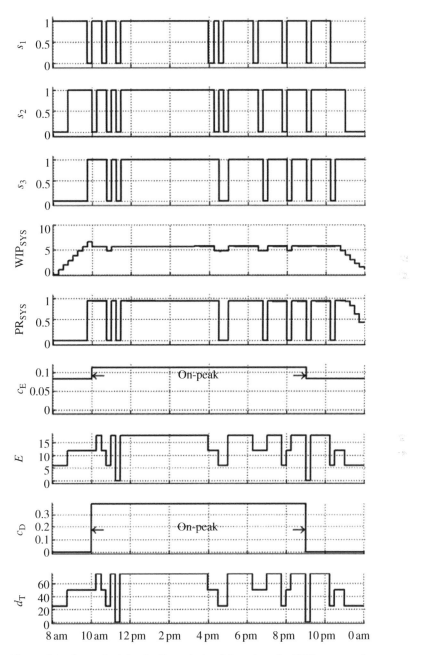

Figure 6.6 Best schedules for Formulation 1 based on the TOU rates in winter. *Source:* [17]/With permission of Elsevier.

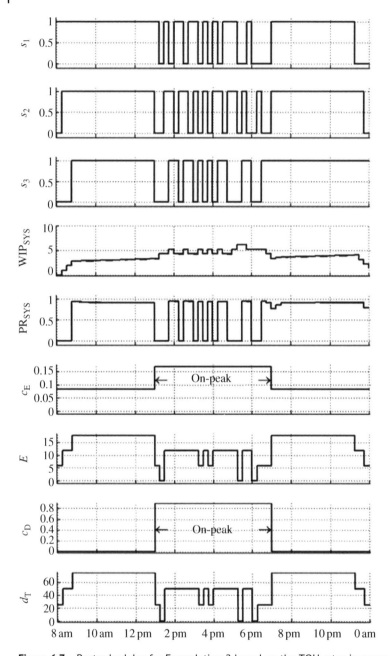

Figure 6.7 Best schedules for Formulation 2 based on the TOU rates in summer. *Source:* [17]/With permission of Elsevier.

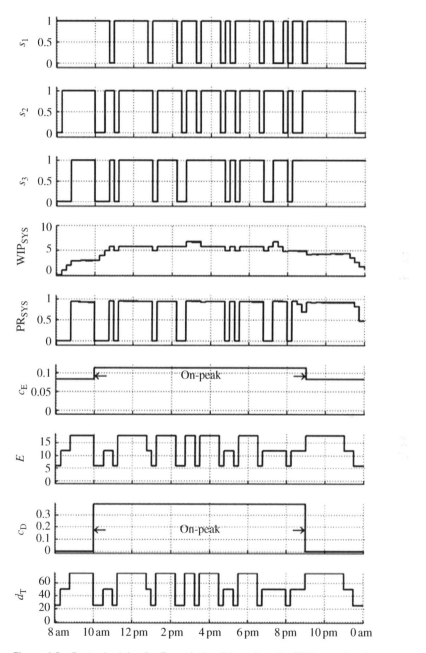

Figure 6.8 Best schedules for Formulation 2 based on the TOU rates in winter. *Source:* [17]/With permission of Elsevier.

Table 6.6 The objective and constraint values corresponding to the best schedules for Formulation 2.

Formulation	E (kWh)	d_T (kW)	C_E ($)	C_D ($)	C ($)	CP
Formulation 2 (summer)	878.750	50	93.359	44.762	138.121	45.027
Formulation 2 (winter)	872.813	75	91.655	29.000	120.655	45.038

6.3 Critical Peak Pricing for Manufacturing Systems

6.3.1 Introduction to Critical Peak Pricing (CPP)

Based on the U.S. FERC survey results, among 11 states providing CPP programs for business customers in the United States, California has the most CPP customers, accounting for more than 75% of the CPP customers nationwide [18]. Hence, the CPP programs offered in California are used to demonstrate the effects of CPP programs on the utility bill and provide insights for manufacturers that are currently enrolled or considering enrolling in CPP programs in other states.

There are three major utilities in California: Pacific Gas and Electric Company (PGE), Southern California Edison (SCE), and San Diego Gas and Electric Company (SDGE). They all operate based on the California Independent System Operator (CAISO) grid and are regulated by the California Public Utilities Commission (CPUC). The CPP programs provided by each utility are designed based on their own pricing model. Although the terms of the CPP programs in the three utility companies are different, they share some key elements. All three utilities put industrial customers on the CPP programs by default but allow customers to opt-out of a TOU rate. However, industrial customers must participate in one of the two types of rates, either TOU or CPP. Tables 3.A.4 and 3.A.5 summarize the detailed pricing information of the default CPP rates and corresponding optional TOU rates offered by PGE, SCE, and SDGE. The features of these rates are further explained as follows.

- For PGE and SCE customers, each CPP event lasts four hours (from 2:00 p.m. to 6:00 p.m.), and for SDGE customers, the event lasts seven hours (from 11:00 a.m. to 6:00 p.m.). During a CPP event, the energy charge may be an order of magnitude higher than the corresponding TOU summer on-peak rate.
- The credits (i.e. the discount pricing) are available on days in the summer months only. The credits may apply to only the on-peak periods, or both on- and mid-peak periods, or all three periods. The credits could be in the form of energy charge discounts, demand charge discounts, or both.

- The number of CPP event days in each year may be a fixed value or a variable. For PGE, the number of event days would be between 9 and 15 days per year. For SCE, this number is fixed at 12 per year. For SDGE, the number of CPP event days shall not be more than 18 days per year. The event day can be any day of the year for PGE and SDGE, and any nonholiday weekdays for SCE.
- Customers of all three utilities would receive a day-ahead notification prior to a CPP event day. Most likely, customers are notified by phone calls or emails. Additional communication tools such as SMS texts, fax, and utility websites are also used by some utilities.

In addition, PGE, SCE, and SDGE are required to regularly report the operation status of their demand response programs to CPUC. The historical CPP event data from 2005 to 2014 of the three utilities are available in [13, 14]. The cumulative total of CPP event days during the data collection period is grouped by month and is shown in Figures 6.9–6.11. It can be observed that the by-month data is approximately normally distributed for all three utilities, and the maximum number of events generally occurs in July, August, and September. The by-weekday data is approximately uniformly distributed for PGE and SCE events, and it is slightly skewed toward Thursday and Friday for SDGE events.

6.3.2 Comparison of Energy Cost Between TOU and CPP Rates

In this section, the annual energy cost of manufacturing systems under CPP and TOU rates listed in Tables 3.A.4 and 3.A.5 are calculated and compared through a case study. Note that the energy cost modeling methods adopted in this section are introduced in Section 5.2.

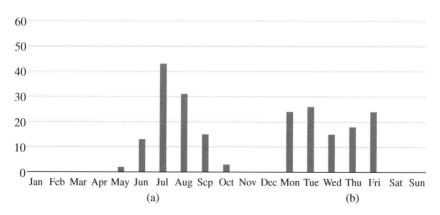

Figure 6.9 The number of historical events in PGE (a) by month and (b) by weekday.

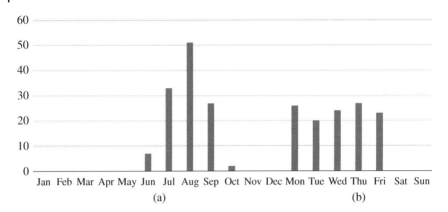

Figure 6.10 The number of historical events in SCE (a) by month and (b) by weekday.

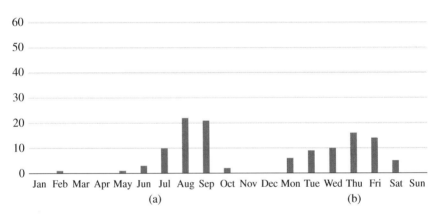

Figure 6.11 The number of historical events in SDGE (a) by month and (b) by weekday.

For each tariff listed in Tables 3.A.4 and 3.A.5, six different scenarios are considered, as shown in Table 6.7. Under each scenario, three different daily production schedules that are usually used in the manufacturing plants are considered: one shift (8 hours), two shifts (16 hours), and three shifts (24 hours).

In Scenario CPP1, the production shift starts at 8:00 a.m. In Scenario CPP2, the customer can reschedule production within a certain time range. More specifically, if there is only one production shift, the start time is between 6:00 a.m. and 6:00 p.m.; if there are two production shifts, the start time is between 6:00 a.m. and 10:00 a.m.; if there are three production shifts, the shift can start at any time of the day. In Scenario CPP3, the shift can start at any time regardless of the work shift. The differences among Scenarios TOU1, TOU2, and TOU3 are

Table 6.7 Six different scenarios.

Scenario	Rate	Production shift starting time
CPP1	CPP	8:00 a.m.
CPP2	CPP	Optimal starting time within the range: 1-shift (6:00 a.m. to 6:00 p.m.); 2-shift (6:00 a.m. to 10:00 a.m.); 3-shift (any time)
CPP3	CPP	Optimal starting time (any time)
TOU1	TOU	8:00 a.m.
TOU2	TOU	Optimal starting time within the range: 1-shift (6:00 a.m. to 6:00 p.m.); 2-shift (6:00 a.m. to 10:00 a.m.); 3-shift (any time)
TOU3	TOU	Optimal starting time (any time)

similar to those among the first three CPP scenarios. In these three TOU scenarios, the customers opt-out from CPP to the corresponding TOU rate. By adjusting the production shift starting time, the customer may take advantage of the lower prices during the off-peak period and avoid (or partially avoid) the electricity use during the critical or on-peak periods.

In the case study, the average number of workdays per month is set as 21 days. The number of CPP events for each utility company is obtained based on the historical data as shown in Figures 6.9–6.11. The power demands used to calculate the energy cost under each tariff are listed in Table 6.8. The comparison results of savings on annual energy costs by switching from one scenario to another are listed in Tables 6.8–6.10.

As shown in Table 6.8, when the manufacturing plant changes from CPP1 to CPP2, the results for all ten tariffs are positive, and the average saving is over 22%, which can be translated into significant savings ranging from several thousand to several hundred thousand US dollars; when the manufacturing plant changes from CPP1 to CPP3, the average electricity cost saving is even more than 30%. This implies that for one shift production, adjusting the production shift can effectively reduce electricity costs. Table 6.8 also shows the cost savings by opting-out of the default CPP to TOU rate. It can be observed that when the manufacturing plant changes from CPP1 to TOU1 and keeps the production shift from 8:00 a.m. to 4:00 p.m., the average savings are negative, which indicates that opting-out of CPP without rescheduling production can lead to higher electricity costs. Comparable results are also observed for the tariff switches with the production rescheduling, such as CPP2→TOU2 and CPP3→TOU3.

Table 6.9 illustrates the comparison results for two-shift production. When the manufacturing plant changes from CPP1 to CPP2 or CPP3, the results remain positive. However, compared with the one-shift production, the savings are

Table 6.8 Savings (%) on the average annual energy costs for one-shift production by switching between scenarios.

Utility	Tariff	System power (kW)	1-shift electric bill saving (%)				
			CPP1– CPP2	CPP1– CPP3	CPP1– TOU1	CPP2– TOU2	CPP3– TOU3
PGE	A-1	100	9.63	14.50	0.67	−2.57	−2.72
	A-6	400	36.99	47.28	−6.98	−2.73	0
	A-10	400	10.59	15.14	−0.81	−5.89	−6.21
	E-19	800	33.54	42.68	4.49	0	0
	E-20	1200	28.68	36.57	−3.64	−2.07	0
SCE	TOU-GS-1	15	16.17	20.69	1.81	−7.62	−8.05
	TOU-GS-2	100	23.18	32.59	−2.04	0	0
	TOU-GS-3	400	22.58	31.84	−2.56	0	0
	TOU-8	1200	25.95	36.02	−2.89	0	0
SDGE	AL-TOU	1200	14.36	27.14	−5.24	0	0
Mean			22.17	30.45	−1.72	−2.09	−1.70

Table 6.9 Savings (%) on the average annual energy costs for two-shift production by switching between scenarios.

Utility	Tariff	Two-shift electric bill saving (%)				
		CPP1– CPP2	CPP1– CPP3	CPP1– TOU1	CPP2– TOU2	CPP3– TOU3
PGE	A-1	1.09	9.55	0.68	0.69	−2.62
	A-6	2.40	34.66	−4.93	−4.64	−2.05
	A-10	1.14	12.15	1.17	1.18	−4.62
	E-19	1.34	31.96	6.31	6.40	0.00
	E-20	1.31	28.21	1.00	1.01	−1.41
SCE	TOU-GS-1	1.15	16.13	1.90	1.92	−7.98
	TOU-GS-2	1.74	24.97	3.21	3.27	0.00
	TOU-GS-3	1.67	24.84	2.83	2.88	0.00
	TOU-8	1.67	27.24	2.51	2.55	0.00
SDGE	AL-TOU	1.57	19.38	−1.78	−1.81	0.00
Mean		1.51	22.91	1.29	1.35	−1.87

Table 6.10 Savings (%) on the average annual energy costs for three-shift production by switching between scenarios

		Three-shift electric bill saving (%)				
Utility	Tariff	CPP1– CPP2	CPP1– CPP3	CPP1– TOU1	CPP2– TOU2	CPP3– TOU3
PGE	A-1	0.00	0.00	−0.35	−0.35	−0.35
	A-6	0.00	0.00	−3.83	−3.83	−3.83
	A-10	0.00	0.00	0.15	0.15	0.15
	E-19	0.00	0.00	5.26	5.26	5.26
	E-20	0.00	0.00	0.82	0.82	0.82
SCE	TOU-GS-1	0.00	0.00	−1.05	−1.05	−1.05
	TOU-GS-2	0.00	0.00	2.67	2.67	2.67
	TOU-GS-3	0.00	0.00	2.37	2.37	2.37
	TOU-8	0.00	0.00	2.11	2.11	2.11
SDGE	AL-TOU	0.00	0.00	−1.47	−1.47	−1.47
Mean		0.00	0.00	0.67	0.67	0.67

significantly smaller, decreased from 22.17% to 1.51% and from 30.45% to 22.91%, respectively. This is because the two-shift production increases the daily operating time from 8 to 16 hours, leaving less flexibility for production rescheduling to take advantage of the lower price during off-peak periods. It can be observed that when the manufacturing plants opt-out of CPP, the switch of CPP3→TOU3 can still lead to an increase in electricity costs; however, switching CPP1→TOU1 and CPP2→TOU2 results in positive average savings.

For three-shift production as shown in Table 6.10, there is no flexibility in production rescheduling as the production must last 24 hours. Therefore, there is no difference among Scenarios CPP1, CPP2, and CPP3. Similarly, there is no difference among Scenarios TOU1, TOU2, and TOU3. Therefore, the results of the three switches, i.e. CPP1→TOU1, CPP2→TOU2, CPP3→TOU3, are the same in all tariffs, and the average saving is 0.67% only. This result indicates that opting out of CPP can slightly reduce electricity costs if the manufacturing plant runs three shifts per day.

In summary, for a manufacturing plant that runs on a one- or two-shift basis, by adjusting the production plan, the electricity cost under the CPP program can be significantly reduced. Whether the customers should continue to register in CPP or transfer to TOU should be decided based on the production schedule, and the program selection decision is usually irrelevant to the starting time of the production.

Problems

6.1 Why does the starting time of the production shift significantly affect the energy costs under TOU tariffs?

6.2 For the case study discussed in Section 6.2, what are the differences between the production schedules for (i) minimizing energy consumption and (ii) minimizing energy cost?

6.3 Based on the case study results illustrated in Section 6.3, do you have any suggestions for a manufacturing plant that considers opting out of the CPP tariff and switch to the TOU tariff?

Appendix 3.A Supplementary Information of Demand Response Tariffs

Table 3.A.1 Surveyed utilities with the corresponding flat and TOU tariffs targeting industrial customers.

| State | Utility name | Flat tariff | | TOU tariff | | Website |
		Name	Demand or energy range	Name	Demand or energy range	
AL	Joe Wheeler Electric Membership Corporation	GSA	[50,1000) kW	TGSA	[50,1000) kW	jwemc.org/rate-forms/
AR	Entergy Arkansas, Inc.	LGS	[100,1000) kW	GST	[0,1000) kW	entergy-arkansas.com/your_business/business_tariffs.aspx
AZ	Salt River Project Agricultural Improvement & Power District	E-36	[5,1500] kW	E-32	[5,1500] kW	srpnet.com/prices/business/
CA	Los Angeles Department of Water and Power	A-1-A	[0,30) kW	A-1-B	[0,30) kW	ladwp.com/ladwp/faces/ladwp/commercial/c-customerservice/c-cs-understandingyourrates/c-cs-uyr-electricrates
CO	Colorado Springs Utilities	E2C	(990,30000) kWh	ETC	(990,30000) kWh	csu.org/pages/electric-tou-b.aspx
CT	Connecticut Light and Power Company	Rate 30	[0,200) kW	Rate 27	[0,350) kW	cl-p.com/rates/rates_and_tariffs

(*Continued*)

Table 3.A.1 (Continued)

State	Utility name	Flat tariff		TOU tariff		Website
		Name	Demand or energy range	Name	Demand or energy range	
DE	Delaware Electric Cooperative, Inc.	GS	[0,50) kW	GS-TOU	[0,50) kW	delaware.coop/about/rules-regulations-rates
FL	Progress Energy Florida	GS-1	[0,∞) kW	GST-1	[0,∞) kW	duke-energy.com/rates/florida-rates-index.asp
GA	Georgia Power	PLM-9	[30,500) kW	TOU-GSD-8	[30,500) kW	georgiapower.com/pricing/business/home.cshtml?bus=prices
HI	Hawaii Electric Light Company, Inc.	P	[200,∞) kW	TOU-P	[200,∞) kW	hawaiielectriclight.com/helco/residential-services/electricrates/hawaii-electric-light-rates
IA	Interstate Power and Light Company	Large GS	[20000,∞) kW	Large GS-TOU	[20000,∞) kW	alliantenergy.com/aboutalliantenergy/companyinformation/tariffs/030242
IL	MidAmerican Energy Company	Rate 22	[0,∞) kW	Rate 22-Rider 15	[0,∞) kW	midamericanenergy.com/rates1.aspx
IN	Indiana Michigan Power Company	LGS	[600,1000) kW	LGS-TOD	[0,1000) kW	indianamichiganpower.com/account/bills/rates/iandmratestariffsin.aspx
KS	Kansas Gas & Electric Company	MGS	[200,∞) kW	OPS	[500,∞) kW	westarenergy.com/wcm.nsf/tariff?openview
KY	Kentucky Power Company	LGS	[100,1000) kW	LGS-TOD	[100,1000) kW	kentuckypower.com/account/bills/rates/kentuckypowerratestariffsky.aspx
LA	Southwestern Electric Power Company	LP	[0,∞) kW	LPTOD	[800,∞) kW	swepco.com/account/bills/rates/swepcoratestariffsla.aspx

State	Company	Rate	Range	Rate	Range	URL
MA	Western Massachusetts Electric Company	G-0	[0,350) kW	T-0	[0,350) kW	wmeco.com/residential/understandbill/ratesrules/ratestariffs.aspx
ME	Central Maine Power Co	MGS-S	[20,400) kW	MGS-S-TOU	[20,400) kW	cmpco.com/yourbusiness/pricing/pricingschedules/default.html
MI	Indiana Michigan Power Company	MGS	[10,1500) kW	LGS	[10,1500) kW	indianamichiganpower.com/account/bills/rates/iandmratestariffsmi.aspx
MN	Otter Tail Power Company	Rate 10.04	[80,∞) kW	Rate 10.05	[80,∞) kW	otpco.com/rates-and-pricing/minnesota/rates,-rules,-andregulations/
MO	Union Electric Company	No.3(M)	[100,∞) kW	No.3(M)-TOD	[100,∞) kW	ameren.com/sites/aue/rates/pages/ratesbundledelecfullsrvmo.aspx
MS	Mississippi Power	LGS-LV-6	[500,∞) kW	LGS-TOU-11	[0,∞) kW	mississippipower.com/my-business/our-pricing/rate-andrider-details
NC	Progress Energy Carolinas	MGS-28	[30,1000) kW	SGS-TOU-8	[30,1000) kW	duke-energy.com/rates/progress-north-carolina.asp
ND	Otter Tail Power Company	Rate 10.03	[80,∞) kW	Rate 10.05	[80,∞) kW	otpco.com/rates-and-pricing/north-dakota/rates,-rules,-andregulations/
NE	Lincoln Electric System	Large L&P 15	(400,∞) kW	Large L&P 27	(400,∞) kW	les.com/business/rates/rate-schedules
NH	Public Service Company of New Hampshire	G	[0,100) kW	G-OTOD	[0,100) kW	psnh.com/ratestariffs/business/small-business-rates.aspx
NJ	Jersey Central Power & Light Company	GS	[0,∞) kW	GST	[750,∞) kW	firstenergycorp.com/content/customer/customer_choice/new_jersey/new_jersey_tariffs.html
NM	Otero County Electric Cooperative, Inc.	Large Power Reg	[50,∞) kW	Large Power Opt	[50,∞) kW	nmprc.state.nm.us/consumer-relations/company-directory/cooperatives/otero-county-electric-coop/

(Continued)

Table 3.A.1 (Continued)

State	Utility name	Flat tariff		TOU tariff		Website
		Name	Demand or energy range	Name	Demand or energy range	
NV	NV Energy (South)	LGS-1	[0,300) kW	OLGS-TOU	[0,300) kW	nvenergy.com/company/rates/index.cfm
NY	Orange & Rockland Utils, Inc.	SC02	[0,∞) kW	SC20	[5,∞) kW	oru.com/aboutoru/ tariffsandregulatorydocuments/newyork/ scheduleforelectricservice.html
OH	Ohio Power Company	GS-3	[10,8000) kW	GS-3(Opt TOD)	[10,8000) kW	aepohio.com/account/bills/rates/ aepohioratestariffsoh.aspx
OK	Public Service Company of Oklahoma	GS	[8000,∞) kWh	GSTOD	[8000,∞) kWh	psoklahoma.com/account/bills/rates/ ratesandtariffs.aspx
PA	Adams Electric Cooperative, Inc.	A-1	[0,50) kW	T-1	[0,50) kW	adamsec.com/content/rates
SC	Progress Energy Carolinas	MGS-29	[30,1000) kW	SGS-TOU-29	[0,1000) kW	duke-energy.com/rates/progress-south-carolina.asp
SD	Otter Tail Power Company	Rate 10.04	[80,∞) kW	Rate 10.05	[80,∞) kW	otpco.com/rates-and-pricing/south-dakota/ rates,-rules,-andregulations/
TN	Kingsport Power Company	MGS	[10,100) kW	MGS-TOD	[10,100) kW	appalachianpower.com/account/bills/ rates/apcoratestariffstn.aspx
TX	Entergy Texas, Inc.	GS	[5,2500) kW	GS-TOD	[5,2500) kW	entergy-texas.com/your_business/tariffs. aspx
VA	Virginia Electric & Power Co	GS-2	[30,500) kW	GS-2T	[30,500) kW	

dom.com/dominion-virginia-power/customer-service/ratesand-tariffs/business-rates-and-tariffs.jsp

State	Company	Schedule	Range	Schedule	Range	URL
VT	Burlington Electric Department	LG	[3000,∞) kWh	LT	[3000,∞) kWh	burlingtonelectric.com/my-business/my-bill/business-ratesand-fees
WA	Snohomish County PUD No 1	Schedule 20	[100,∞) kW	Schedule 24	[500,∞) kW	snopud.com/aboutus/rates.ashx?p=1166
WI	Wisconsin Power and Light Company	GS-1	[0,75) kW	GS-3	[0,75) kW	alliantenergy.com/aboutalliantenergy/companyinformation/tariffs/030306
WV	Appalachian Power Company	GS	[10],1000) kW	GS (Opt TOD)	[10],1000) kW	appalachianpower.com/account/bills/rates/apcoratestariffswv.aspx
WY	Rocky Mountain Power	Schedule 28	[20,∞) kW	Schedule 46	[10]00,∞) kW	rockymountainpower.net/about/rar/wri.html

Table 3.A.2 Details of the TOU tariffs (Note: Empty cells are interpreted as not applicable or the values are zero).

State	Months	On-peak	Mid-peak	Off-peak	Energy charge ($/kWh) On-peak	Mid-peak	Off-peak	Demand charge ($/kW) On-peak	Mid-peak	Off-peak	Customer charge ($)	Demand metering (min)
AL	May–Oct	11–21		21–11	0.09715		0.07457	14.1		3.73	46.46	30
	Nov–Apr	2–12		12–2	0.08738		0.07711	13.22		3.73		
AR	Jun–Sep	13–20		20–13	0.01779		0.0127	14.49		4.27	89.92	15
	Oct–May	7–18		18–7	0.00978		0.00838	12.21		3.70		
AZ	May–Jun	14–19	11–14; 19–23	23–11	0.1649	0.099	0.0518	3.33	3.33	1	24.87	30
	Sep–Oct Nov–Apr	5–9	17–21	9–17; 21–5	0.1098	0.0982	0.0513	3.33	3.33	1		
	Jul–Aug	14–19	11–14; 19–23	23–11	0.2064	0.1101	0.0524	3.33	3.33	1		
CA	Jun–Sep	13–17	10–13; 17–20	20–10	0.17089	0.1096	0.03826	5.36	5.36	5.36	15	15
	Oct–May	13–17	10–13; 17–20	20–10	0.06558	0.06558	0.03826	5.36	5.36	5.36		
CO	Apr–Sep	16–22		22–16	0.1176		0.0472				20.307	
	Oct–Mar	11–18		18–11	0.1176		0.0472					
CT	Whole year	12–20		20–12	0.19759		0.12123	9.06		9.06	38.5	30
DE	May–Sep	15–18		18–15	0.374538		0.063648				10.7	
	Oct–Apr	6–8; 17–21		8–17; 21–6	0.175858		0.063648					

FL	Apt–Oct	12–21		21–12	0.17218		0.02714				19.01	
	Nov–Mar	6–10; 18–22		22–6; 10–18	0.17218		0.02714					
GA	Jun–Sep	14–19	12–14; 19–21	21–12; 14–19	0.122372	0.060059	0.022592	15.03	5.02	5.02	209	30
	Oct–May			Whole day			0.022592			5.02		
HI	Whole year	17–21	7–17	21–7	0.288184	0.268184	0.168184	24.5	19.5		410	15
IA	Jun16–Sep15	7–20		20–7	0.02483		0.01586	First 200 kW: 15.61; Next 800 kW: 15.48	First 200 kW: 15.61; Next 800 kW: 15.48		0	15
	Sep16–Jun15	7–20		20–7	0.01586		0.00687	First 200kW: 8.21; Next 800kW: 7.49	First 200kW: 8.21; Next 800kW: 7.49			
IL	Jun–Sep	8–20		20–8	0.0924		0.0779				12.85	
	Oct–May	8–20		20–8	First 6000 kWh: 0.0756; Over: 0.0650		First 6000 kWh: 0.0611; Over: 0.0505					

(Continued)

Table 3.A.2 (Continued)

State	Months	On-peak	Mid-peak	Off-peak	Energy charge ($/kWh)			Demand charge ($/kW)			Customer charge ($)	Demand metering (min)
					On-peak	Mid-peak	Off-peak	On-peak	Mid-peak	Off-peak		
IN	Whole year	7–21		21–7	0.0871		0.02903	3.727		3.727	35.3	15
KS	Jun–Sep	14–20		20–14	0.01815		0.01815	10.65		2.1	100	15
	Oct–May			Whole day			0.01815			2.1		
KY	Whole year	7–21		21–7	0.09778		0.04116	7.64		7.64	85	15
LA	Jul–Sep	13–19		19–13	First 500 MWh: 0.0346; Next 4500 MkW: 0.0142; Over: 0.0130		First 500 MWh: 0.0179; Next 4500 MkW: 0.0144; Over: 0.0110	18.36		5.17	0	15
	Oct–Jun			Whole day			First 500 MWh: 0.0179; Next 4500 MkW: 0.0144; Over: 0.0110			5.17		

MA	Whole year	12–20		20–12	0.10205		0.09997	17.24		17.24	30000	30
ME	Apr–Nov	7–12; 16–20		20–7	0.00579	0.005215	0.003647	9.31	1.98		40.11	15
	Dec–Mar	7–12; 16–20		20–7	0.00579	0.005215	0.003647	10.81	1.98			
MI	Whole year	7–21		21–7	0.08699		0.01796	11.1		11.1	44	15
MN	Jun–Sep	13–19	9–13; 19–22	22–9	0.07319	0.05397	0.02437	5.54	1.68		208.5	6
	Oct–May	7–12; 17–21	6–7; 12–17; 21–22	22–6	0.06507	0.04917	0.03665	5.13	0.94			
MO	Jun–Sep	10–22		22–10	First 150 kWh/kW: 0.1114; Next 200 kWh/kW: 0.0869; Over: 0.0625		First 150 kWh/kW: 0.0931; Next 200 kWh/kW: 0.0686; Over: 0.0442	4.62		2.31	108.32	15
	Oct–May	10–22		22–10	First 150 kWh/kW: 0.0663; Next 200 kWh/kW: 0.0502; Over: 0.0403		First 150 kWh/kW: 0.0608; Next 200 kWh/kW: 0.0447; Over: 0.0308	1.71		0.855		

(Continued)

Table 3.A.2 (Continued)

State	Months	On-peak	Mid-peak	Off-peak	Energy charge ($/kWh) On-peak	Mid-peak	Off-peak	Demand charge ($/kW) On-peak	Mid-peak	Off-peak	Customer charge ($)	Demand metering (min)
MS	Jun–Sep	12–20	10–12; 20–22	22–10; 10–18; 22–6	0.06488	0.02769	0.0091	13.9	2.7	2.7	1060	15
	Nov–Mar	6–10; 18–22		10–18; 22–6	0.03699		0.0091	8.25		2.1		
	Apr, May, Oct	12–20		20–12	0.03699		0.0091	8.25		2.1		
NC	Jun–Sep	10–22		22–10	0.05812		0.04636	9.9		1.51	58.69	15
	Oct–Mar	6–13; 16–21		13–16; 21–6	0.05812		0.04636	7.34		1.51		
ND	Apr–May	10–22		22–10	0.05812		0.04636	7.34		1.51		
	Jun–Sep	13–19	9–13; 19–22	22–9	0.0815	0.06247	0.03721	5.75	1.59		195	60
	Oct–May	7–12; 17–21	6–7; 12–17; 21–22	22–6	0.07314	0.05949	0.04199	4.42	1.22			
NE	Jun–Sep	14–20		20–14	0.0285		0.0285	19.95		9.1	305	30
	Oct–May	14–20		20–14	0.0237		0.0237	19.95		9.1		
NH	Whole year	7–20		20–7	0.14838		0.10805	14.88			53.79	30
NJ	Jun–Sep	8–20		8–20	0.004815		0.004815	7.22		7.22	35.25	15

	Oct–May	8–20	0.004815	8–20	0.004815	6.75	6.75		
NM	Whole year	12–22	0.07825	22–12	0.06804	11	5.25	85	15
NV	Jun–Sep	13–19	0.21007	19–13	0.06401	12.47	4.22	24.62	15
	Oct–May			Whole day	0.05314		4.38		
NY	Jun–Sep	13–19	0.18815	19–13	0.10551	19.41	0	51.32	15
	Oct–May	10–21	0.13065	21–10	0.10551	8.38	0		
OH	Whole year	7–21	0.0012605	21–7	0.0012605	13.06	6.97	22.79	30
OK	Jun–Oct	14–19	0.1189	19–14	0.0477			58.63	
	Nov–May			Whole day	First 150 kWh/kW: 0.0640; Next 150 kWh/kW: 0.0579; Over: 0.0477				
PA	Jun–Aug	7–19	0.181	19–7	0.06			28.5	
	Sep–May	7–19	0.121	19–7	0.06				
SC	Jun–Sep	10–22	0.06458	22–10	0.05123	10.49	1	19.75	15
	Oct–Mar	6–13; 16–21	0.0628	13–16; 21–6	0.05123	7.77	1		
	Apr–May	10–22	0.0628	22–10	0.05123	7.77	1		

(Continued)

Table 3.A.2 (Continued)

State	Months	On-peak	Mid-peak	Off-peak	Energy charge ($/kWh)			Demand charge ($/kW)			Customer charge ($)	Demand metering (min)
					On-peak	Mid-peak	Off-peak	On-peak	Mid-peak	Off-peak		
SD	Jun–Sep	13–19	9–13; 19–22	22–9	0.04649	0.02761	0.00292	5.59	1.7		218.5	60
	Oct–May	7–12; 12–21	6–7; 12–17; 21–22	22–6	0.03851	0.02289	0.01059	3.91	0.72			
TN	Whole year	6–21		21–6	0.08847		0.02755				23.45	
TX	May–Oct	13–21		21–13	0.04882		0.01682	9.8		9.8	34.95	30
	Nov–Apr	6–10; 18–22		10–18; 22–6	0.01943		0.01682	5.07		5.07		
VA	Jun–Sep	10–22		22–10	0.03172		0.00541	10.484		2.647	26.17	30
	Oct–May	7–22		22–7	0.03172		0.00541	9.075		2.647		
VT	Jun–Sep	12–18		18–12	0.107754		0.076418	25.47		3.53		
	Dec–Mar	6–22		22–6	0.115459		0.076418	25.47		3.53		
	Apr–May; Oct–Nov			Whole day			0.076418			3.53		
WA	Apr–Sep	7–11		11–7	First 30 MWh: 0.0787; Over: 0.0607		First 30 MWh: 0.0787; Over: 0.0607	First 100 kW: 0; Over: 7.36		0	9.9	60

	Oct–Mar	7–11	11–7	First 30 MWh: 0.0873; Over: 0.0708		First 100 kW: 0; Over: 7.36	0		
WI	Jun–Sep	8–20	20–8	0.20739	0.05505			15.339	
	Oct–May	8–20	20–8	0.19598	0.05505				
WV	Whole year	7–21	21–7	First 350 kWh/kW: 0.06553; Over: 0.03729	First 350 kWh/kW: 0.06553; Over: 0.03729	4.24	2.46	21	15
WY	Whole year	7–23	23–7	0.00826	0.00826	16.78	2.26	625	15

Table 3.A.3 Details of the flat tariffs (Note: Empty cells are interpreted as not applicable or the values are zero).

State	Months	Energy charge ($/kWh)	Demand charge ($/kW)	Customer charge ($)	Demand metering interval (min)
AL	Jun–Sep	First 15000 kWh: 0.10791; Over: 0.06310	First 50 kW: 0; Over: 13.97	46.46	30
	Dec–Mar	First 15000 kWh: 0.10778; Over: 0.06310	First 50 kW: 0; Over: 13.10		
	Apr–May	First 15000 kWh: 0.10710; Over: 0.06310	First 50 kW: 0; Over: 13.10		
AR	Oct–Nov				
	Jun–Sep	0.0244	10.56	89.92	15
AZ	May–Jun	First 350 kWh: 0.0989; Over: 0.0988		24.84	
	Sep–Oct				
	Nov–Apr	First 350 kWh: 0.0785; Over: 0.0778			
	Jul–Aug	First 350 kWh: 0.1211; Over: 0.1203			
CA	Jun–Sep	0.07219	5.36	6.5	15
	Oct–May	0.04929	5.36		
CO	Whole year	0.0801		18.972	
CT	Whole year	0.1314	11.96	38.5	30
DE	Jun–Sep	0.098438		8.7	
	Oct–May	First 700 kWh: 0.093488; Over: 0.078088		8.7	
FL	Whole year	0.07286		11.59	
GA	Whole year	First 200 kWh/kW: 0.085880; Next 200 kWh/kW: 0.011052; Next 200 kWh/kW: 0.008317; Over: 0.007235		19	
HI	Whole year	0.218184	19.5	400	15

Table 3.A.3   (Continued)

State	Months	Energy charge ($/kWh)	Demand charge ($/kW)	Customer charge ($)	Demand metering interval (min)
IA	Jun 16–Sep 15	0.01971	First 200 kW: 15.61; Next 800 kW: 15.48	0	15
	Sep 15–Jun 16	0.01073	First 200 kW: 8.21; Next 800 kW: 7.49		
IL	Jun–Sep	0.0845		8.85	
	Oct–May	First 6000 kWh: 0.0677; Over: 0.0571			
IN	Whole year	First 300 kWh/kW: 0.06217; Over: 0.04216	4.695	35.3	15
KS	Jun–Sep	0.019261	12.506021	100	15
	Oct–May	0.014627	12.506021		
KY	Whole year	0.07795	4.02	85	15
LA	Whole year	First 500 MWh: 0.0196; Next 4500 MWh: 0.0126; Over: 0.0118	8.56	0	15
MA	Whole year	0.10099	15.27	30	30
ME	Apr–Nov	0.004398	10.23	37.22	15
	Dec–Mar	0.004398	11.33		
MI	Whole year	0.06615	5.69	17.45	15
MN	Jun–Sep	0.04618	7.22	188.5	15
	Oct–May	0.05	6.07		
MO	Jun–Sep	First 150 kWh/kW: 0.0997; Next 200 kWh/kW: 0.0752; Over: 0.0508	4.62	88.82	15
	Oct–May	First 150 kWh/kW: 0.0628; Next 200 kWh/kW: 0.0467; Over: 0.0368	1.71		
MS	Whole year	First 200 kWh/kW: 0.02774; Next 200 kWh/kW: 0.02425; Over: 0.00921	7.85	725	15

(Continued)

Table 3.A.3 (Continued)

State	Months	Energy charge ($/kWh)	Demand charge ($/kW)	Customer charge ($)	Demand metering interval (min)
NC	Whole year	0.06763	4.81	58.69	15
ND	Jun–Sep	0.05115	7.29	175	15
	Oct–May	0.05165	5.61		
NE	Jun–Sep	0.0285	19.95	275	30
	Oct–May	0.0237	19.95		
NH	Whole year	First 500 kWh: 0.18525; Next 1000 kWh: 0.12358; Over: 0.10977	13.08	29.04	30
NJ	Jun–Sep	First 1000 kWh: 0.061999; Over: 0.004958	6.94	11.65	15
	Oct–May	First 1000 kWh: 0.057366; Over: 0.004958	6.47		
NM	Whole year	0.07188	10.5	75	15
NV	Whole year	0.07165	8.57	21.1	15
NY	Jun–Sep	0.18015		33.15	
	Oct–May	0.16052			
OH	Whole year	0.0012605	13.06	22.79	30
OK	Jun–Oct	First 150 kWh/kW: 0.0785; Next 150 kWh/kW: 0.0716; Over: 0.0493		58.63	
	Nov–May	First 150 kWh/kW: 0.0640; Next 150 kWh/kW: 0.0579; Over: 0.0477			
PA	Whole year	0.0939		28.5	
SC	Whole year	0.07353	4.91	18.1	15
SD	Jun–Sep	0.01696	7.29	198.5	15
	Oct–May	0.02046	4.63		

Table 3.A.3 (Continued)

State	Months	Energy charge ($/kWh)	Demand charge ($/kW)	Customer charge ($)	Demand metering interval (min)
TN	Whole year	First 200 kWh/kW: 0.07374; Over: 0.03689		21.5	
TX	Whole year	0.01965	6.58	34.95	30
VA	Jun–Sep	First 150 kWh/kW: 0.04634; Next 150 kWh/kW: 0.02605; Next 150 kWh/kW: 0.01136; Over: 0.00289	7.477	21.17	30
	Oct–May	First 150 kWh/kW: 0.04634; Next 150 kWh/kW: 0.02605; Next 150 kWh/kW: 0.01136; Over: 0.00289	6.036		
VT	Whole year	0.083003	20.03	41.04	15
WA	Apr–Sep	First 300 MWh: 0.0772; Over: 0.0592	First 100 kW: 0; Over: 4.2	9.9	60
	Oct–Mar	First 300 MWh: 0.0860; Over: 0.0689	First 100 kW: 0; Over: 4.2		
WI	Jun–Sep	0.1219		15.339	
	Oct–May	0.11092			
WV	Whole year	First 350 kWh/kW: 0.06553; Over: 0.03729	4.24	21	15
WY	Whole year	0.0125	14.29	37	15

Table 3.A.4 Detailed CPP rates of PGE, SCE, and SDGE (Note: empty cells are interpreted as zero or not applicable).

Utility	Tariff	Discount months	Energy and demand charge credits	Adder during CPP events	Event days per year	Event time period	Notification time
PGE	A-1	May–Oct	Energy charge ($/kWh): all period 0.00977	Energy charge ($/kWh): 0.60	9–15	14–18, year-round, any day of the week	By 14:00 day ahead
	A-6	May–Oct	Energy charge ($/kWh): On-peak 0.10830; Mid-peak 0.02166	Energy charge ($/kWh): 1.20			
	A-10	May–Oct	Energy charge ($/kWh): all period 2.6	Energy charge ($/kWh): 0.90			
	E-19	May–Oct	Energy charge ($/kWh): On-peak 6.37; Mid-peak 1.38	Energy charge ($/kWh): 1.20			
	E-20	May–Oct	Energy charge ($/kWh): On-peak 6.10; Mid-peak 1.24	Energy charge ($/kWh): 1.20			
SCE	TOU-GS-1	Jun–Sep	Energy charge ($/kWh): all period 0.03776	Energy charge ($/kWh): 1.37453	12	14–16, year-round, non-holiday weekdays	By 15:00 day ahead
	TOU-GS-2	Jun–Sep	Demand charge ($/kW): On-peak 10.75	Energy charge ($/kWh): 1.37453			
	TOU-GS-3	Jun–Sep	Demand charge ($/kW): On-peak 11.44	Energy charge ($/kWh): 1.37453			
	TOU-8	Jun–Sep	Demand charge ($/kW): On-peak 11.93	Energy charge ($/kWh): 1.37453			
SDGE	ALL-TOU	May–Oct	Demand charge ($/kW): On-peak 11.30	Energy charge ($/kWh): 1.39243	0–18	11–18, year-round, any day of the week	By 15:00 day ahead

Table 3.A.5 Detailed TOU rates (Note: empty cells are interpreted as zero or not applicable)

Utility	Tariff	Months	On-peak	Mid-peak	Off-peak	Energy charge ($/kWh) On-peak	Mid-peak	Off-peak	Demand charge ($/kW) On-peak	Mid-peak	Off-peak	Base (any time)	Other monthly charge ($)
PGE	A-1	May–Oct	12–18	8:30–12; 18–21:30	21:30–8:30	0.2547	0.24562	0.21801					19.975232
		Nov–Apr		8:30–21:30	21:30–8:30		0.17359	0.15381					
	A-6	May–Oct	12–18	8:30–12; 18–21:30	21:30–8:30	0.58648	0.2723	0.15207					26.08776
		Nov–Apr		8:30–21:30	21:30–8:30								
	A-10	May–Oct	12–18	8:30–12; 18–21:30	21:30–8:30	0.17989	0.17198	0.14792				14.28	169.790688
		Nov–Apr		8:30–21:30	21:30–8:30		0.1308	0.11017				6.47	
	E-19	May–Oct	12–18	8:30–12; 18–21:30	21:30–8:30	0.16594	0.11342	0.07903	18.37	4.25		12.99	629.236224
		Nov–Apr		8:30–21:30	21:30–8:30		0.10645	0.08296		0.22		12.99	
	E-20	May–Oct	12–18	8:30–12; 18–21:30	21:30–8:30	0.15115	0.10722	0.07807	17.91	3.89		12.65	1028.73114
		Nov–Apr		8:30–21:30	21:30–8:30		0.10094	0.07925		0.25		12.65	

(Continued)

Table 3.A.5 (Continued)

Utility	Tariff	Months	On-peak	Mid-peak	Off-peak	Energy charge ($/kWh) On-peak	Mid-peak	Off-peak	Demand charge ($/kW) On-peak	Mid-peak	Off-peak	Base (any time)	Other monthly charge ($)
SCE	TOU-GS-1	Jun–Sep	12–18	8–12; 18–23	23–8	0.24438	0.19989	0.17031					27.208
		Oct–May		8–12	21–8		0.16766	0.15631					
	TOU-GS-2	Jun–Sep	12–18	8–12; 18–23	23–8	0.1457	0.09447	0.06933	19.9	5.82		12.71	267.53
		Oct–May		8–12	21–8		0.09973	0.07503				12.71	
	TOU-GS-3	Jun–Sep	12–18	8–12; 18–23	23–8	0.14656	0.09156	0.06538	20.36	5.97		15.77	432.01
		Oct–May		8–12	21–8		0.09337	0.07077				15.77	
	TOU-8	Jun–Sep	12–18	8–12; 18–23	23–8	0.15267	0.09289	0.06592	25.16	7.11		14.99	596.11
		Oct–May		8–12	21–8		0.09454	0.07165				14.99	
SDGE	AL-TOU	May–Oct	11–18	6–11; 18–22	22–6	0.12849	0.11807	0.08777	21.1			20.77	[20,500) kW: 87.83
		Nov–Apr	17–20	6–17; 20–22	22–6	0.11845	0.10184	0.07896	7.14			20.77	[500,12000) kW: 349.31

References

1 Zhang, G., Wang, G., Chen, C.-H. et al. (2021). Augmented Lagrangian coordination for energy-optimal allocation of smart manufacturing services. *Robot. Comput. Integr. Manuf.* 71: 102161.

2 Finn, P., Fitzpatrick, C., and Connolly, D. (2012). Demand side management of electric car chargingb Benefits for consumer and grid. *Energy* 42 (1): 358–363.

3 Lv, J., Peng, T., Zhang, Y., and Wang, Y. (2021). A novel method to forecast energy consumption of selective laser melting processes. *Int. J. Prod. Res.* 59 (8): 2375–2391.

4 Kim, J. and Shcherbakova, A. (2011). *Energy, and Undefined 2011, Common Failures of Demand Response*. Elsevier.

5 U.S. Federal Energy Regulatory Commission (2010). National Action Plan on Demand Response.

6 U.S. Federal Energy Regulatory Commission (2011). Implementation proposal for the national action plan on demand response.

7 Lee, M.P., Aslam, O., Foster, B. et al. (2014). Assessment of demand response and advanced metering. *Fed. Energy Regul. Comm. Tech. Rep.* 74: 1895–1902.

8 Wu, J., Wu, L., Xu, Z. et al. (2022). Dynamic pricing and prices spike detection for industrial park with coupled electricity and thermal demand. *IEEE Trans. Autom. Sci. Eng.* 1–12.

9 Wan, Y., Qin, J., Yu, X. et al. (2022). Price-based residential demand response management in smart grids: a reinforcement learning-based approach. *IEEE/CAA J. Autom. Sin.* 9 (1): 123–134.

10 Zheng, W. et al. (2017). Percentile performance estimation of unreliable IaaS clouds and their cost-optimal capacity decision. *IEEE Access* 5: 2808–2818.

11 Dong, Q., Yu, L., Song, W. et al. (2017). Fast distributed demand response algorithm in smart grid. *IEEE/CAA J. Autom. Sin.* 4 (2): 280–296.

12 Consumer Experiences Of Time of Use Tariffs | Ipsos MORI. (2012). https://www.ipsos.com/ipsos-mori/en-uk/consumer-experiences-time-use-tariffs (accessed 15 January 2020).

13 Barbose, G., Goldman, C., Bharvirkar, R. et al. (2005). *Real Time Pricing as a Default or Optional Service for C&I Customers: A Comparative Analysis of Eight Case Studies*. Lawrence Berkeley National Laboratory: Berkeley, CA.

14 Barbose, G., Goldman, C., and Neenan, B. (2004). *A Survey of Utility Experience with Real Time Pricing*. Lawrence Berkeley National Laboratory: Berkeley, CA.

15 Wang, Y. and Li, L. (2015). Time-of-use electricity pricing for industrial customers: a survey of U.S. utilities. *Appl. Energy* 149: 89–103.

16 Orange and Rockland Utilities. Service classification No. 20, (2012). https://www.oru.com/documents/tariffsandregulatory%0Adocuments/ny/electrictariff/electricsc20.pdf (accessed 15 January 2020).

17 Wang, Y. and Li, L. (2013). Time-of-use based electricity demand response for sustainable manufacturing systems. *Energy* 63: 233–244.

18 U.S. Federal Energy Regulatory Commission (2012). Assessment of demand response and advanced metering.

7

Energy Control and Optimization for Manufacturing Systems Utilizing Combined Heat and Power System

In Chapter 2, a combined heat and power (CHP) system is introduced as one of the most important contributors in the indirect end-use energy consumption at the plant level, and the energy flow in the CHP system is demonstrated in Section 2.2.3. As an on-site energy generation method, the CHP system can provide electricity to the plant while providing heat, leading to a reduction in the grid power demand of the manufacturing plant. Therefore, schedule arrangements of a CHP system and its supported manufacturing system toward minimized overall energy costs have become the demand of many manufacturers. In this chapter, the energy control and optimization for manufacturing systems utilizing CHP systems are introduced. More precisely, in Section 7.1, the knowledge related to a CHP system is reviewed. In Section 7.2, an energy cost optimization model is demonstrated, which is specifically established for the combined CHP and manufacturing systems. Meanwhile, four constraints are proposed based on the characteristics of a manufacturing system and CHP system. In Section 7.3, a particle swarm optimization (PSO) approach is adopted to solve the proposed optimization problem. In accordance with the workflow of PSO introduced in Chapter 3, the initialization of position and velocity matrices, fitness function formulation and evaluation, as well as the updating processes of the matrices are illustrated. Finally, a comparative case study is presented in Section 7.4 to demonstrate the effectiveness of the proposed method. The results indicate that the energy control with the CHP system can effectively reduce the energy cost of a manufacturing system.

7.1 Introduction to Combined Heat and Power System

CHP generation systems have been considered as a promising method to implement energy management and, thus, have been widely applied in commercial and residential building sectors. Compared with traditional separated heat and power (SHP) generation systems, less greenhouse gas (GHG) emissions and higher

Sustainable Manufacturing Systems, First Edition. Lin Li and MengChu Zhou.
© 2023 The Institute of Electrical and Electronics Engineers, Inc.
Published 2023 by John Wiley & Sons, Inc.

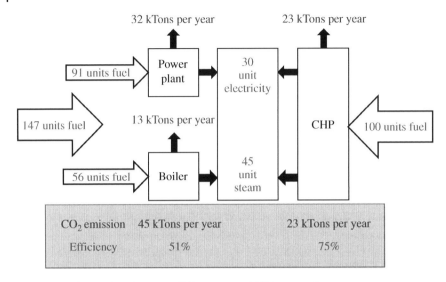

Figure 7.1 Comparison between CHP system and SHP generation systems.

energy efficiency are two advantages of CHP systems [1]. Figure 7.1 shows the comparison between the CHP and SHP systems concerning the CO_2 emission and energy efficiency [2]. The results indicate that the CHP system can reduce 48.9% of the CO_2 emission and increase energy efficiency by 24% over SHP.

Research efforts have been devoted to investigating the benefits of the CHP system implementation in industrial systems with special emphases on GHG emission reduction and energy efficiency improvement [2, 3]. However, the investigation on the detailed CHP application methods in industrial systems toward reduced energy cost has received little attention. In this chapter, a production scheduling problem for the combined manufacturing and CHP systems is formulated, with the objective of identifying a schedule for both manufacturing and CHP systems that can minimize the energy cost under a time-of-use (TOU) tariff with the production yield constraint.

7.2 Problem Definition and Modeling

The system layout and energy flow of a combined manufacturing and CHP system studied in this chapter are shown in Figure 7.2. The manufacturing facility contains a manufacturing system, a CHP system, and an auxiliary boiler. The manufacturing system consists of N machines and $N - 1$ buffers, and electricity and heat are two requisites for system operations. In particular, the CHP system consumes

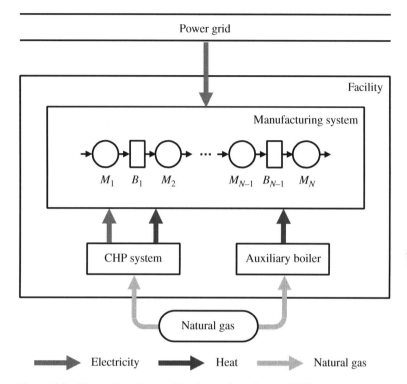

Figure 7.2 Illustration of a combined manufacturing and CHP system.

natural gas and generates both electricity and heat for the manufacturing system. If the electricity generated by the CHP system cannot meet the production requirement, additional electricity needs to be purchased from the grid. Similarly, the auxiliary natural gas boiler can function as an alternative to the CHP system in replenishing the heat demand in the manufacturing system.

In the modeling, the production horizon is slotted into a set of identical intervals with the length of H. Let $t = \{[1, 2, ..., T]\}$ be the index of a time interval. Depending on the TOU tariff, a time interval belongs to either the on-peak or off-peak period. The output electrical power of the CHP system keeps constant in a time step and may vary at the end of a time step.

The scheduling problem discussed in this chapter can be formulated as an optimization problem with the following three decision variables.

A binary decision variable $x_i(t)$ denotes whether machine M_i ($i = [1, ..., N]$) is turned on or off at time step t.

$$x_i(t) = \begin{cases} 1, & \text{machine } i \text{ is turned on} \\ 0, & \text{machine } i \text{ is turned off} \end{cases} \tag{7.1}$$

A binary decision variable $y(t)$ denotes whether the CHP system is turned on or off at time step t.

$$y(t) = \begin{cases} 1, & \text{CHP system is turned on} \\ 0, & \text{CHP system is turned off} \end{cases} \tag{7.2}$$

A nonnegative real decision variable $p(t) \in (0, \hat{p}]$ denotes the output electrical power of the CHP system at time step t, and $\hat{p}$ is the output power capacity of the CHP system.

The objective function and corresponding constraints of the optimization problem are presented as follows.

7.2.1 Objective Function

The objective of the optimization problem is to minimize the total energy cost of the combined manufacturing and CHP system, i.e.

$$\min C = C_G + C_C \tag{7.3}$$

The total energy cost, denoted by C, includes two cost elements: the cost of electricity purchased from the power grid (denoted by C_G), and the operational costs of the CHP system and auxiliary boiler (denoted by C_C).

7.2.1.1 Electricity Cost

The electricity-related cost consists of three parts: electricity consumption cost (C_E), demand cost (C_D), and fixed cost (C_F).

$$C_{ELE} = C_E + C_D + C_F \tag{7.4}$$

Since the CHP system can satisfy part of the electricity requirement in the manufacturing system, the electricity (kWh) purchased from the power grid in each time interval can be calculated as

$$E(t) = \sum_{i=1}^{N} P_i x_i(t) - p(t)y(t) \tag{7.5}$$

where P_i is the power (kW) of machine i; the summation of $P_i x_i(t)$ represents the electricity consumption of the manufacturing system at time t, and $p(t)y(t)$ denotes the electricity generated by the CHP system at time t. The difference between these two values is the electricity that should be purchased from the grid.

Hence, the electricity consumption cost can be calculated as:

$$C_E = c_E(t) \sum_{t=1}^{T} E(t) \tag{7.6}$$

where $c_E(t)$ is the electricity consumption rate ($/kWh) under TOU tariff at time slot t.

Similarly, the demand cost can be calculated by:

$$C_{\mathrm{D}} = \max_t \left(\sum_{i=1}^{N} P_i x_i(t) c_{\mathrm{D}}(t) - p(t) y(t) c_{\mathrm{D}}(t) \right) \tag{7.7}$$

where $c_{\mathrm{D}}(t)$ is the power demand rate (\$/kW) under TOU tariff at time t. Note that the power demand rate during off-peak periods is 0.

7.2.1.2 Operation Cost for the CHP System and Boiler

The operation-related cost consists of four parts: natural gas costs of the CHP system ($C_{\mathrm{C}}^{\mathrm{gas}}$), and the auxiliary boiler ($C_{\mathrm{A}}^{\mathrm{gas}}$), as well as the startup costs for the CHP system (C_{C}^0), and the auxiliary boiler (C_{A}^0).

$$C_{\mathrm{OP}} = C_{\mathrm{C}}^{\mathrm{gas}} + C_{\mathrm{A}}^{\mathrm{gas}} + C_{\mathrm{C}}^0 + C_{\mathrm{A}}^0 \tag{7.8}$$

The CHP system generates two types of energy: electricity and heat. Suppose that the electricity to heat ratio of the CHP system is ε, i.e. when the CHP system generates ε kWh of electricity, it generates 1 kWh of heat. Hence, the natural gas cost needed to operate a CHP system can be calculated by:

$$C_{\mathrm{C}}^{\mathrm{gas}} = c_{\mathrm{gas}} \sum_{t=1}^{T} \left(\frac{p(t)y(t) \cdot \left(1 + \frac{1}{\varepsilon}\right)}{\delta \eta_{\mathrm{C}}} \right) \tag{7.9}$$

where η_{C} is the efficiency of the CHP system (%); δ is the calorific value of the natural gas (kWh/Nm^3), and c_{gas} is the price of the natural gas (\$/Nm^3). Note that the unit Nm^3 stands for normal cubic meter, which is used to define the gas volume under the temperature of 0°C and the pressure of 1.013 25 bar.

When the heat generated by the CHP system cannot meet the heat demand of the manufacturing system, the auxiliary boiler starts working as a heat generator. The cost of natural gas consumed to operate the boiler can be calculated as:

$$C_{\mathrm{A}}^{\mathrm{gas}} = c_{\mathrm{gas}} \sum_{t=1}^{T} \left(\frac{\left(D_{\mathrm{H}}(t) - \frac{p(t)y(t)}{\varepsilon}\right) z_{\mathrm{A}}(t)}{\delta \eta_{\mathrm{A}}} \right) \tag{7.10}$$

where η_{A} is the efficiency of the auxiliary boiler (%); $D_{\mathrm{H}}(t)$ is the heat demand of the manufacturing system at time t, and $z_{\mathrm{A}}(t)$ is the operation status of the auxiliary boiler, which equals 1 if the CHP system cannot meet the heat demand at time t, and equals 0 otherwise. Mathematically, $z_{\mathrm{A}}(t)$ can be defined as

$$z_{\mathrm{A}}(t) = \begin{cases} 1, & D_{\mathrm{H}}(t) - \dfrac{p(t)y(t)}{\varepsilon} > 0 \\ 0, & \text{otherwise} \end{cases} \tag{7.11}$$

The startup costs of the CHP system and the auxiliary boiler are proportional to the number of startups. Hence, the startup costs can be calculated as follows.

$$C_C^0 = c_C^0 \sum_{t=1}^{T} \max\left(y(t) - y(t-1), 0\right) \tag{7.12}$$

$$C_A^0 = c_A^0 \sum_{t=1}^{T} \max\left(z_A(t) - z_A(t-1), 0\right) \tag{7.13}$$

where c_C^0 and c_A^0 are the unit startup cost (\$/startup) of the CHP system and the auxiliary boiler, respectively. Note that $y(0)$ and $z_A(0)$ are assumed zeros as initial conditions for the CHP system and the auxiliary boiler.

7.2.2 Constraints

Constraint 1 The first constraint is the production yield constraint, i.e.

$$Y \geq Y^{\circ} \tag{7.14}$$

where Y is the production yield of the manufacturing system during the production horizon, and Y° is the predefined production target. The production yield can be calculated by:

$$Y = \sum_{t=1}^{T} (x_N(t) \cdot \rho_N \lambda_N) \tag{7.15}$$

where ρ_N is the production rate of the last machine M_N (with the unit of the number of products produced per time interval), and λ_N is the efficiency of the last machine M_N (%).

Constraint 2 The second constraint is the buffer capacity constraint. More specifically, the buffer content during each time interval should be maintained in the range between zero and the maximum buffer capacity, which is

$$0 \leq c_i(t) \leq \hat{C}_i \tag{7.16}$$

where $c_i(t)$ is the buffer content in buffer B_i at time t, and $\hat{C}_i$ is the maximum capacity of buffer B_i.

According to the conservation of material flow, the buffer content of buffer B_i in time interval t equals the buffer content in previous time interval $t-1$ plus the incoming parts in the past interval due to the operation of the upstream machine M_i and the outgoing parts in the past interval to the downstream machine M_{i+1}. Therefore, $c_i(t)$ can be calculated as

$$c_i(t) = c_i(t-1) + x_i(t-1) \cdot \rho_i \lambda_i - x_{i+1}(t-1) \cdot \rho_{i+1} \lambda_{i+1} \tag{7.17}$$

where ρ_i is the production rate of machine M_i, and λ_i is the efficiency of machine M_i.

Constraint 3 The third constraint is the CHP system operation constraint. Owing to its operational principles, the CHP system cannot be turned on and off frequently. It means that the CHP system cannot be activated (or shut down) until the consecutive off (or on) time is not less than the minimum off (on)-time. Mathematically, this constraint can be expressed as

$$y(t) = \begin{cases} 1, & 1 \le U(t) < U_1 \\ 0, & -1 \ge U(t) > -U_o \end{cases} \tag{7.18}$$

where U_1 is the minimum on-time (i.e. the number of time intervals that the CHP is on) after startup; U_o is minimum off-time after shutdown, and $U(t)$ is the consecutive on/off time at time t. In particular, a positive $U(t)$ denotes the consecutive on-time, and a negative $U(t)$ indicates the consecutive off-time:

$$U(t) = \begin{cases} \max\left(U(t-1), 0\right) + 1, & y(t-1) = 1 \\ \min\left(U(t-1), 0\right) - 1, & y(t-1) = 0 \end{cases} \tag{7.19}$$

Note that in the first time interval, it is assumed that $U(1) = 0$.

Constraint 4 The fourth constraint is the CHP system output power constraint. More specifically, the CHP system output power $p(t)$ must be not larger than the output power capacity of the CHP system ($\hat{p}$) or the electricity demand of the manufacturing system. This constraint can be formulated as:

$$p(t) \le \min\left(\sum_{i=1}^{N} P_i x_i(t), \hat{p}\right) \tag{7.20}$$

7.3 Solution Approach

In reference to Section 3.3, PSO is suitable for solving nonlinear and non-differentiable problems in high-dimension space. Hence, it is adopted to solve the proposed problem. The initialization, evaluation, and updating process are demonstrated as follows.

7.3.1 Initialization

In this problem, the dimension of the solution space is $(N + 2) \times T$. Hence, the solutions for each particle can be formulated as an $(N + 2) \times T$ matrix, i.e. the position matrix X in PSO, as shown in Figure 7.3. The position matrix consists of three submatrices. First, a $N \times T$ production scheduling submatrix indicates the production scheduling $x_i(t)$ of N machines in the manufacturing system throughout the

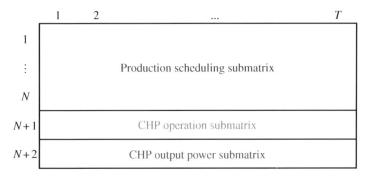

Figure 7.3 Encoding scheme of position and velocity matrices of each particle.

production horizon with T time intervals. Next, at the $N+1$ row, an $1 \times T$ CHP operation submatrix indicates the on/off decisions of the CHP system, i.e. the decision variable $y(t)$, throughout the production horizon with T intervals. Finally, at the $N+2$ row, an $1 \times T$ CHP output power submatrix indicates the decisions for the CHP power output, i.e. decision variable $p(t)$, throughout the production horizon with T intervals. The velocity matrix V has the same dimension and is encoded in the same way as the position matrix.

The swarm is initialized by a given number of matrices for all particles. For each position matrix, the initial values of the elements in the production scheduling submatrix and CHP operation submatrix are randomly selected from the set $\{0, 1\}$. The initial values of elements in the CHP output power submatrix are randomly selected from the set $\{\Delta p, 2\Delta p, ..., PC - \Delta p, PC\}$, where $\Delta p = PC/z$ and z is a positive integer. For each velocity matrix, the initial values of the elements in the production scheduling submatrix and CHP operation submatrix are randomly selected from the set $\{-1, 0, 1\}$, and the initial values of elements in the CHP output power submatrix are randomly selected from the set $\{-\Delta p, 0, \Delta p\}$.

7.3.2 Evaluation

The fitness function used for the evaluation step in the PSO is given as:

$$
\begin{aligned}
&C + \sigma \min\left(Y - Y^\circ, 0\right)^2 + \sigma \sum_{t=1}^{T} \sum_{i=1}^{N-1} \min\left(\hat{C}_i - c_i(t), 0\right)^2 \\
&+ \sigma \sum_{t=1}^{T} \sum_{i=1}^{N-1} \min\left(c_i(t), 0\right)^2 + \sigma \sum_{t \in T^1} \min\left(U(t) - U_1, 0\right)^2 \\
&+ \sigma \sum_{t \in T^0} \max\left(U(t) + U_0, 0\right)^2 + \sigma \min\left(\min\left(\sum_{i=1}^{N} P_i x_i(t), \hat{p}\right) - p(t), 0\right)^2
\end{aligned}
\tag{7.21}
$$

In (7.21), constraints 1–4 are integrated as penalty terms into the fitness function, where σ is penalty coefficient; T^1 denotes the set of time intervals where the decision is to turn off the CHP from an on state (i.e. $y(t) = 0$ and $U(t) > 0$), and T^o denotes the set of time intervals where the decision is to turn on the CHP from an off state (i.e. $y(t) = 1$ and $U(t) < 0$). Note that the first term C represents the objective function in the fitness function.

7.3.3 Updating Process

Recall the matrices updating introduced in Section 3.3. The updating process from iteration s to iteration $s + 1$ for continuous variables is shown as follows:

$$V_r(s + 1) = \omega V_r(s) + c_1 r_1 (p_r(s) - X_r(s)) + c_2 r_2 (g(s) - X_r(s)) \qquad (7.22)$$

$$X_r(s + 1) = X_r(s) + V_r(s + 1) \qquad (7.23)$$

where $V_r(s)$ and $X_r(s)$ are, respectively, the velocity and position matrices of particle r at iteration s; $p_r(s)$ is the personal best position of particle r at iteration s; $g(s)$ is the global best position at iteration s; ω is the inertia weight; c_1 is the cognitive parameter; c_2 is the social parameter, and r_1 and r_2 are two random numbers drawn between 0 and 1.

Since decision variables $x_i(t)$ and $y(t)$ are binary variables, and the above updating process is for the continuous variable, additional steps for converting a real number to an integer number are required. In particular, (7.24) converts the real number velocity to the set $\{-1, 0, 1\}$.

$$V_{i,j}(s+1) = \begin{cases} -1, & V_{i,j}(s+1) < -0.5 \\ 0, & -0.5 \le V_{i,j}(s+1) \le 0.5 \\ 1, & V_{i,j}(s+1) > 0.5 \end{cases} \qquad (7.24)$$

The next one updates the position using the integer velocity.

$$X_{i,j}(s+1) = \begin{cases} 0, & X_{i,j}(s) + V_{i,j}(s+1) \le 0 \\ 1, & X_{i,j}(s) + V_{i,j}(s+1) > 0 \end{cases} \qquad (7.25)$$

$V_{i,j}$ and $X_{i,j}$ in these two equations denote the respective element at the ith row and jth column of the velocity and position matrices. Since Eqs. (7.24) and (7.25) are only applied to binary variables $x_i(t)$ and $y(t)$, subscript i is in the range of $\{1, 2, ..., N+1\}$ and j is in the range of $\{1, 2, ..., T\}$.

Since the position and velocity of the CHP output power submatrix elements are also discrete, the following equations are adopted to convert real number to integers.

$$V_{i,j}(s+1) = \begin{cases} -\Delta p, & V_{i,j}(s+1) < -0.5\Delta p \\ 0, & -0.5\Delta T \leq V_{i,j}(s+1) \leq 0.5\Delta T \\ \Delta p, & V_{i,j}(s+1) > 0.5\Delta T \end{cases} \tag{7.26}$$

$$X_{i,j}(s+1) = \begin{cases} \Delta p, & X_{i,j}(s) + V_{i,j}(s+1) < \Delta p \\ X_{i,j}(s) + V_{i,j}(s+1), & \Delta p \leq X_{i,j}(s) + V_{i,j}(s+1) \leq PC \\ CP, & X_{i,j}(s) + V_{i,j}(s+1) > PC \end{cases}$$
$$\tag{7.27}$$

Specifically, Eq. (7.26) converts the real number velocity to a number in $\{-\Delta p, 0, \Delta p\}$, and Eq (7.27) updates the position using the discrete velocity. Since Eqs. (7.26) and (7.27) are only applied to variable $p(t)$, subscript i is equal to $N + 2$ and j is in the range of $\{1, 2, ..., T\}$.

7.4 Case Study

In this section, a manufacturing facility including a five-machine-four-buffer manufacturing system and a CHP system with an auxiliary boiler is employed to illustrate the effectiveness of the proposed method. In addition, a separate heat and power system is adopted as a comparison to demonstrate the benefits associated with the CHP system. More specifically, the following two scenarios are considered in the case study:

Scenario I: The electricity demand of the manufacturing system is satisfied by the CHP system and power grid, and the heat demand of the manufacturing system is satisfied by the CHP system and auxiliary boiler.

Scenario II: The electricity demand of the manufacturing system is satisfied by the power grid, and the heat demand of the manufacturing system is satisfied by the auxiliary boiler.

The problem formulation and modeling method for Scenario I is discussed in Section 7.2. For Scenario II, the objective function is slightly adjusted to minimize the total cost of the manufacturing and auxiliary boiler. By revising the original formulation in Section 7.2, the objective function in Scenario II can be expressed as:

$$\min C = C_G + C_C \tag{7.28}$$

$$C_E = c_E(t) \sum_{t=1}^{T} \sum_{i=1}^{N} P_i x_i(t) + c_D(t) \max_t \left(\sum_{i=1}^{N} P_i x_i(t) \right) + C_F \tag{7.29}$$

$$C_C = c_{gas} \sum_{t=1}^{T} \left(\frac{D_H(t) \cdot z_A(t)}{\delta \eta_A} \right) + c_A^0 \sum_{t=1}^{T} \max (z_A(t) - z_A(t-1), 0) \tag{7.30}$$

where

$$z_A(t) = \begin{cases} 1, & D_H(t) > 0 \\ 0, & \text{otherwise} \end{cases} \qquad (7.31)$$

7.4.1 Case Study Settings

In the case study, the production horizon is an eight-hour shift from 7:00 a.m. to 3:00 p.m. The length of each time interval H is set to be 15 minutes. The production target Y° is set as 220. The parameters of the machines, i.e. machine power P_i, production rate ρ_i, and efficiency λ_i are listed in Table 7.1. The parameters of the buffers, i.e. the initial buffer contents $c_i(0)$ and the maximum capacity $\hat{C}_i$, are listed in Table 7.2.

The parameters related to the operation of the CHP system and auxiliary boiler are shown in Table 7.3. These parameters include the calorific value of the natural gas δ; the efficiencies of CHP system η_C and auxiliary boiler η_A; the unit startup cost of the CHP system c_C^0 and auxiliary boiler c_{AB}^{st}; the power to heat ratio of CHP system ε; the minimum on-/off-time of CHP system U_1 and U_0; the power output capacity $\hat{p}$ and power adjustment interval Δp of CHP system; and the cost of natural gas c_{gas}.

A winter day in Chicago, USA, is used to size the heat demand of the entire facility using EnergyPlus [4], which is a whole building energy simulation program provided by the U.S. Department of Energy. In this case, the facility is set as a one-story

Table 7.1 Parameters of the machines.

	Power (kW)	Production rate (units/time interval)	Efficiency (%)
Machine 1	13	10	90
Machine 2	15	10	90
Machine 3	21	10	90
Machine 4	16	10	90
Machine 5	20	10	90

Table 7.2 Parameters of the buffers.

	Initial contents (units)	Capacity (units)
Buffer 1	90	180
Buffer 2	80	160
Buffer 3	75	150
Buffer 4	80	180

Table 7.3 CHP operation-related parameters.

Parameter	Value	Unit
δ	9.649	kWh/Nm^3
η_C	82	%
η_A	80	%
c_C^0	15	$/startup
c_A^0	15	$/startup
ε	0.65	No unit
U_1	2	Number of intervals
U_0	2	Number of intervals
$\hat{p}$	80	kW
Δp	8	kW
c_{gas}	0.173	$/Nm^3

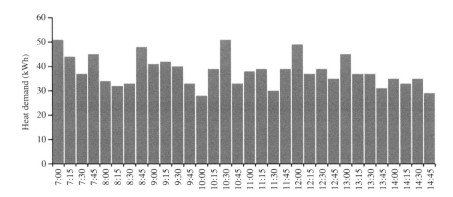

Figure 7.4 Heat demand (kWh) of the facility.

and one-thermal-zone building. The floor area of the building is assumed to be 20 m × 20 m, and the height is assumed to be 10 m. Various materials are used to represent different parts of the facility in the simulation program to best replicate the practical real-world setting. The walls and roof are made of fiberglass and plasterboard, and the floor is made of heavy concrete. The allowable range of the indoor temperature of the facility is 20–24 °C. These parameters are used as input in EnergyPlus to obtain the simulation results of the heat demand, as shown in Figure 7.4.

The on-peak and off-peak periods under the TOU tariff and the pricing for on-peak and off-peak periods are illustrated in Table 7.4.

Table 7.4 Electricity consumption rate and power demand rate.

Type	Time of day	Consumption rate ($/kWh)	Demand rate ($/kW)
On-peak	12 p.m. to 7 p.m.	0.17	18.8
Off-peak	7 p.m. to 12 p.m.	0.08	0

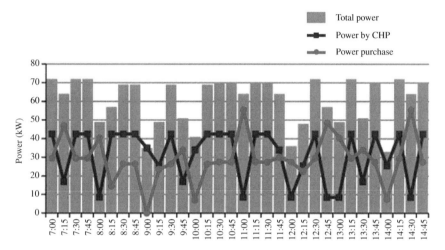

Figure 7.5 Power consumption of each interval in Scenario I. *Source:* [1]/With permission of Elsevier.

The population size of the PSO algorithm is set as 8000; the number of iterations is set as 1000, and the penalty coefficient is set as 10 000.

7.4.2 Results and Discussions

In Scenario I, the production yield throughout the planning horizon is 255 units. The power consumption of the manufacturing system, the output power of the CHP system, and the electricity purchased from the grid in the best solution under Scenario I are shown in Figure 7.5. In the best CHP operation schedule, the CHP system is turned on throughout the entire production horizon and covers part of electricity demand.

Figure 7.6 reflects the heat demand and supply in Scenario I. The results indicate that the auxiliary boiler is still the major heat provider to the manufacturing system.

In addition, the cost comparisons between the two scenarios are presented in Table 7.5. Compared with Scenario II, the combined manufacturing and CHP system needs an additional $14.28 of CHP system operation under Scenario I, while

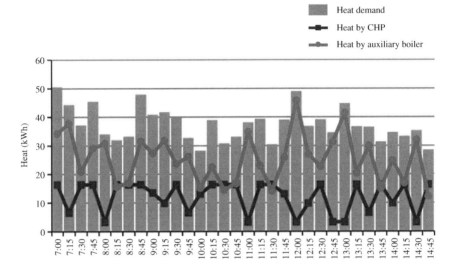

Figure 7.6 Heat demand and supply of each interval in Scenario I. *Source:* [1]/With permission of Elsevier.

Table 7.5 Cost comparison between Scenario I and Scenario II.

	Scenario I	Scenario II	Saving
Electricity bill cost ($)	1071.61	1409.75	338.14
CHP cost ($)	14.28	–	−14.28
Auxiliary boiler cost ($)	17.77	26.21	8.44
Total cost ($)	1103.66	1435.96	332.3
Total saving percentage			23.14%

the electricity cost of the combined system drops significantly from $1409.75 to $1071.61. The overall cost associated with the combined system is about 23.14% lower than that in Scenario II, where the CHP system is not used.

Problems

7.1 What are the advantages of implementing the combined heat and power system in manufacturing systems?

7.2 What parameters affect the operation cost for the combined heat and power system in a manufacturing system?

7.3 What are the constraints in the energy cost minimization problem in the combined manufacturing and CHP system? Why are those constraints necessary for the optimization problem?

7.4 Please use other PSO variants [5–14] to solve Scenarios I and II and discuss the solution speed and performance.

References

1 Sun, Z., Li, L., Bego, A., and Dababneh, F. (2015). Customer-side electricity load management for sustainable manufacturing systems utilizing combined heat and power generation system. *Int. J. Prod. Econ.* 165: 112–119. https://doi.org/10.1016/j.ijpe.2015.04.002.

2 U.S. Environmental Protection Agency (2015). Fuel and carbon dioxide emissions savings calculation methodology for combined heat and power systems. https://www.epa.gov/sites/production/files/2015-07/documents/fuel_and_carbon_dioxide_emissions_savings_calculation_methodology_for_combined_heat_and_power_systems.pdf.

3 Anderson, J.O. and Toffolo, A. (2013). Improving energy efficiency of sawmill industrial sites by integration with pellet and CHP plants. *Appl. Energy* https://doi.org/10.1016/j.apenergy.2013.05.066.

4 U.S. Department of Energy (2018). Getting started with EnergyPlus. The Board of Trustees of the University of Illinois and the Regents of the University of California through the Ernest Orlando Lawrence Berkeley National Laboratory, 2010. http://www.vs.inf.ethz.ch/edu/HS2011/CPS/papers/energyplus_gettingstarted.pdf (accessed 15 January 2020).

5 Bi, J., Yuan, H., Duanmu, S. et al. (2021). Energy-optimized partial computation offloading in mobile-edge computing with genetic simulated-annealing-based particle swarm optimization. *IEEE Internet Things J.* 8 (5): 3774–3785. https://doi.org/10.1109/JIOT.2020.3024223.

6 Cao, Y., Zhang, H., Li, W. et al. (2019). Comprehensive learning particle swarm optimization algorithm with local search for multimodal functions. *IEEE Trans. Evol. Comput.* 23 (4): 718–731. https://doi.org/10.1109/TEVC.2018.2885075.

7 Dong, W. and Zhou, M. (2017). A supervised learning and control method to improve particle swarm optimization algorithms. *IEEE Trans. Syst. Man, Cybern. Syst.* 47 (7): 1135–1148. https://doi.org/10.1109/TSMC.2016.2560128.

8 Tian, G., Ren, Y., and Zhou, M. (2016). Dual-objective scheduling of rescue vehicles to distinguish forest fires via differential evolution and particle swarm optimization combined algorithm. *IEEE Trans. Intell. Transp. Syst.* 17 (11): 3009–3021. https://doi.org/10.1109/TITS.2015.2505323.

9 Li, J., Zhang, J., Jiang, C., and Zhou, M. (2015). Composite particle swarm optimizer with historical memory for function optimization. *IEEE Trans. Cybern.* 45 (10): 2350–2363. https://doi.org/10.1109/TCYB.2015.2424836.

10 Liang, X., Li, W., Zhang, Y., and Zhou, M. (2015). An adaptive particle swarm optimization method based on clustering. *Soft Comput.* 19 (2): 431–448.

11 Tang, J., Liu, G., and Pan, Q. (2021). A review on representative swarm intelligence algorithms for solving optimization problems: applications and trends. *IEEE/CAA J. Autom. Sin.* 8 (10): 1627–1643. https://doi.org/10.1109/JAS.2021.1004129.

12 Wang, Y. and Zuo, X. (2021). An effective cloud workflow scheduling approach combining PSO and idle time slot-aware rules. *IEEE/CAA J. Autom. Sin.* 8 (5): 1079–1094. https://doi.org/10.1109/JAS.2021.1003982.

13 Gao, K., Cao, Z., Zhang, L. et al. (2019). A review on swarm intelligence and evolutionary algorithms for solving flexible job shop scheduling problems. *IEEE/ CAA J. Autom. Sin.* 6 (4): 904–916. https://doi.org/10.1109/JAS.2019.1911540.

14 Yun, L., Li, L., and Ma, S. (2022). Demand response for manufacturing systems considering the implications of fast-charging battery powered material handling equipment. *Appl. Energy* 310: 118550. https://doi.org/10.1016/j.apenergy.2022.118550.

8

Plant-level Energy Management for Combined Manufacturing and HVAC System

In Chapter 2, a heating, ventilation, and air-conditioning (HVAC) system is introduced as one of the primary contributors to the direct non-process end-use energy consumption at the plant level. On the one hand, the HVAC system maintains the indoor temperature of plant buildings, and the associated energy consumption can be affected by a variety of environmental factors, such as outdoor temperature, heat capacity of the buildings, and other building parameters. On the other hand, the thermal energy generated during manufacturing operations can also significantly affect the energy consumption of the HVAC system. Therefore, it is necessary to establish an integrated electricity demand response model capable of capturing the characteristics of both manufacturing and HVAC systems. In this chapter, the energy control and optimization method for manufacturing systems coupled with HVAC systems are introduced. In particular, Section 8.1 illustrates the formulation of an energy cost optimization problem. More specifically, the energy cost model for the combined manufacturing and HVAC system is presented in Section 8.1.1. Four constraints are demonstrated in Section 8.1.2 based on the characteristics of manufacturing systems and HVAC systems. In Section 8.2, particle swarm optimization (PSO) is adopted as the solution approach to solve a proposed cost optimization problem. The details about PSO initialization, evaluation, and matrices updating are also illustrated in this section. Finally, in Section 8.3, a comparative case study is used to demonstrate the effectiveness of the proposed method in identifying the optimal demand response strategy for a combined manufacturing and HVAC system.

8.1 Definition and Modeling

In this chapter, an integrated electricity demand response model for a combined manufacturing and HVAC system is demonstrated. Factors such as production capacity, electricity price, power demand restrictions during demand response

Sustainable Manufacturing Systems, First Edition. Lin Li and MengChu Zhou.
© 2023 The Institute of Electrical and Electronics Engineers, Inc.
Published 2023 by John Wiley & Sons, Inc.

events, and ambient temperature are incorporated in the model to determine the optimal demand response strategy considering both the production schedule and HVAC control.

The electricity demand response model is established upon an incentive-based demand response program. More specifically, the manufacturers enrolled in this program promise to limit their power demand during the demand response event. If the manufacturers do not violate the committed power demand limitation, they are rewarded; otherwise, they face a penalty. In this program, the notification of a demand response event is sent to manufacturers one day in advance, including the starting time and duration of the event. Therefore, manufacturers need to determine an optimal production schedule for their manufacturing system and the optimal control strategy for their HVAC system to minimize the electricity cost while maintaining their desired production yield, acceptable indoor temperature, and commitment with the power demand limitations.

The manufacturing system is assumed to be a typical serial production line with N machines and $N-1$ buffers. In the model, the production horizon is slotted into a set of identical intervals with the length of H. Let $t = [1, ..., T]$ be the index of these time intervals. Depending on the demand response program, these time intervals belong to either the on-peak or off-peak periods.

The scheduling problem discussed in this chapter can be formulated as an optimization problem with the following two decision variables.

- A binary decision variable $x_i(t)$ denotes whether machine M_i ($i = [1, ..., N]$) is turned on or off at time step t.

$$x_i(t) = \begin{cases} 1, & \text{machine } M_i \text{ is turned on} \\ 0, & \text{machine } M_i \text{ is turned off} \end{cases} \tag{8.1}$$

- A real decision variable $y(t) \in [T_L, T_U]$ denotes the target indoor temperature set by the HVAC at time step t. T_L and T_U are the lower and upper bounds of the acceptable indoor temperature.

The objective function and constraints of the optimization problem are presented as follows.

8.1.1 Objective Function

The objective of the optimization problem is to minimize the total electricity cost of the combined manufacturing and HVAC system. The total electricity cost, denoted by C, consists of three elements: electricity consumption cost (C_E), demand cost (C_D), and fixed cost (C_F).

$$\min C = C_{\mathrm{E}} + C_{\mathrm{D}} + C_{\mathrm{F}} \tag{8.2}$$

In addition to the manufacturing system, the HVAC system also consumes electricity. Hence, the electricity consumption cost can be calculated as

$$C_{\mathrm{E}} = \sum_{t=1}^{T} \left(\sum_{i=1}^{N} HP_i x_i(t) + E_{\mathrm{HVAC}}(t) \right) c_{\mathrm{E}}(t) \tag{8.3}$$

where P_i is the power (kW) of machine M_i; the summation of $HP_i x_i(t)$ represents the electricity consumption of the manufacturing system at time t; $E_{\mathrm{HVAC}}(t)$ is the electricity consumption of the HVAC system (kWh) at time t, and $c_{\mathrm{E}}(t)$ is the electricity consumption rate ($/kWh) at time t.

Similarly, the demand cost can be calculated as

$$C_{\mathrm{D}} = \max_t \left(\sum_{i=1}^{N} P_i x_i(t) c_{\mathrm{D}}(t) + \frac{E_{\mathrm{HVAC}}(t) c_{\mathrm{D}}(t)}{H} \right) \tag{8.4}$$

where $c_{\mathrm{D}}(t)$ is the power demand rate ($/kW) at time t. Note that the power demand rate during off-peak periods is 0 $/kW.

The electricity consumption of the HVAC system can be formulated as

$$E_{\mathrm{HVAC}}(t) = k_1 |\mathrm{TEL}(t) - y(t)| \tag{8.5}$$

where k_1 is the heat capacity of the entire plant (kWh/$^{\circ}$F), and $\mathrm{TEL}(t)$ is the actual indoor temperature of the plant building at time t. In practice, k_1 is related to two factors: the heat capacity of the environmental medium (e.g. air) in a plant building, and the performance coefficient of the HVAC system. The heat capacity of the environmental medium can usually be considered as a constant. The performance coefficient of the HVAC system is used to describe its energy efficiency and quantify how much energy is needed to add or remove one unit of heat. The performance coefficient is a complex function of all factors influencing the HVAC performance (i.e. outdoor temperature and required HVAC load), which is hard to estimate through an analytical approach. Hence, to simplify the model as a demonstration for exploring the control strategies of a combined manufacturing and HVAC system in electricity demand response, the performance coefficient of HVAC is also assumed to be a constant. Consequently, k_1 is assumed to be a constant that describes the heat capacity of the entire power plant, and this kind of assumption has also been used in the previous literature [1].

To obtain $\mathrm{TEL}(t)$ at each time interval, the Newton's law of cooling considering an internal heat source due to a manufacturing operation is derived as follows. Note that the time is continuous in the following explanation of Newton's law. Let t' denote the continuous time to distinguish it from the discrete time interval t.

8.1.1.1 Calculate TEL(t)

The temperature changing rate in the plant building due to the difference between the indoor and outdoor temperatures can be formulated as

$$\frac{dT}{dt'}\bigg|_{\Delta T(out-in)} = -k(T(t') - T_{out}) \tag{8.6}$$

where $T(t')$ is the indoor temperature of the plant building ($^\circ$F), which is a differentiable function of time t'; T_{out} is the outdoor temperature ($^\circ$F), and k is a positive constant of the Newton's law of cooling.

The temperature changing rate in the plant building due to the manufacturing operation can be formulated as

$$\frac{dT}{dt'}\bigg|_{manufacturing\ operation} = \frac{q}{k_1} \tag{8.7}$$

where q is the rate of the thermal energy (kW) transferred from the manufacturing system to the plant indoor environment.

Considering both effects described in (8.6) and (8.7), the net temperature changing rate of the plant building can be formulated as

$$\frac{dT}{dt'} = \frac{q}{k_1} - k(T(t') - T_{out})$$
$$= -k\left(T(t') - \left(T_{out} + \frac{q}{kk_1}\right)\right) \tag{8.8}$$

Note that Eq. (8.8) is a separatable equation. After separating the variables, the following equation can be obtained.

$$\frac{dT}{T(t') - \left(T_{out} + \dfrac{q}{kk_1}\right)} = -kdt' \tag{8.9}$$

Integrating both sides of Eq. (8.9) yields

$$\ln\left(T(t') - \left(T_{out} + \frac{q}{kk_1}\right)\right) = -kt' + C \tag{8.10}$$

where C is a constant. With the initial temperature $T(0)$, C can be calculated as

$$C = \ln\left(T(0) - \left(T_{out} + \frac{q}{kk_1}\right)\right) \tag{8.11}$$

Substituting Eq. (8.11) into Eq. (8.10) leads to the following equation:

$$\ln\frac{\left(T(t') - \left(T_{out} + \dfrac{q}{kk_1}\right)\right)}{\left(T(0) - \left(T_{out} + \dfrac{q}{kk_1}\right)\right)} = -kt' \tag{8.12}$$

Figure 8.1 Schematic diagram of the discrete time and continuous time.

Hence, the solution of the plant building temperature at time t' can be expressed as

$$T(t') = \left(T_{\text{out}} + \frac{q}{kk_1}\right) + \left(T(0) - \left(T_{\text{out}} + \frac{q}{kk_1}\right)\right) e^{-kt'} \tag{8.13}$$

Now, switching back to a discrete time system, as shown in Figure 8.1, at the beginning of time interval t, continuous time t' is reset as zero. At the end of time interval t (also referred to as the beginning of time interval $t + 1$), t' equals to the length of the time interval H.

During the time interval t, the outdoor temperate, denoted by TEO(t), is assumed to be a constant, and the rate of the thermal energy transferred from the manufacturing system to the indoor environment of the manufacturing plant, denoted by $q(t)$, is also assumed to be a constant. Note that TEO(t) and $q(t)$ do not change within a time interval but can have different values in different time intervals, and thus they can be expressed as functions of the discrete time t. Applying Eq. (8.13) in the time interval t yields

$$T(H) = \left(\text{TEO}(t) + \frac{q(t)}{kk_1}\right) + \left(T(0) - \left(\text{TEO}(t) + \frac{q(t)}{kk_1}\right)\right) e^{-kH} \tag{8.14}$$

The actual indoor temperature at time t, i.e. TEL(t), is defined as the temperature at the beginning of time interval t. Therefore, TEL($t + 1$) represents the temperature at the beginning of time interval $t + 1$, which is also the temperature at the end of time interval t, and it can be used to replace $T(H)$ in Eq. (8.14). In addition, it is assumed that the temperature change by HVAC is instantaneous, and thus $y(t)$ can be used to represent $T(0)$. Consequently, the relationship between TEL($t + 1$) and decision variable $y(t)$ can be expressed as

$$\text{TEL}(t + 1) = \left(\text{TEO}(t) + \frac{q(t)}{kk_1}\right) + \left(y(t) - \left(\text{TEO}(t) + \frac{q(t)}{kk_1}\right)\right) e^{-kH} \tag{8.15}$$

In Eq. (8.15), the values of TEO(t) and $q(t)$ are required. The outdoor temperature TEO(t) can be obtained from the historical weather data. The rate of the thermal energy transferred from a manufacturing system to the plant indoor environment $q(t)$ can be estimated by using the following method described.

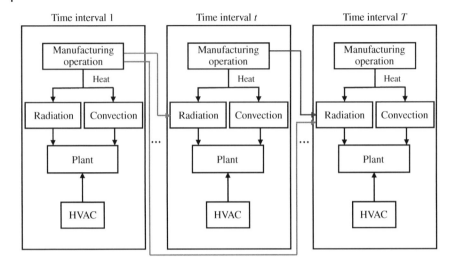

Figure 8.2 Illustrations of the convective and radiant heat transfer due to manufacturing operations at different time intervals.

8.1.1.2 Estimate $q(t)$

The thermal energy is transferred from a manufacturing system to the surroundings in two ways: convection and radiation (thermal conduction is assumed negligible). Figure 8.2 illustrates the behaviors of the convective and radiant heat transfer at each time interval. The convective portion of the thermal energy is added to the surroundings as an instantaneous load [2], which indicates that the convective portion of thermal energy at time t is only related to the thermal energy generated from the manufacturing system at time t. Conversely, radiative heat is absorbed by surfaces in the room and then dissipated to the surroundings over time [2], which indicates that the radiant portion of thermal energy at time t is not only related to the heat generated at time t but also the heat generated at time $t - 1$, $t - 2$, etc.

Let $q_G(t)$ denote the rate of the thermal energy generated from a manufacturing system at time t, which can be calculated as

$$q_G(t) = \sum_{i=1}^{N} (1 - \eta_i) P_i x_i(t) \tag{8.16}$$

where η_i is the motor efficiency of machine M_i. Assuming the ratio between the thermal energy transferred from a manufacturing system to the surroundings through convection and radiation is $c : (1 - c)$, the instantaneous convective heat transferred to the surroundings at time t, denoted by $q_C(t)$, can be calculated as

$$q_C(t) = cq_G(t) \tag{8.17}$$

Radiant heat transfer with a ratio of $(1 - c)$ is considered non-instantaneous with accumulative effects. It is assumed that $s_0\%$ of the radiant heat generated at time t is transferred to the surroundings immediately at time t; $s_1\%$ of radiant heat is transferred to the surroundings at time $t + 1$; $s_2\%$ of radiant heat is transferred to the surroundings at time $t + 2$, and so forth. The radiant heat fraction series s_k ($k = 0, 1, ..., j$) should have the following properties.

$$\sum_{k=0}^{j} s_k = 100\% \tag{8.18}$$

$$s_0 > s_1 > ... > s_j \tag{8.19}$$

Equation (8.18) illustrates that the total radiant heat transferred to the surroundings cannot exceed the total radiant heat generated from the manufacturing system. Equation (8.19) represents the fact that the radiant heat gradually attenuates. Let $q_R(t)$ denote the total radiant heat transferred to the surroundings at time t, which can be calculated as:

$$q_R(t) = (1 - c) \sum_{t_1=1}^{t} s_{t-t_1} \cdot q_G(t_1) \tag{8.20}$$

The rate of the thermal energy transferred from the manufacturing system to the plant indoor environment $q(t)$ can be calculated as

$$q(t) = q_C(t) + q_R(t) \tag{8.21}$$

8.1.2 Constraints

- Constraint 1:

The first constraint is the production yield constraint, i.e.

$$\text{TP} \geq \text{TA} \tag{8.22}$$

where TP is the production yield of the manufacturing system during the production horizon, and TA is the predefined production target. The production yield can be calculated as

$$\text{TP} = \sum_{t=1}^{T} (x_N(t) \cdot \text{PR}_N \text{EFF}_N) \tag{8.23}$$

where PR_N is the production rate of the last machine M_N (with the unit of the number of products produced per time interval), and EFF_N is the efficiency of the last machine M_N (%).

- Constraint 2:

  The second constraint is the buffer content constraint, i.e.

  $$0 \leq c_i(t) \leq C_i \tag{8.24}$$

  where $c_i(t)$ is the buffer content in buffer B_i at time t, and C_i is the maximum capacity of buffer B_i. Similar to the buffer content constraint demonstrated in Chapter 7, according to the conservation of material flow, $c_i(t)$ can be calculated as

  $$c_i(t) = c_i(t-1) + x_i(t-1) \cdot PR_i EFF_i - x_{i+1}(t-1) \cdot PR_{i+1} EFF_{i+1} \tag{8.25}$$

  where PR_i is the production rate of machine M_i, and EFF_i is the efficiency of machine M_i.

- Constraint 3:

  The third constraint is the demand response event constraint. The manufacturer promises to limit its power demand during the demand response event under a limitation, which is

  $$D_p \leq R_p \tag{8.26}$$

  where R_p is the committed limitation of power demand during a demand response event, D_p is the peak demand of the combined manufacturing and HVAC system during the demand response event, which can be expressed as

  $$D_p = \max_{t \in T_{dr}} \left(\sum_{i=1}^{N} P_i x_i(t) + \frac{E_{HVAC}(t)}{H} \right) \tag{8.27}$$

  where T_{dr} is the set of intervals that belong to the demand response event.

- Constraint 4:

  The fourth constraint is the temperature constraint. The actual indoor temperature during the entire production horizon should be within a comfortable range $[T_L, T_U]$, i.e.

  $$T_L \leq TEL(t) \leq T_U \tag{8.28}$$

8.2 Solution Approach

In reference to Section 3.3, the PSO algorithm is adopted to solve the proposed problem. The initialization, evaluation, and updating process are demonstrated as follows.

8.2.1 Initialization

In this problem, the dimension of the solution space is $(N + 1) \times T$. Hence, the solutions for each particle can be formulated as a $(N + 1) \times T$ matrix, i.e. the position matrix X, as shown in Figure 8.3. The position matrix consists of two submatrices: a $N \times T$ production scheduling submatrix indicates the production scheduling $x_i(t)$ of N machines in the manufacturing system, and a $1 \times T$ temperature control submatrix indicates the HVAC temperature settings $y(t)$ throughout the production horizon. The velocity matrix V has the same dimension and is encoded in the same way as the position matrix.

The swarm is initialized by a given number of matrices for all particles. For each position matrix, the initial values of decision variables $x_i(t)$ are randomly selected from the set $\{0, 1\}$. The initial values of decision variables $y(t)$ are randomly selected from the set $\{T_L, T_L + \Delta T, T_L + 2\Delta T, ..., T_U - \Delta T, T_U\}$, where $T_U - T_L = z\Delta T$ and z is a positive integer. For each velocity matrix, the initial values of the decision variables $x_i(t)$ are randomly selected from the set $\{-1, 0, 1\}$, and the initial values of decision variables $y(t)$ are randomly selected from the set $\{-\Delta T, 0, \Delta T\}$.

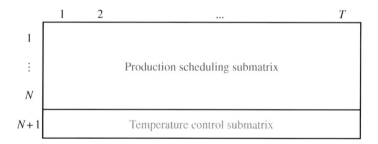

Figure 8.3 Encoding scheme of position and velocity matrices of each particle.

8.2.2 Evaluation

The fitness function used for the evaluation step in PSO is given as

$$
\begin{aligned}
C + \sigma_1 \min(\mathrm{TP} - \mathrm{TA}, 0)^2 &+ \sigma_2 \sum_{t=1}^{T} \sum_{i=1}^{N-1} \min(C_i - c_i(t), 0)^2 \\
&+ \sigma_3 \sum_{t=1}^{T} \sum_{i=1}^{N-1} \min(c_i(t), 0)^2 + \sigma_4 \min(R_\mathrm{p} - D_\mathrm{p}, 0)^2 \\
&+ \sigma_5 \min(\mathrm{TEL}(t) - T_\mathrm{L}, 0)^2 + \sigma_6 \min(T_\mathrm{U} - \mathrm{TEL}(t), 0)^2
\end{aligned}
\tag{8.29}
$$

In Eq. (8.29), σ_1, σ_2, σ_3, σ_4, σ_5, and σ_6 are six large real numbers. Constraints 1–4 are integrated as penalty terms into the fitness function. Note that the first term C represents the objective function in the fitness function. The term with σ_1 represents the penalty for constraint 1. The terms with σ_2 and σ_3 represent the penalty for constraint 2. The term with σ_4 represents the penalty for constraint 3. The terms with σ_5 and σ_6 represent the penalty for constraint 4.

8.2.3 Updating Process

Recall the matrices updating demonstrated in Section 3.3. The updating process from iteration s to iteration $s + 1$ for continuous variables is shown as follows:

$$
V_r(s + 1) = \omega V_r(s) + c_1 r_1 (\mathrm{pbest}_r(s) - X_r(s)) + c_2 r_2 (\mathrm{gbest}(s) - X_r(s))
\tag{8.30}
$$

$$
X_r(s + 1) = X_r(s) + V_r(s + 1)
\tag{8.31}
$$

where $V_r(s)$ and $X_r(s)$ are the velocity and position matrices of the particle r at the iteration s; $\mathrm{pbest}_r(s)$ is the personal best position of the particle r at the iteration s; $\mathrm{gbest}(s)$ is the global best position at the iteration s; ω is the inertia weight; c_1 is the cognitive parameter; c_2 is the social parameter, and r_1 and r_2 are two random numbers drawn between 0 and 1.

Since decision variables $x_i(t)$ are binary variables, and the above updating process is for the continuous variable, additional steps are required to convert a real number to an integer. In particular, the following equation converts the real number velocity to the set $\{-1, 0, 1\}$, the following one updates the position using the integer velocity. The $V_{i,j}$ and $X_{i,j}$ in these two equations denote the respective element at the ith row and jth column of the velocity and position matrices. Since Eqs. (8.32) and (8.33) are only applied to binary variables $x_i(t)$, subscript i is in the range of $[1, N]$ and j is in the range of $[1, T]$.

$$
V_{i,j}(s + 1) =
\begin{cases}
-1, & V_{i,j}(s + 1) < -0.5 \\
0, & -0.5 \le V_{i,j}(s + 1) \le 0.5 \\
1, & V_{i,j}(s + 1) > 0.5
\end{cases}
\tag{8.32}
$$

$$X_{i,j}(s+1) = \begin{cases} 0, & X_{i,j}(s) + V_{i,j}(s+1) \leq 0 \\ 1, & X_{i,j}(s) + V_{i,j}(s+1) > 0 \end{cases} \tag{8.33}$$

Since the position and velocity of the temperature control submatrix elements are also discrete, the following equations are adopted to convert real number to integer. More specifically, Eq. (8.34) converts the real number velocity to the set $\{-\Delta p, 0, \Delta p\}$, and Eq. (8.35) updates the position using the discrete velocity. Since Eqs. (8.34) and (8.35) are only applied to variable $y(t)$, subscript i is equal to $N+1$ and j is in the range of $[1, T]$.

$$V_{i,j}(s+1) = \begin{cases} -\Delta T, & V_{i,j}(s+1) < -\Delta T \\ 0, & -\Delta T \leq V_{i,j}(s+1) \leq \Delta T \\ \Delta T, & \Delta T, \ V_{i,j}(s+1) > \Delta T \end{cases} \tag{8.34}$$

$$X_{i,j}(s+1) = \begin{cases} T_{\mathrm{L}}, & X_{i,j}(s) + V_{i,j}(s+1) < T_{\mathrm{L}} \\ X_{i,j}(s) + V_{i,j}(s+1), & T_{\mathrm{L}} \leq X_{i,j}(s) + V_{i,j}(s+1) \leq T_{\mathrm{U}} \\ T_{\mathrm{U}}, & X_{i,j}(s) + V_{i,j}(s+1) > T_{\mathrm{U}} \end{cases} \tag{8.35}$$

8.3 Case Study

Based on the knowledge introduced above, an optimal case of a combined manufacturing and HVAC systems is illustrated, and the associated electricity cost is compared with the situation where a manufacturing system and an HVAC system are controlled separately.

In this section, a manufacturing facility including a five-machine-four-buffer manufacturing system and an HVAC system is employed to illustrate the effectiveness of the proposed method. In addition to the integrated control strategy for the combined manufacturing and HVAC system, separated control strategies for the manufacturing system and HVAC system are adopted as a comparison. More specifically, the following two scenarios are considered in the case study:

- Scenario I: The integrated control strategy is applied for the combined manufacturing and HVAC system to minimize the energy cost while satisfying the production yield and temperature constraints.
- Scenario II: First, an optimal production schedule for the manufacturing system is obtained, which aims to minimize the production-related energy cost while satisfying the production yield constraint. Meanwhile, the HVAC system is controlled to meet the temperature constraint based on the production schedule.

The problem formulation and modeling method for Scenario I is discussed in Section 8.1. For Scenario II, the objective function is slightly adjusted to minimize the energy cost of the manufacturing system. By revising the original formulation in Section 8.1, the objective function in Scenario II can be expressed as

$$\min C = C_E + C_D + C_F \tag{8.36}$$

$$C_E = c_E(t) \sum_{t=1}^{T} \sum_{i=1}^{N} HP_i x_i(t) \tag{8.37}$$

$$C_D = \max_t \left(\sum_{i=1}^{N} P_i x_i(t) c_D(t) \right) \tag{8.38}$$

In addition, the third constraint, i.e. the demand response event constraint, should be updated by

$$D_p^{mfg} \leq \alpha R_p \tag{8.39}$$

$$D_p^{mfg} = \max_{t \in T_{dr}} \left(\sum_{i=1}^{N} P_i x_i(t) \right) \tag{8.40}$$

where D_p^{mfg} is the peak demand of the manufacturing operation during the demand response event; α is the ratio between the manufacturing power and the total power (i.e. the power of the combined manufacturing and HVAC system), which can be obtained from the historical data. In Eq. (8.39), αR_p represents the committed power demand limitation of manufacturing operations during the demand response event.

In Scenario II, after solving the optimal production schedule, the corresponding power curve of manufacturing operations can be obtained and used as input in EnergyPlus to calculate the energy consumption of the HVAC system [3]. In EnergyPlus, the HVAC target temperature is set based on the lower or upper bounds of the acceptable temperature range.

The case study settings and results are presented next.

8.3.1 Model Settings

In the case study, the production horizon is an eight-hour shift from 7:00 a.m. to 3:00 p.m. The length of each time interval H is set as 15 minutes. The production target TA is set as 280. The parameters of the machines, i.e. machine power P_i, production rate PR_i, and efficiency EFF_i are listed in Table 8.1, followed by the parameters of the buffers, i.e. the initial buffer contents $c_i(0)$ and the maximum capacity C_i.

To calculate the thermal energy transferred from the manufacturing system to the surroundings, the motor efficiency of each machine (denoted by η_i) is set as 0.85. The ratio between convective and radiant heat, i.e. $c : (1 - c)$, is set as $0.7 : 0.3$ [4]. The radiant heat fraction series s_k is shown in Figure 8.4.

At the beginning of the production horizon, the indoor temperature is assumed to be 68 °F. The upper bound and lower bounds of the acceptable indoor

Table 8.1 Parameters of the machines and the buffers in the system.

	Machine 1	Machine 2	Machine 3	Machine 4	Machine 5
Power (kW)	13	15	21	16	20
Production rate (units/time interval)	10	10	10	10	10
Efficiency (%)	90	90	90	90	90

	Buffer 1	Buffer 2	Buffer 3	Buffer 4
Initial contents (units)	90	80	75	80
Capacity (units)	180	160	150	180

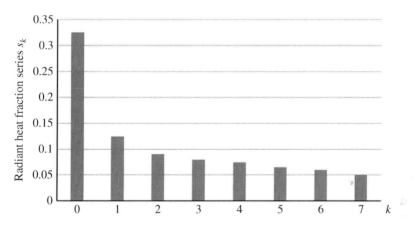

Figure 8.4 Radiant heat fraction series. *Source:* Adopted from [5].

temperature are 70 °F and 76 °F, respectively. The heat capacity of the entire plant k and the constant of the Newton's law of cooling k_1 are assumed to be 1.5 and 1.2, respectively. ΔT is assumed to be 0.5 °F, and the target temperature $y(t)$ should be selected from the set {70 °F, 70.5 °F, 71 °F, 71.5 °F, 72 °F, 72.5 °F, 73 °F, 73.5 °F, 74 °F, 74.5 °F, 75 °F, 75.5 °F, 76 °F}.

The outdoor temperature data is obtained from the weather report of a summer day in Chicago, USA [6], and the temperature profile is shown in Figure 8.5.

Furthermore, a warmer day with higher temperatures is used as the input to evaluate the robustness of the proposed method, as shown in Figure 8.6.

The demand response event occurs between 12:00 p.m. and 1:00 p.m., and the committed limitation of the power demand (R_p) is 95 kW. In Scenario II, the committed power demand limitation of manufacturing operation (αR_p) is set as 70 kW. Table 8.2 reflects the rates of electricity consumption and the power demand during on-peak and off-peak periods.

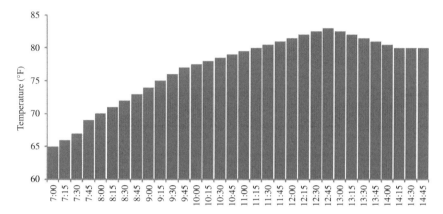

Figure 8.5 Outdoor temperature during each interval.

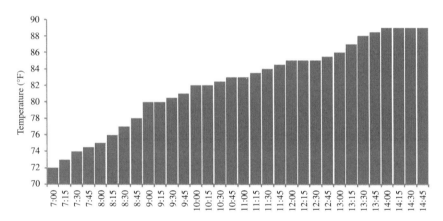

Figure 8.6 Outdoor temperature on a warmer day during each interval.

Table 8.2 Electricity consumption rate and power demand rate.

Type	Time of day	Consumption rate ($/kWh)	Demand rate ($/kW)
On-peak	12 p.m. to 7 p.m.	0.17	18.8
Off-peak	7 p.m. to 12 p.m.	0.08	0

The population size of the PSO algorithm is set as 8000, and the number of iterations is set as 1000.

8.3.2 Results and Discussions

In Scenario I, the production yield throughout the planning horizon is 288 units. The optimal target temperature set by the HVAC system and the corresponding indoor temperature are shown in Figure 8.7.

The total power consumption of the combined system throughout the planning horizon can be calculated based on the optimal production scheduling and HVAC system control results. The results are shown in Figure 8.8.

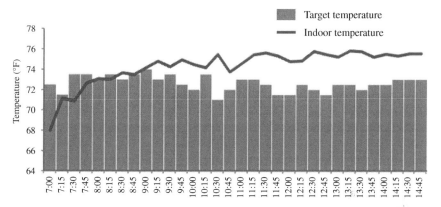

Figure 8.7 Target temperature set by HVAC and indoor temperature evolution in Scenario I.

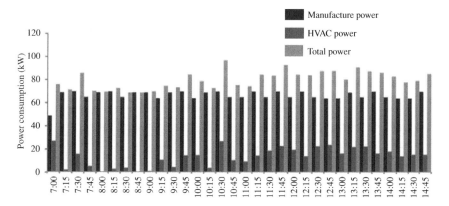

Figure 8.8 Power consumption of each interval in Scenario I.

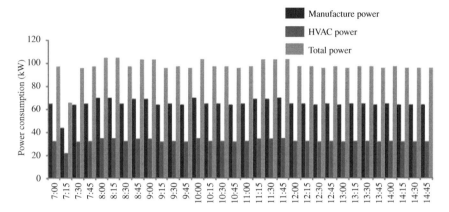

Figure 8.9 Power consumption of each interval in Scenario II.

In Scenario II, the production yield throughout the planning horizon is also 288 units. After obtaining the optimal production schedule, EnergyPlus is used to find the power consumption of the HVAC system. The total power consumption in Scenario II is shown in Figure 8.9.

It can be observed that the HVAC power and total power consumption in Scenario II are generally higher than those in Scenario I. The total energy consumption of Scenario II is 18% higher than that of Scenario I.

In addition, during the on-peak period, the power demand in Scenario I is 91 kW, which is relatively lower than the power demand of 97.5 kW in Scenario II. During the off-peak period, the power demand in Scenario I is 97 kW, and the power demand is 107 kW in Scenario II. In both scenarios, the electricity demand during the on-peak period is lower than the one during the off-peak period. During the demand response event, the power demand in Scenario I is 88 kW, which is lower than the promised limitation of 95 kW. The power demand in Scenario II is 97 kW, which is slightly higher than the promised limitation. These results indicate that the separated management strategies can neither effectively reduce the power demand during the on-peak period nor ensure the power demand during the demand response event to be controlled below the committed level. The comparison of electricity demand per unit product in these two scenarios is shown in Table 8.3. It can be observed that the electricity demand per unit product in Scenario I is much lower than that in Scenario II.

Table 8.4 illustrates the power demand on a normal summer day (the corresponding outdoor temperature profile is shown in Figure 8.5) and a hot summer day (referring to the temperature profile shown in Figure 8.6). The integrated control strategy in Scenario I is applied to calculate the power demand. The power

Table 8.3 Comparison of the power demand per unit product between Scenarios I and II.

Power demand per unit product	Scenario I (kW/unit)	Scenario II (kW/unit)	Reduction (%)
On-peak period	0.32	0.34	5.9
Off-peak period	0.34	0.36	5.5
Demand response event	0.30	0.34	11.8

Table 8.4 Comparison of power demand on normal and hot summer days.

Power demand	Normal day (kW)	Hot day (kW)
On-peak period	91	95
Off-peak period	97	101
Demand response event	88	90

demand during the on-peak period, off-peak period, and demand response event of a hot day is 95 kW, 101 kW, and 90 kW, respectively. These values are slightly higher than the results on a normal day since the HVAC system needs to consume more energy for generating cooling on a warmer day. It can be observed that even on a sizzling summer day, the power demand during the demand response event can be controlled to be lower than the promised limitation.

Problems

8.1 What are the advantages of determining the demand response strategy with joint consideration of production schedule and HVAC control?

8.2 How does a manufacturing operation affect the energy consumption of an HVAC system?

8.3 In the combined manufacturing and HVAC system, what parameters affect the energy consumption of an HVAC system?

8.4 What are the constraints in the energy cost minimization problem in the combined manufacturing and HVAC system? Why are those constraints necessary for the optimization problem?

References

1 Liang, Y., Levine, D.I., and Shen, Z.J.M. (2012). Thermostats for the smart grid: models, benchmarks, and insights. *Energy J.* 33 (4): 61–95. https://doi.org/10.5547/01956574.33.4.4.

2 Hosni, M.H., Jones, B.W., and Xu, H. (1999). Experimental results for heat gain and radiant/convective split from equipment in buildings. *ASHRAE Trans.* 105: 527.

3 US Department of Energy (2018). Getting started with EnergyPlus.

4 Dababneh, F., Li, L., and Sun, Z. (2016). Peak power demand reduction for combined manufacturing and HVAC system considering heat transfer characteristics. *Int. J. Prod. Econ.* 177: 44–52. https://doi.org/10.1016/j.ijpe.2016.04.007.

5 Bruning, S.F. (2004). A new way to calculate cooling loads. *Ashrae J.* 46 (1): 20.

6 Weather Underground (2013). Weather history. https://www.wunderground.com/history/daily/us/il/chicago/KMDW/date/2013-7-21 (accessed 2 March 2021).

Part IV

Energy Management in Advanced Manufacturing Systems

9

Energy Analysis of Stereolithography-based Additive Manufacturing

In Chapter 5, a component-based energy modeling method is introduced. In this chapter, stereolithography (SL), one of the most commonly used additive manufacturing (AM) technologies, is used as the demonstrative process, and its process-level energy consumption is modeled by decomposing the total energy consumption into the energy consumed by major process components. More precisely, in Section 9.1, the basics and technical advantages of the AM technologies are introduced, followed by a detailed description of a projection-based SL process. In Section 9.2, an energy consumption model is demonstrated, which is established based on three major energy-consuming components in the SL machine. In Section 9.3, the experimental validation of the established energy model is illustrated. Meanwhile, the design of experiments (DOE) methodology is employed to investigate the impacts of different parameters on the overall energy consumption, considering the potential interactions among process parameters. In Section 9.4, a case study is presented by comparing the energy consumption under the default parameter settings and the optimal combination of control parameters. The results indicate that by adopting the optimal parameter settings, the energy consumption of an SL-based AM process can be significantly reduced without observable product quality decay.

9.1 Introduction to Additive Manufacturing

Additive manufacturing, also known as three-dimensional (3D) printing, has emerged as a promising manufacturing technology for fabricating parts with a high degree of design freedom. In general, 3D geometry is initially designed in computer-aided design (CAD) software and imported into control software for building file generation. The generated file is subsequently transferred to an AM machine, and a 3D part is fabricated in a layer-by-layer fashion [1]. Attributable to the innovative layer-wise part fabrication, AM technologies are suitable for

Sustainable Manufacturing Systems, First Edition. Lin Li and MengChu Zhou.
© 2023 The Institute of Electrical and Electronics Engineers, Inc.
Published 2023 by John Wiley & Sons, Inc.

manufacturing structures with complex geometries and customizable printing materials [2]. Consequently, AM has gained significant public interest since its emergence and has been widely adopted in many fields, such as aerospace [3], healthcare [4], automotive [5], electronics [6], and construction [7].

AM technologies possess many advantages over traditional manufacturing processes, and some representative features are summarized as follows [8–10].

- The unique characteristic of AM is the part fabrication via the addition of printing material layer by layer. This feature eliminates the need for tooling, lubricants, and cutting fluids.
- Given the layer-wise part fabrication scheme, an AM process is often associated with a relatively shorter product development cycle, easier process control, and higher product quality than a traditional manufacturing process, e.g. milling and drilling.
- Since the entire production process can be completed on one machine, integrating regional and localized production into an AM process can lead to a shortened supply chain, and the associated energy consumption can also be reduced.

To date, multiple types of AM technologies have been developed to achieve differing production goals with different types of printing materials. Some AM processes build parts from the powder materials, such as selective laser sintering (SLS), electron beam melting (EBM), and binder jetting (BJ). Some use filament materials, such as fused deposition modeling (FDM). Some other AM processes, such as SL and Polyjet, fabricate parts by solidifying the photo-curable liquid materials through a light source.

In this chapter, a mask image projection (MIP) SL-based AM process is selected as a model process, and a component-based modeling approach is adopted in energy model formulation. Additionally, the DOE is used to explore the potential interactions among different process parameters and determine the optimal combination of process parameters toward the minimal energy consumption.

9.1.1 Illustration of MIP SL-based AM Process

An MIP SL-based AM process is mostly used in indoor environment as a desktop 3D printer. Owing to the innovative image projection approach, where the liquid material is solidified by exposing two-dimensional (2D) cross-sectional images instead of the line-by-line scanning with a laser beam, MIP SL-based AM permits faster production than other AM processes [11]. In addition, it can produce a wide variety of parts with different shapes while ensuring good quality and maintaining reasonable production cost [12]. Considering its process characteristics, it has great potential for energy efficiency improvement from several aspects. For example, parameters associated with the layer image projection, such as the layer thickness and curing time, can be configured in building files [13, 14]. In addition, parameters related to part geometry, such as part orientation and position, can be adjusted in its control [15, 16].

As shown in Figure 9.1, an MIP SL-based AM process consists of a control computer, a material tank with a transparent bottom, a building platform, a Z stage, a digital micromirror device (DMD), an ultraviolet (UV) light source, and a lens. The liquid resin is stored in the material tank, which can be solidified upon exposure to the UV light based on the layer image projected by DMD. During a part fabrication process, all components are stationary except the movement of the building platform along the Z stage (in the vertical direction), which allows the layer-wise part fabrication on a building platform.

The entire production process consists of the following steps:

1) The 3D geometry built in CAD software is imported into the control software and sliced into 2D layer images with uniform thickness. Then, the control software generates a building file, which compiles material-specific and part design-dependent parameter settings, such as layer thickness, layer image, and curing time.
2) After the building file is transferred to the SL machine, the building platform moves down along the Z stage and descends into the material tank, leaving a single layer of liquid material between the bottom of the material tank and the building platform.
3) To start the production, DMD projects the first layer image on the bottom of the building platform through the transparent material tank. The liquid resin turns to the non-tacky solid upon exposure to the UV light [17].

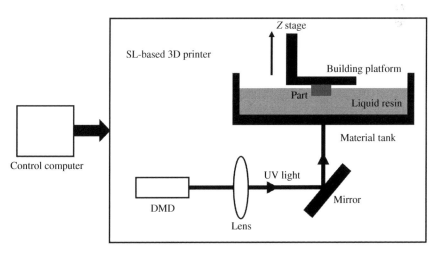

Figure 9.1 A schematic diagram of the MIP SL-based AM process. *Source:* Adapted from [8].

4) After the first layer is solidified, the building platform ascends along the Z stage by a distance of the layer thickness. Subsequently, DMD automatically projects the next layer image on the new building surface.
5) The whole part can be fabricated by repeating the above two steps. At the end of the printing process, the building platform moves up along the Z stage to its original position.

9.2 Energy Consumption Modeling

As shown in Figure 9.2, the MIP SL-based AM process includes multiple subsystems: layer image projection, UV curing, building platform movement, lighting, and cooling. The energy consumption of the entire system is related to the energy consumed in each subsystem. Hence, the component-based modeling approach (in reference to Section 5.1) can be employed to model the energy consumption in the production process. Note that the subsystems with minor contributions to the total energy consumption are not included in the model, such as layer image projection and lighting.

The total energy consumption includes three parts: energy consumption of the UV curing process (E_{curing}), energy consumption of building platform movement ($E_{platform}$), and energy consumption of cooling system ($E_{cooling}$), as formulated as follows:

$$E = E_{curing} + E_{platform} + E_{cooling} \tag{9.1}$$

The energy consumption for each subsystem can be modeled next.

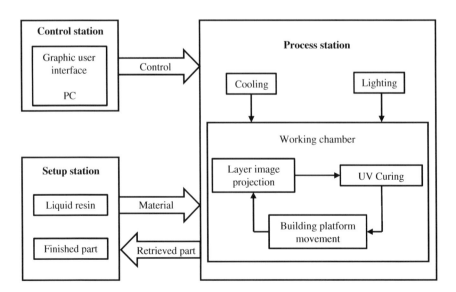

Figure 9.2 The major components in MIP SL-based AM process. *Source:* Adapted from [8].

9.2.1 Energy Consumption of UV Curing Process

For a certain part geometry with a total height of h and layer thickness of d, the number of layers K can be calculated as

$$K = \frac{h}{d} \tag{9.2}$$

The UV curing energy consumption for the kth layer, denoted by e_k, can be calculated as

$$e_k = \frac{P_{UV}t_k}{a} \tag{9.3}$$

where P_{UV} is the power output of the UV light source, t_k is the curing time for the kth layer, and a is a constant determined by the UV source characteristics, which can be calculated as

$$a = \eta_1\eta_2\eta_3 \tag{9.4}$$

where η_1, η_2, and η_3 denote the lighting efficiency, ratio of effective wavelength over the total wavelength, and material absorptivity for a specific UV source, respectively. These values can be obtained from the AM machine documentation.

In Eq. (9.3), the curing time t_k is dependent on the layer index k, which can be estimated as

$$t_k = \begin{cases} t_1, & 1 \leq k \leq k_b \\ t_1 - s(k - k_b), & k_b + 1 \leq k \leq k_c \\ t_2, & k_c + 1 \leq k \leq K - 3 \\ t_1, & K - 2 \leq k \leq K \end{cases} \tag{9.5}$$

Equation (9.5) describes the relationships between curing time and the layer index. From the first layer to layer k_b, the curing time is a constant t_1. Note that in order to ensure that the first several layers of a part can be fully cured, t_1 is usually longer than the time required for solidifying a single layer of material. From layer $(k_b + 1)$ to layer k_c, curing time decreases from t_1 to t_2 with a linear rate $s = (t_2 - t_1)/(k_c - k_b)$. Hence, layers from the $(k_b + 1)$ to k_c are defined as transition layers. Then, from layer $(k_c + 1)$ to layer $(K - 3)$, the curing time maintains at the constant value of t_2. These layers are defined as stable layers as they have a uniform curing time. The curing time of the last three layers is normally switched back to t_1 to ensure a fully cured part with good quality.

The total UV curing energy consumption can be calculated as

$$E_{\text{curing}} = \sum_{k=1}^{K} e_k \tag{9.6}$$

9.2.2 Energy Consumption of Building Platform Movement

During a production process, the building platform moves along the Z stage. An electric stepper motor provides power for the movement of the building platform, and the associated energy consumption can be calculated as

$$E_{\text{platform}} = \sum_{k=1}^{K} P_m t_k \tag{9.7}$$

where P_m is the power output of the stepper motor power.

9.2.3 Energy Consumption of Cooling System

The energy consumption of a cooling system can be calculated as

$$E_{\text{cooling}} = P_{\text{cooling}} t_{\text{cooling}} \tag{9.8}$$

where P_{cooling} is the power output of a cooling fan, and t_{cooling} is the cooling time, which is slightly longer than the production time.

9.3 Experimentation

In the previous section, a mathematical model is established by analyzing the energy consumption contributed by each subsystem of an MIP SL-based AM process. Different process parameters, such as the number of layers and curing time for each layer, are integrated into the model. In order to find the optimal combination of process parameters that can reduce the total energy consumption while maintaining satisfactory product quality, the interactions among different parameters need to be investigated, which mean that the energy consumption may vary when changing one parameter while keeping other parameters at different levels. For example, the curing time can be changed from a lower value to a higher value when the layer thickness is set at different levels, while the associated energy consumption may either remain the same or change accordingly.

9.3.1 Experiment Design Methodology

To investigate such interactions, it is necessary to conduct multiple experiments under different combinations of input parameters. By adopting the DOE method, one can study the influences of input parameters with the smallest number of experiments, which is more efficient than the one-factor-at-a-time experimental strategy [18]. Therefore, in addition to the mathematical model of energy consumption, the DOE approach is applied to design experiments with different

Table 9.1 Description of experimental control parameters.

Symbol	Control parameter	High level (+1)	Low level (−1)	Center point
A	Layer thickness, d (mm)	0.05	0.025	0.0375
B	Curing time for stable layers, t_2 (s)	6.5	4	5.25
C	Curing time transition rate, s (s/layer)	2.7	1.125	1.9125
D	Orientation	90°	0°	45°

combinations of input parameters. The experimental results can be used to examine potential interactions among different parameters, identify the optimal parameter settings toward reduced energy consumption, and validate the proposed energy model. The detailed configurations of the experiment designed by the DOE are illustrated as follows.

As shown in Table 9.1, a two-level factorial design is used to establish the experiments with four input factors, including layer thickness d, curing time for stable layers t_2, curing time transition rate s, and build orientation. The first three factors are process-related parameters, whereas the last one is a geometry-related factor. Two levels of each parameter, i.e. low and high, are considered. Two replicate experiments are conducted for each possible combination of all four parameters with different levels. Four center points are added to the experiments to provide a measure of process stability and inherent variability, and to check for curvature. A 2^4 factorial design combined with four center points results in a total of 36 experiments.

For the MIP SL-based AM machine in the experiments, the default value for layer thickness is 0.025 mm, and the high level of this parameter is set as 0.05 mm. The curing time for stable layers is set initially as 6.5 seconds, and the low level is set as four seconds. The high and low levels of the transition rate from longer curing time to shorter one are set as 1.125 s/layer (i.e. the default value) and 2.7 s/layer, respectively. The orientation is investigated with the initial value of 0° and a high level of 90°.

The software is used for experimental data analysis and result visualization in this chapter. For the convenience of data recording in software, the high levels of input factors are coded as +1, lower levels are coded as −1, and center points are coded as 0. Besides, four input factors are signified by A, B, C, and D.

9.3.2 Experiment Apparatus

The MIP SL-based AM machine used in experiments is a desktop 3D printer (Perfactory Micro EDU 3D) with the highest resolution of 150 microns (μm) for the X and Y axes and 50–100 μm for the Z axis [19]. The 3D printer is equipped with state-of-the-art direct light projection technology from Texas Instruments, and it can achieve a precise layer image projection by DMD. A light-emitting diode UV light source is used for material solidification. The 3D printer can achieve a building speed of up to 20 mm/h for full building capacity ($100 \times 75 \times 100$ mm^3). In addition, the material used in the experiments is LS600M from EnvisionTEC [20]. In addition, the real power consumption is measured using a clamp-on power meter, with the maximum alternating current/direct current (AC/DC) measurement of 600 Å and the maximum AC/DC voltage of 1000 V. The measured data are recorded with an interval of five seconds.

Due to the existence of the power factor, the measured current and voltage data cannot be directly used to calculate the power consumption. The power factor, denoted by PF, can be measured by the watt-ampere-voltmeter method [21]. Hence, the real power of the AM machine can be calculated as:

$$P_{\text{real power}} = \text{PF}(U_{\text{measured}} I_{\text{measured}}) \tag{9.9}$$

where U_{measured} and I_{measured} represent the measured voltage and current. The experimentally determined value of the PF is 0.85.

9.4 Results and Discussions

In this section, under the default parameter settings, the proposed energy model is validated by comparing the calculated results with the experimental measurements. The influences of selected input parameters and the significance of interactions among factors are illustrated through the factorial analysis of experimental results. The optimal combination of process parameters is determined by using the response surface optimization method. Note that a bolt with a height of 1 cm is used as the demonstrative part in the model validation and factorial experiments.

9.4.1 Baseline Case Results Using Default Conditions

Under the default conditions (i.e. layer thickness 0.025 mm, curing time for stable layers 6.5 seconds, curing time transition rate 1.125 s/layer, and orientation 0°), the measured energy consumption is 278 707.35 J for fabricating a bolt. The result calculated by using the energy model is 264 531.672 J, which indicates an absolute percentage error of 5.36%. The difference may come from the use of rated power of each subsystem provided in the AM machine documentation rather than the actual power in energy consumption calculation using the proposed model. Due to the limitation of measuring equipment and the complexity of subsystems,

Table 9.2 Comparison of energy consumption in different AM processes.

Case	AM process	Material	Capacity utilization (%)	Layer thickness (mm)	Specific energy consumption (kWh/kg)	Reference
1	SLS	PA2200	3.41	0.12	39.20	[22]
2		PA2200	3.02	0.15	40.30	
3		PA3200	2.50	0.15	36.00	
4	SLS	Polymer	–	0.15	40.10	[23]
5	SL	Epoxy resin	–	0.15	32.47	
6	FDM	ABS	–	0.4	115.20	
7	SL	LS600M	0.05	0.025	175.95	

the real power of each subsystem cannot be directly measured. Hence, the rated power is used for approximate calculation.

The energy consumption measurement under the default conditions is also compared with other processes in the existing literature. Table 9.2 illustrates the comparison of energy consumption in different AM processes. In particular, the energy consumption of the SLS process with different layer thicknesses and material types is presented in Cases 1–3 [22]. Cases 4–6 depict the energy consumption of three AM processes, including SLS, SL, and FDM [23]. The energy consumption of the MIP SL-based AM process discussed in this chapter is demonstrated in Case 7.

As shown in Table 9.2, the calculated energy consumption using the proposed energy model (Case 7) is relatively higher than the results in the literature (Cases 1–6). This can be explained by the following three main reasons.

- Different types of AM processes adopt diverse manufacturing technologies and produce dissimilar types of materials, which result in different energy consumption characteristics.
- The energy consumption of an AM machine can be affected by the capacity utilization rate, i.e. the ratio of printed part volume to the maximum building volume. It has been reported that the lower capacity utilization rate may result in higher specific energy consumption (SEC) for some AM processes [24]. Compared with the capacity utilization rates used in Cases 1–3, a lower capacity utilization rate is used in Case 7 (i.e. 0.05%), leading to higher SEC.
- The layer thickness used in Case 7 is much smaller than the others, leading to better product quality and increased energy consumption.

9.4.2 Factorial Analysis Results

The experimental results are imported for further factorial analysis to obtain the statistical model with a significance level of 0.05. As shown in Table 9.3, according to the factorial analysis results, factors A, B, and D and interactions AB, AD, BD, CD, ABC, ABD, and BCD are of great significance. Factor C and interactions AC, BC, ACD, and ABCD have minor influences on energy consumption.

More specifically, factors A, B, and D can affect the production time, which is also the reason for the high significance of interactions among these three factors. Although factor C also affects the production time, the effect only lasts for a relatively shorter period in part fabrication (usually 25 layers). Therefore, the influence of factor C is negligible. From the results of DOE, although the impacts of factor C and other related interactions such as AC, BC, ACD, and ABCD are not significant, the interactions CD, ABC, and BCD exert substantial influences. Based on the SL process observations, the duration of factor C changes with factor D. The duration of factor C is longer when the part is built horizontally, and it becomes shorter when the part is built vertically. Hence, the effect of the interaction between C and D is significant. The impacts of three-order interactions ABC and BCD are substantial because of similar reasons where factor C changes according to the changes of factors A and B and factors B and D.

Table 9.3 Factorial analysis results.

Factor	Sum of squares	p Value
A	2.12×10^{11}	0.000
B	5.83×10^{9}	0.000
C	2.20×10^{6}	0.556
D	6.27×10^{9}	0.000
AB	1.41×10^{8}	0.000
AC	9.89×10^{5}	0.692
AD	3.29×10^{8}	0.000
BC	1.37×10^{7}	0.150
BD	7.95×10^{7}	0.002
CD	6.78×10^{7}	0.005
ABC	4.38×10^{7}	0.015
ABD	2.27×10^{8}	0.000
ACD	3.82×10^{5}	0.805
BCD	5.80×10^{7}	0.006
ABCD	1.38×10^{7}	0.640

The Pareto chart of the standardized effects is shown in Figure 9.3a, and the normal plot of the standardized effects is presented in Figure 9.3b. It can be observed that factor A has a negative impact on the total energy consumption,

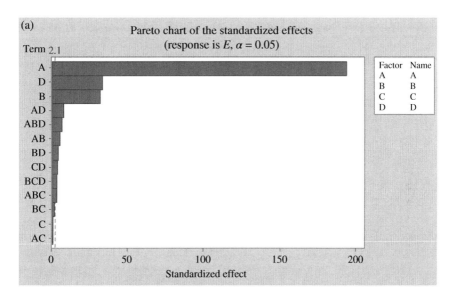

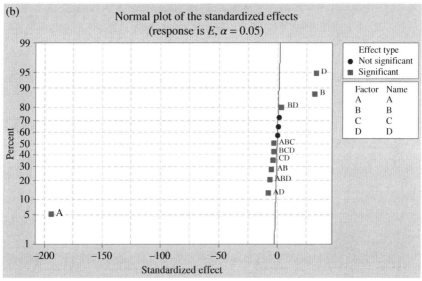

Figure 9.3 Factorial analysis results: (a) Pareto chart and (b) normal plot. *Source:* [8]/John Wiley & Sons/Public Domain CC BY 4.0.

whereas factors B and D have positive effects. The interactions AD, ABD, BD, CD, BCD, ABC, and BC have relatively minor influences, whereas factor C and interaction AC have ignorable effects.

The model adequacy checking is demonstrated in Figure 9.4. The normality plot and the plot of residuals versus observation order indicate that the residuals are normally and independently distributed. The plot of residuals versus fitted value implies that the assumption of constant variance is satisfied.

The corresponding refined statistical model can be expressed as in Eq. (9.10), where $C_t P_t$ is the center point.

$$
\begin{aligned}
E = {} & 259\,101 - 81\,321 \times A + 13\,492 \times B + 13\,993 \times D - 2096 \times A \times B \\
& - 3207 \times A \times D + 1576 \times B \times D - 1456 \times C \times D - 1170 \times A \times B \times C \\
& - 2662 \times A \times B \times D - 1346 \times B \times C \times D + 34\,970 \times C_t P_t
\end{aligned}
$$

$$(9.10)$$

The regression model adequacy checking is shown in Figure 9.5.

Based on the factorial analysis, the optimal levels of the input factors that can minimize energy consumption can be identified by using the response surface method. The results are shown in Figure 9.6a. A higher level of factor A and the lower level of factors B and D would reduce the energy consumption of the

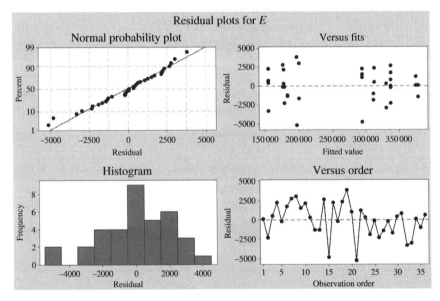

Figure 9.4 Adequacy checking for factorial design model. *Source:* [8]/John Wiley & Sons/ Public Domain CC BY 4.0.

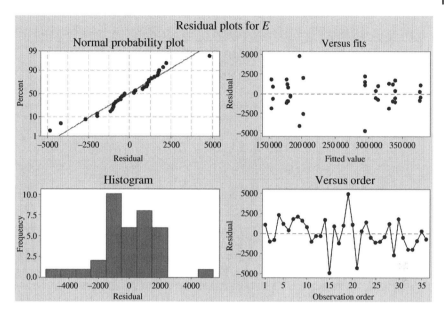

Figure 9.5 Adequacy checking for the refined statistical model. *Source:* [8]/John Wiley & Sons/Public Domain CC BY 4.0.

MIP SL-based AM process. In addition, Figure 9.6b shows the surface plots with respect to the factors A, B, and D and the total energy consumption.

The measured energy consumption using the optimal combination of control parameters is 127 707.35 J. An approximate 54.16% of reduction in energy consumption can be achieved compared with the default working condition where the measured energy consumption is 278 707.35 J.

9.4.3 Product Quality Comparison

As demonstrated in the previous section, the adjustments of process parameters can reduce the total energy consumption; however, different combinations of parameters may lead to possible decay of part quality. To evaluate the influence of process parameter settings on part quality, a Micro-Vu vision system can be used to obtain the surface images of an AM fabricated part, and a 3D optical profiler can be employed for surface roughness measurements. The variations in part quality with respect to different parameter combinations are shown in Figure 9.7.

More specifically, Figure 9.7a reflects the part quality under default working conditions (referring to the parameter settings used in Section 9.4.1). The result indicates good quality for the screw thread of the printed bolt. To evaluate the impact of layer thickness on part quality, the same geometry is printed under

(a)

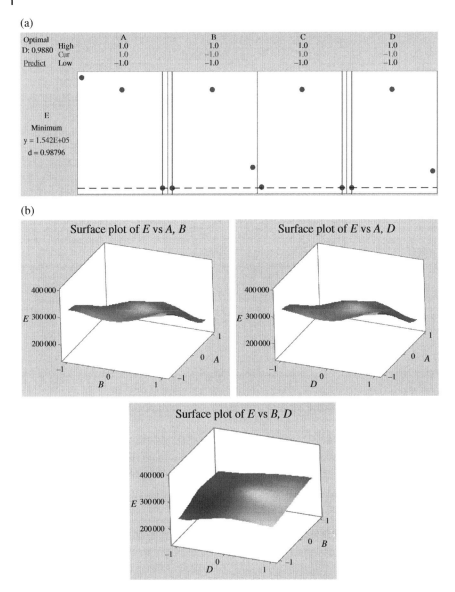

(b)

Figure 9.6 Response optimization results and surface plots: (a) response optimization results; and (b) surface plot of energy consumption versus A and B, A and D, B and D. *Source:* [8]/John Wiley & Sons/Public Domain CC BY 4.0.

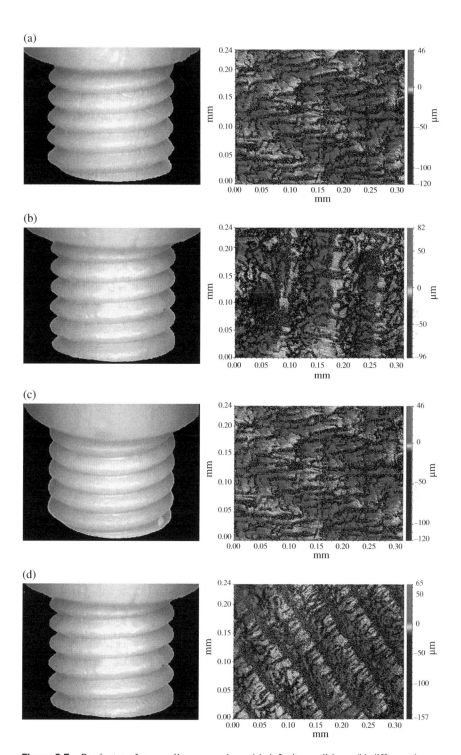

Figure 9.7 Product surface quality comparison: (a) default conditions, (b) different layer thickness, (c) different curing time, and (d) optimized conditions. *Source:* [8]/John Wiley & Sons/Public Domain CC BY 4.0.

the conditions where the layer thickness is changed to 0.05 mm, while the other parameters remain the same as the default conditions. The corresponding surface image is shown in Figure 9.7b. Similarly, Figure 9.7c shows the surface image of the bolt fabricated with the curing time of the stable layer being changed to four seconds. In addition, the surface quality of the part printed under optimized working conditions is demonstrated in Figure 9.7d, where the layer thickness is 0.05 mm; curing time for stable layers is four seconds; curing time transition rate is 1.125 s/layer, and orientation is 0°.

The variations in the surface roughness of the screw thread on the parts can be quantified using the arithmetic mean surface roughness R_a. As shown in Figure 9.7, although the four parts are built under different combinations of parameters, the corresponding R_a values are within 10 μm magnitude (from 2.599 to 4.946 μm). Therefore, it can be concluded that energy consumption can be reduced by adopting the optimal parameter settings while maintaining acceptable surface quality.

Problems

9.1 What are the advantages of additive manufacturing over traditional manufacturing?

9.2 Briefly describe the working principle of an MIP SL-based AM process.

9.3 Demonstrate the major subsystems of an MIP SL-based AM process. Which need to be considered in the energy consumption modeling?

9.4 Why is it important to consider the interactions among different process parameters in analyzing the energy consumption in an AM process?

References

1 Yang, Y., He, M., and Li, L. (2020). Power consumption estimation for mask image projection stereolithography additive manufacturing using machine learning based approach. *J. Clean. Prod.* 251: 119710. https://doi.org/10.1016/j.jclepro.2019.119710.

2 Han, M., Yang, Y., and Li, L. (2021). Techno-economic modeling of 4D printing with thermo-responsive materials towards desired shape memory performance. *IISE Trans.* 1–13. https://doi.org/10.1080/24725854.2021.1989093.

3 Uriondo, A., Esperon-Miguez, M., and Perinpanayagam, S. (2015). The present and future of additive manufacturing in the aerospace sector: a review of important aspects. *Proc. Inst. Mech. Eng. Part G J. Aerosp. Eng.* 229 (11): 2132–2147. https://doi.org/10.1177/0954410014568797.

4 Gibson, I. and Srinath, A. (2015). Simplifying medical additive manufacturing: making the surgeon the designer. *Procedia Technol.* 20: 237–242. https://doi.org/10.1016/j.protcy.2015.07.038.

5 Leal, R., Barreiros, F.M., Alves, L. et al. (2017). Additive manufacturing tooling for the automotive industry. *Int. J. Adv. Manuf. Technol.* 92 (5–8): 1671–1676. https://doi.org/10.1007/s00170-017-0239-8.

6 Saengchairat, N., Tran, T., and Chua, C.-K. (2017). A review: additive manufacturing for active electronic components. *Virtual Phys. Prototyp.* 12 (1): 31–46. https://doi.org/10.1080/17452759.2016.1253181.

7 Lim, S., Buswell, R.A., Le, T.T. et al. (2012). Developments in construction-scale additive manufacturing processes. *Autom. Constr.* 21: 262–268. https://doi.org/10.1016/j.autcon.2011.06.010.

8 Yang, Y., Li, L., Pan, Y., and Sun, Z. (2017). Energy consumption modeling of stereolithography-based additive manufacturing toward environmental sustainability. *J. Ind. Ecol.* 21 (S1): S168–S178. https://doi.org/10.1111/jiec.12589.

9 Simon, T., Yang, Y., Lee, W.J. et al. (2019). Reusable unit process life cycle inventory for manufacturing: stereolithography. *Prod. Eng.* 13 (6): 675–684. https://doi.org/10.1007/s11740-019-00916-0.

10 Zhao, J., Han, M., and Li, L. (2021). Modeling and characterization of shape memory properties and decays for 4D printed parts using stereolithography. *Mater. Des.* 203: 109617. https://doi.org/10.1016/j.matdes.2021.109617.

11 Pan, Y., Zhou, C., and Chen, Y. (2012). A fast mask projection stereolithography process for fabricating digital models in minutes. *J. Manuf. Sci. Eng.* 134 (5): 51011, 1–9. https://doi.org/10.1115/1.4007465.

12 Yang, Y. and Li, L. (2018). Cost modeling and analysis for Mask Image Projection Stereolithography additive manufacturing: simultaneous production with mixed geometries. *Int. J. Prod. Econ.* 206: 146–158. https://doi.org/10.1016/j.ijpe.2018.09.023.

13 Han, M., Yang, Y., and Li, L. (2020). Energy consumption modeling of 4D printing thermal-responsive polymers with integrated compositional design for material. *Addit. Manuf.* 34: 101223. https://doi.org/10.1016/j.addma.2020.101223.

14 Zhao, J., Yang, Y., and Li, L. (2020). Efficiency-aware process planning for mask image projection stereolithography: leveraging dynamic time of exposure. *Addit. Manuf.* https://doi.org/10.1016/j.addma.2020.101407.

15 Yang, Y., Li, L., and Zhao, J. (2019). Mechanical property modeling of photosensitive liquid resin in stereolithography additive manufacturing: bridging degree of cure with tensile strength and hardness. *Mater. Des.* 162: 418–428. https://doi.org/10.1016/j.matdes.2018.12.009.

16 Zhao, J., Yang, Y., and Li, L. (2020). A comprehensive evaluation for different post-curing methods used in stereolithography additive manufacturing. *J. Manuf. Process.* 56: 867–877.

17 Bajpai, M., Shukla, V., and Kumar, A. (2002). Film performance and UV curing of epoxy acrylate resins. *Prog. Org. Coatings* 44 (4): 271–278. https://doi.org/10.1016/S0300-9440(02)00059-0.

18 Montgomery, D.C. (2012). *Design and Analysis of Experiments*. New York: Wiley.

19 EnvisionTEC (2015). Perfactory micro EDU. https://envisiontec.com/envisiontec/wp-content/uploads/MK-MCS-MicroEDU-V01-FN-EN.pdf (accessed 11 January 2015).

20 EnvisionTEC. Tachnical data: envisiontec LS600 resin. https://envisiontec.com/wp-content/uploads/2016/09/2020-LS600.pdf (accessed 6 March 2020).

21 Prasanna Kumar, C.S., Sabberwal, S.P., and Mukharji, A.K. (1995). Power factor measurement and correction techniques. *Electr. Power Syst. Res.* 32 (2): 141–143. https://doi.org/10.1016/0378-7796(94)00906-K.

22 Kellens, K., Renaldi, R., Dewulf, W. et al. (2014). Environmental impact modeling of selective laser sintering processes. *Rapid Prototyp. J.* 20 (6): 459–470. https://doi.org/10.1108/RPJ-02-2013-0018.

23 Luo, Y., Ji, Z., Leu, M.C., and Caudill, R. (1999). Environmental performance analysis of solid freedom fabrication processes. *Proceedings of the 1999 IEEE International Symposium on Electronics and the Environment (Cat. No.99CH36357)*, Danvers, Massachusetts, USA (11–13 May 1999) pp. 1–6, https://doi.org/10.1109/ISEE.1999.765837.

24 Baumers, M., Tuck, C., Wildman, R. et al. (2011). Energy inputs to additive manufacturing: does capacity utilization matter. *Eos (Washington. DC)* 1000 (270): 30–40.

10

Energy Efficiency Modeling and Optimization of Cellulosic Biofuel Manufacturing System

In Chapter 5, a system-level energy consumption modeling approach is demonstrated, where both the intra-process and inter-process variables are taken into consideration to establish an energy model. In this chapter, the cellulosic biofuel manufacturing system consisting of several processes is used as an example to demonstrate the formulation of a system-level energy model. More specifically, in Section 10.1, the knowledge related to cellulosic biofuel manufacturing is introduced, followed by a detailed description of the major processes involved in cellulosic biofuel manufacturing. In Section 10.2, the formulation of the energy consumption model for cellulosic biofuel manufacturing is illustrated, which involves the discussions on four energy-related factors in cellulosic biofuel manufacturing, including heating energy, energy loss, reaction energy, and energy recovery. In Section 10.3, particle swarm optimization (PSO) is adopted to solve the proposed energy consumption minimization problem. Additionally, the comparison of the system-level energy efficiency under a baseline case and an optimal case is conducted in Section 10.4.

10.1 Introduction to Cellulosic Biofuel Manufacturing

Along with the widespread use of fossil fuels, there are growing concerns about the sustainability issues associated with the irreversible depletion of fossil fuels, such as climate change, environmental degradation, and energy supply challenges. The adoption of clean, renewable, and sustainable energy sources has drawn significant attention over recent decades. As one of the most popular renewable energy sources, biofuel is expected to play a critical role in future global energy infrastructure. Based on the biomass feedstocks used in biofuel production, biofuels can be divided into several categories, such as sugarcane, cellulosic, and algae ones. Among these categories, cellulosic biofuel is considered the most promising alternative energy source due to its widely available feedstock resources and relatively

Sustainable Manufacturing Systems, First Edition. Lin Li and MengChu Zhou.
© 2023 The Institute of Electrical and Electronics Engineers, Inc.
Published 2023 by John Wiley & Sons, Inc.

low life-cycle greenhouse gas (GHG) emissions [1]. Compared with petroleum fuels, cellulosic ethanol can achieve up to 90% of GHG emission reduction [2].

The block diagram of a typical cellulosic biofuel production system is shown in Figure 10.1, which consists of four major processes: size reduction, pretreatment, enzymatic hydrolysis, and fermentation.

- Size reduction: The biomass size reduction process usually involves grinding or cutting the cellulosic feedstock into small pieces. After the size reduction process, the specific surface area (i.e. the total surface area of the material per unit of mass) of biomass increases significantly, which facilitates the subsequent biomass hydrolysis process. The energy consumption of this process mainly comes from the electricity consumption of size reduction machines.

- Pretreatment: In the pretreatment process, the steam-assisted diluted-acid pretreatment method is adopted [4]. The cellulosic material and water are fully mixed in the reactor under the high temperature (~158 °C) and high pressure (5 atm) reaction conditions. During the reaction, the crystallinity of cellulose is reduced, which indicates that the cellulose becomes more amorphous and suitable for enzymatic hydrolysis. Most of the hemicellulose contained in the feedstock can

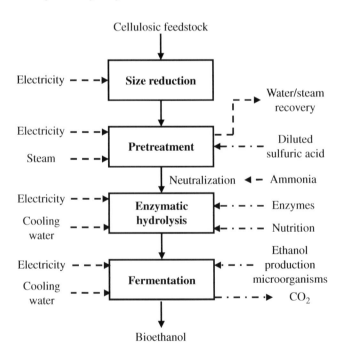

Figure 10.1 Block diagram of biofuel production from cellulosic biomass. *Source:* [3]/With permission of Elsevier.

Table 10.1 Solid compositions of corn-stover.

Component	Dry weight percentage (%)
Glucan	39.1
Xylan	23.1
Lignin	16.3
Galactan	1.8
Arabinan	3.4
Mannan	0.3
Acetyl	2.7
Structural inorganics	4.8
Protein	4.5
Ash	4

Source: Adapted from [7].

be converted into soluble pentoses, such as xylose, which can be used as substrates in fermentation. In this process, energy is mainly consumed by heating the feedstock–water mixture and maintaining the reaction temperature. The heat in the water/steam can be partially recovered after the pretreatment process.

- Enzymatic hydrolysis: In the hydrolysis process, the cellulose chains are broken down into glucose for fermentation. Currently, enzymatic hydrolysis is the most recognized bioethanol hydrolysis technology because no degradation products of glucose are formed [5]. After the hydrolysis process, most of the cellulosic contents (hemicellulose and cellulose) are converted into the fermentable sugar. Since the hydrolysis reaction is exothermic, energy is consumed to cool down the hydrolysis reactor.
- Fermentation: In the fermentation process, all soluble sugars (mainly xylose and glucose) are assumed to be simultaneously fermented into ethanol using recombinant *Zymomonas mobilis* strain ZM4 (pZB5) (*Z. mobilis* ZM4 (pZB5)) [6]. In this process, energy is mainly consumed to maintain the reaction temperature.

In this chapter, the corn-stover is assumed to be the biomass feedstock. The solid compositions of corn-stover are listed in Table 10.1.

10.2 Energy Modeling of Cellulosic Biofuel Production

In this section, a mathematical energy consumption model of cellulosic biofuel production is introduced. The energy modeling of a biomass size reduction process is demonstrated in Section 10.2.1, and the energy modeling of biofuel chemical conversion processes, i.e. pretreatment, enzymatic hydrolysis, and fermentation, is presented in Section 10.2.2.

10.2.1 Energy Modeling of Biomass Size Reduction Process

The energy consumption in the size reduction process can be generally described by three empirical equations, i.e. Rittinger's law, Bond's law, and Kick's law. In particular, the first two laws are employed to model the energy consumption in processing small particles ($<$1 cm), while the last law can be used for particles that are larger than 1 cm [8]. Since the size of the original biomass feedstock particles is usually larger than 1 cm, Kick's law is adopted to calculate the energy consumption in the size reduction process.

According to the Kick's law [9], the energy consumed in the size reduction process can be calculated as

$$E_{\text{size reduction}} = C_k \ln\left(\frac{L_1}{L_2}\right) \tag{10.1}$$

where $E_{\text{size reduction}}$ is the energy consumed in this process; C_k is the Kick's constant; L_1 is the particle size before size reduction, and L_2 is the particle size after size reduction. Note that the particle size is defined as the screen size that the particles can pass through regardless of the particles' shape.

In Eq. (10.1), the Kick's constant is an experimental constant that depends on the biomass type and size reduction machines' properties. The size reduction process is assumed to use a grinding machine with a 500-rpm rotor, and the feeding rate is assumed to be 5 kg/min. According to [10], the grinding machine consumes 0.02 kWh when grinding 150 mm corn-stover into 25.4 mm; thus, the Kick's constant can be calculated as

$$C_k = E_{\text{size reduction}}/\ln\left(L_1/L_2\right) = 72\,(\text{kJ/kg})/\ln\left(150/25.2\right) = 40.55 \tag{10.2}$$

10.2.2 Energy Modeling of Biofuel Chemical Conversion Processes

In this section, the energy modeling of the three biochemical processes, i.e. pretreatment, enzymatic hydrolysis, and fermentation, is illustrated. The energy consumed in these processes is mainly used to maintain the desired reaction temperature. In addition, the energy that can be partially recovered during the cooling stages is also considered in the modeling.

Figure 10.2 shows the reaction temperature profile during the pretreatment, enzymatic hydrolysis, and fermentation processes. The entire timeline can be divided into eight stages according to the temperature changes, which includes one heating stage (i.e. stage H), four heat preservation stages (i.e. stages

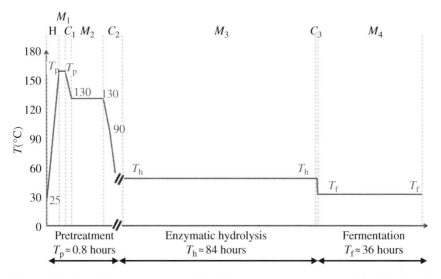

Figure 10.2 Reaction temperature profile during pretreatment, enzymatic hydrolysis, and fermentation processes. *Source:* [3] With permission of Elsevier.

M_1–M_4), and three cooling stages (i.e. stages C_1–C_3). In the heating stage, biomass is mixed with water at room temperature (25 °C) and heated up to the reaction temperature T_p (>155 °C). Subsequently, in the M_1 stage, the reaction temperature is maintained at T_p for 5–10 minutes, and oligomers are formed in this stage [11]. Then, the reactor cools down to 130 °C in stage C_1 and maintains at that temperature for approximately 25 minutes in stage M_2 to allow the conversion of oligomers to monomers [7]. The reactor then cools down to T_h in stage C_2 and maintains at T_h for a period of time t_h ($t_h \approx 84$ hours) in stage M_2. Energy is consumed in maintaining the temperature in stage M_2 to ensure the efficiency of the cellulase enzyme. After enzymatic hydrolysis, the saccharified slurry is cooled to T_f and kept for a period of time t_f ($t_f \approx 36$ hours) for fermentation to ensure the ethanol yield from the strain of *Z. mobilis* ZM4 (pZB5).

10.2.2.1 Heating Energy

As shown in Figure 10.2, the heating stage is within the pretreatment process. The heating energy represents the energy required to heat the mixture of biomass, water, and sulfuric acid. The generic equation of heating energy can be written as

$$E_{\text{heating}} = m_w \int_{25}^{T_p} C_{p_w} dT + m_b \int_{25}^{T_p} C_{p_b} dT + m_a C_{p_a} \left(T_p - 25 \right) \tag{10.3}$$

where m_w, m_b, and m_a are the respective mass (kg) of water, biomass, and diluted acid; C_{p_w} and C_{p_b} are the specific heat capacity (kJ/mol/K) of water and biomass, which are functions of temperature T; C_{p_a} is the average specific heat capacity of diluted sulfuric acid, which is a constant; and T_p is the pretreatment temperature (°C).

In the heating stage, the initial pressure in the reactor is set as 557 kPa (5.5 atm), and the boiling point of water under this pressure is 155 °C. The latent heat of water vaporization at 155 °C is H_L. It is assumed that 5% of water evaporates at this temperature, and the rest 95% of water is heated from 155 °C to T_p. According to [12], the heat capacity of water (C_{p_w}) and steam (C_{p_s}) can be calculated as

$$C_{p_w} = 0.00001T^2 - 0.0013T + 4.2085 \tag{10.4}$$

$$C_{p_s} = 0.000\,000\,8T^2 - 0.0002T + 1.8572 \tag{10.5}$$

In addition, according to [13], the heat capacity of biomass (C_{p_b}) can be calculated as

$$C_{p_b} = 0.000\,04T^2 - 0.0015T + 0.9325 \tag{10.6}$$

Consequently, the heating energy can be calculated as

$$
\begin{aligned}
E_{\text{heating}} = {}& m_w \int_{25}^{155} \left(0.000\,01T^2 - 0.0013T + 4.2085\right) dT \\
& + 0.95 m_w \int_{155}^{T_p} \left(0.000\,01T^2 - 0.0013T + 4.2085\right) dT \\
& + 0.05 m_w \int_{155}^{T_p} \left(0.000\,000\,8T^2 - 0.0002T + 1.8572\right) dT \\
& + 0.05 m_w H_L + m_b \int_{25}^{T_p} \left(0.000\,04T^2 - 0.0015T + 0.9325\right) dT \\
& + m_a C_{p_a} \left(T_p - 25\right)
\end{aligned}
\tag{10.7}
$$

10.2.2.2 Energy Loss

Since the reaction temperatures during all the stages (as shown in Figure 10.2) are higher than the room temperature (25 °C), heat escapes from the reactor to the surrounding environment by conduction and convection, which causes energy loss. The reactor is assumed to be a cylindrical nature [7], and thus the energy loss can be calculated as the heat loss of a pipe:

$$E_{\text{lost}} = UA\Delta Tt \tag{10.8}$$

where A is the reactor surface area (m^2); ΔT is the temperature difference between the inside of the reactor and the ambient temperature (°C); t is the duration time (h), and U is the overall heat transfer coefficient, which can be formulated as

$$U = \cfrac{1}{\cfrac{R_{\text{out}} \ln \frac{R_{\text{out}}}{R_{\text{in}}}}{k_{\text{p}}} + \cfrac{1}{h_{\text{out}}}} \tag{10.9}$$

where R_{in} and R_{out} are the inner and outer radiuses of the reactor (m); k_{p} is the thermal conductivity of the reactor (W/m/°C), and h_{out} is the convection coefficient at the outside surface (W/m^2/°C).

Since the duration of the heating stage is relatively short, it can be assumed that the temperature in the reactor increases linearly. Therefore, the total energy required to compensate for the energy loss and maintain the reaction temperature can be calculated as

$$E_{\text{heatlost}} = U^p A^p \int_0^{t_H} t\left(\frac{T_p - 25}{t_H}\right) dt + U^p A^p \Delta T^{M_1} t_{M_1} + U^p A^p \Delta T^{M_2} t_{M_2}$$
$$+ U^h A^h \Delta T^{M_3} t_{M_3} + U^f A^f \Delta T^{M_4} t_{M_4}$$

$$\tag{10.10}$$

where the superscript/subscript p, h, f, and M_i ($i = 1, 2, 3, 4$) signify the pretreatment, hydrolysis, fermentation processes, and stages M_1 to M_4, respectively, and t_H is the duration of the heating stage.

10.2.2.3 Reaction Energy

The reaction energy model is established based on the conservation of energy in a closed system, i.e. the internal energy change of a system is equal to the heat provided by the outside world to the system minus the work done by the system. Since all the reactions occur in the isochoric reactor, the work done by the system is zero. Hence, the energy absorbed/released during the reaction is equal to the change of internal energy between the input and output materials in each process. Consequently, the mass of the products in each process needs to be calculated.

Table 10.2 summarizes nine major chemical reactions involved in cellulosic biofuel production. It is assumed that arabinan, mannan, and galactan have the same reactions and conversion pathways as xylan. The heat of formation (HoF) represents the energy required/released for converting 1 kg reactant into the product. The positive values represent endothermic reactions, and the negative values indicate exothermic reactions.

Table 10.2 Major reactions involved in cellulosic biofuel production.

Process	Reaction	Heat of formation (kJ/kg)[a]
Pretreatment	1: $(Xylan)_n + nH_2O \rightarrow n$ Xylose	248.8
	2: $(Xylan)_n + mH_2O \rightarrow m$ Xylose Oligomer	1903.2
	3: $(Glucan)_n + nH_2O \rightarrow n$ Glucose	279.1
	4: $(Arabinan)_n + nH_2O \rightarrow n$ Arabinose	2102.2
	5: $(Mannan)_n + nH_2O \rightarrow n$ Mannose	2941.5
	6: $(Galactan)_n + nH_2O \rightarrow n$ Galactose	2941.5
Enzymatic hydrolysis	7: $(Glucan)_n + nH_2O \rightarrow n$ Glucose	−1463.1
Fermentation	8: Glucose $\rightarrow$ 2Ethanol + 2CO$_2$	−1836.9
	9: 3Xylose $\rightarrow$ 5Ethanol + 5CO$_2$	−1610.3

[a]Calculated based on the data in [7].

Catalyzed by diluted acid, xylan is converted into xylose and xylose oligomer (which can be finally converted into xylose) [14]. The reaction rate of xylan is expressed as

$$
\begin{aligned}
\frac{dp_x}{dt} &= -(k_{xo} + k_{xl})p_x C_{acid} \frac{dc_{xo}}{dt} \\
&= k_{xo}\left(\frac{p_x}{M_x}\right)\left(\frac{M_x}{M_{xo}}\right)C_{acid}\frac{1}{\left(\dfrac{m_w}{\rho_w}\right)} - k_{x2}c_{xo}C_{acid}\frac{dc_{xl}}{dt} \\
&= k_{xl}\left(\frac{p_x}{M_x}\right)\left(\frac{M_x}{M_{xl}}\right)C_{acid}\frac{1}{\left(\dfrac{m_w}{\rho_w}\right)} + k_{x2}c_{xo}\left(\frac{M_{xo}}{M_{xl}}\right)C_{acid}, \quad t\epsilon[0, t_{M_1}]
\end{aligned}
\tag{10.11}
$$

where p_x, c_{xo}, and c_{xl} represent xylan mass fraction in biomass, xylose oligomers concentration (kg/l), and xylose concentration (kg/l); C_{acid} represents active acid concentration (w/w %); M_x, M_{xo}, and M_{xl} denote the respective molecular weight (g/mol) of xylan, xylose oligomer, and xylose; m_w is the mass of water (kg); ρ_w is the density of water (kg/m^3); k_{xo} and k_{xl} are the reaction rates (1/s) of the conversion from xylan to xylose oligomers and xylan to xylose, respectively. The reaction rates are assumed to have an Arrhenius dependence on temperature [15], which can be expressed as

$$
k = Ae^{-E_a/RT}
\tag{10.12}
$$

where A is the pre-exponential factor; E_α is the activation energy, and R is the gas constant.

In Eq. (10.11), the acid concentration C_{acid} is represented by the concentration of acid that has infiltrated into feedstock particles. According to [16], the acid infiltrated into a spherical particle can be described by the following equation:

$$C_{\text{acid}}(t) = 1 - \sum_{n=1}^{\infty} \left(\frac{6\alpha(1+\alpha)}{9 + 9 + q_n^2 \alpha} \right) \exp\left(-\frac{q_n^2 D_e t}{(L_2/2)^2} \right) \tag{10.13}$$

where $C_{\text{acid}}(t)$ denotes the acid concentration (w/w %) at time t; q_n is a reaction constant; D_e is the diffusion coefficient (m^2/s), and α is the ratio of the liquid volume to the volume of solid spheres, which can be expressed as

$$\alpha = \frac{3\left(\frac{m_w}{\rho_w}\right)}{\left[4\pi\left(\frac{L_2}{2}\right)^2\right]} \tag{10.14}$$

In the pretreatment process, i.e. $t \in [0, t_{M_1}]$, a fraction of glucan is converted into glucose, and the reaction rate can be formulated by the following equation according to [14].

$$\frac{dp_g}{dt} = -k_g p_g C_{\text{acid}}$$
$$\frac{dc_g}{dt} = k_g \left(\frac{p_g}{M_g}\right)\left(\frac{M_g}{M_{gl}}\right) C_{\text{acid}} \frac{1}{(m_w/\rho_w)} \tag{10.15}$$

where p_g and c_g represent glucan mass fraction in biomass and glucose concentration, respectively; M_g represents the molecular weight (g/mol) of glucose, and k_g is the reaction rates (1/s) of the conversion from glucan to glucose.

In the enzymatic hydrolysis process, i.e. $t \in [0, t_h]$, glucose formation can be simulated by a substrate inhibition model [17], which can be expressed as

$$\frac{dp_g}{dt} = -\frac{k_h E_b p_g}{1 + \left(\frac{c_g}{K_g}\right) + \left(\frac{c_{xl}}{K_{xl}}\right)}$$
$$\frac{dc_g}{dt} = 1.111 \frac{k_h E_b p_g \frac{1}{(m_w/\rho_w)}}{1 + \left(\frac{c_g}{K_g}\right) + \left(\frac{c_{xl}}{K_{xl}}\right)} \tag{10.16}$$

where E_b is the enzyme loading, and K_g and K_{xl} represent the substrate inhibition coefficient of glucose and xylose, respectively.

In a fermentation process, i.e. $t \in [0, t_f]$, the xylose and glucose generated from the pretreatment and enzymatic hydrolysis process are converted into ethanol. The model in the following equation is established based on substrate inhibition and product inhibition [18], which can be used to calculate ethanol yield:

$$
\begin{aligned}
\frac{dc_{xl}}{dt} &= -(1-\alpha)\frac{q_{smax,x}}{m_w}\left(1 - \frac{c_e - P_{i,x}}{P_{m,x} - P_{i,x}}\right)\left(\frac{K_{i,x}}{K_{i,x} + c_{xl}}\right)c_Z\frac{dc_g}{dt} \\
&= -\alpha\frac{q_{smax,g}}{m_w}\left(1 - \frac{c_e - P_{i,g}}{P_{m,g} - P_{i,g}}\right)\left(\frac{K_{i,g}}{K_{i,g} + c_g}\right)c_Z\frac{dc_e}{dt} \\
&= -\left(\frac{q_{emax,x}}{q_{smax,x}}\frac{dc_{xl}}{dt} + \frac{q_{emax,g}}{q_{smax,g}}\frac{dc_g}{dt}\right)
\end{aligned}
\tag{10.17}
$$

where c_e is the concentration of ethanol (kg/l); c_Z is the concentration of ethanol production microbial (kg/l); q_{smax} is the maximum substrate uptake rate (g/l); q_{emax} is the maximum ethanol production rate (g/l); P_i is the production inhibition coefficient; P_m is the maximum production inhibition coefficient, and K_i is the substrate inhibition coefficient.

The total reaction energy is equal to the changes of internal energy in all processes. The internal energy change can be calculated by multiplying the conversion rate of the reactant with the heat of formation of the corresponding reactions, which can be expressed as

$$
\begin{aligned}
E_{reaction} &= \Delta Q_{i,p} + \Delta Q_{i,h} + \Delta Q_{i,f} \\
&= \sum_{i=1}^{9} CR_i\text{HoF}_i = \sum_{i=1}^{9}\left(\frac{R_i^0 + R_i^E}{R_i^0}\right)\text{HoF}_i
\end{aligned}
\tag{10.18}
$$

where R_i^0 and R_i^E are the mass of the reactant at the beginning and end of the ith reaction listed in Table 10.2, and HoF_i represents the heat of formation of the ith reaction.

10.2.2.4 Energy Recovery
During the cooling stages, i.e. stages C_1 and C_2 in Figure 10.2, the partially recovered energy can be calculated based on the enthalpy changes of water and biomass at each stage:

$$
\begin{aligned}
E_{recovery} &= \sum_{C_1, C_2} m_{b,remain}\int_{T_1}^{T_2}\left(0.00004T^2 - 0.0015T + 0.9325\right)dT \\
&\quad + m_w\int_{T_1}^{T_2}\left(0.00001T^2 - 0.0013T + 4.2085\right)dT
\end{aligned}
\tag{10.19}
$$

10.2.2.5 Total Energy Consumption

In summary, the total energy required to produce ethanol from the size reduction process to the end of fermentation can be expressed as

$$E_{total} = E_{size\ reduction} + E_{heating} + E_{heatloss} + E_{reaction} - E_{recovery} \qquad (10.20)$$

10.3 Energy Consumption Optimization Using PSO

10.3.1 Problem Formulation

In the optimization problem of energy consumption in cellulosic biofuel production, the following nine decision variables are considered based on the proposed energy model:

- Feedstock particle diameter L_2
- Water loading (per kg feedstock) m_w
- Diluted acid loading C_{acid}
- Pretreatment temperature T_p
- Pretreatment time t_p
- Enzymatic hydrolysis temperature T_h
- Enzymatic hydrolysis time t_h
- Fermentation temperature T_f
- Fermentation time t_f

The objective function and constraints are given as follows:
Objective function

$$\min E_{total} \qquad (10.21)$$

Constraints

$$2.54\,\text{mm} \leq L_2 \leq 25.4\,\text{mm} \qquad (10.22)$$

$$1.5\,\text{kg} \leq m_w \leq 3\,\text{kg} \qquad (10.23)$$

$$1\%\text{w/w} \leq C_{acid} \leq 3\%\text{w/w} \qquad (10.24)$$

$$150\,^{\circ}\text{C} \leq T_p \leq 170\,^{\circ}\text{C} \qquad (10.25)$$

$$4\,\text{minutes} \leq t_p \leq 6\,\text{minutes} \qquad (10.26)$$

$$44\,^{\circ}\text{C} \leq T_h \leq 52\,^{\circ}\text{C} \qquad (10.27)$$

$$72\,\text{hours} \leq t_h \leq 96\,\text{hours} \qquad (10.28)$$

Table 10.3 Values of decision variables in the baseline case.

Decision variables	Values	Unit
Feedstock particle diameter, L_2	6.67	mm
Water loading (per kg feedstock), m_w	2.2	kg
Diluted acid loading, C_{acid}	1.0	w/w %
Pretreatment temperature, T_p	158	°C
Pretreatment time, t_p	5	min
Enzymatic hydrolysis temperature, T_h	48	°C
Enzymatic hydrolysis time, t_h	84	h
Fermentation temperature, T_f	32	°C
Fermentation time, t_f	36	h

$$28\,^{\circ}\mathrm{C} \leq T_f \leq 36\,^{\circ}\mathrm{C} \tag{10.29}$$

$$28\,\text{hours} \leq t_f \leq 36\,\text{hours} \tag{10.30}$$

$$ce_f \geq ce_{f,\text{baseline}} \tag{10.31}$$

Equations (10.22)–(10.30) represent the feasible ranges of nine variables. Equation (10.31) describes the ethanol yield constraint, where ce_f is the ethanol yield, and $ce_{f,\,baseline}$ is the ethanol yield in the baseline case. Note that the values of decision variables in the baseline case are adopted from the National Renewable Energy Laboratory in the United States, as shown in Table 10.3 [7]. The final ethanol yield $ce_{f,\,baseline}$ in the baseline case is 0.2411 kg ethanol/kg feedstock.

10.3.2 Solution Procedures

In reference to Section 3.3, the PSO algorithm is adopted to solve the proposed optimization problem. The initialization, evaluation, and updating process are demonstrated as follows.

10.3.2.1 Initialization
In this problem, the dimension of solution space is 1×9. Hence, the solutions for each particle can be formulated as a vector, i.e. a position vector X. A velocity vector V has the same dimension as X and is encoded in the same way as X.

$$X = \begin{bmatrix} L_2, m_w, C_{acid}, T_p, t_p, T_h, t_h, T_f, t_f \end{bmatrix} \tag{10.32}$$

The particle swarm is initialized by a given number of vectors for all particles. For each position vector, all decision variables are randomly selected from their feasible ranges.

10.3.2.2 Evaluation

For the evaluation step in the PSO, the ethanol yield constraint is integrated as a penalty term into the fitness function, as formulated as

$$E_{\text{total}} + \sigma_1 \min \left(ce_f - ce_{f,\text{baseline}}, 0 \right)^2 \tag{10.33}$$

where σ_1 is a large real number.

10.3.2.3 Updating Process

Since all decision variables are continuous, the updating process demonstrated in Section 3.3 can be directly applied. More specifically, the updating process from iteration s to iteration $s + 1$ for continuous variables can be expressed as

$$V_r(s + 1) = \omega V_r(s) + c_1 r_1 (\text{pbest}_r(s) - X_r(s)) + c_2 r_2 (\text{gbest}(s) - X_r(s)) \tag{10.34}$$

$$X_r(s + 1) = X_r(s) + V_r(s + 1) \tag{10.35}$$

where $V_r(s)$ and $X_r(s)$ are the velocity and position vectors of the particle r at iteration s; $\text{pbest}_r(s)$ is the personal best position of the particle r at iteration s; $\text{gbest}(s)$ is the global best position at iteration s; ω is the inertia weight; c_1 is the cognitive parameter; c_2 is the social parameter; r_1 and r_2 are two random numbers drawn between 0 and 1.

In this optimization problem, c_1 and c_2 are set as 2, and the maximum iteration number is set as 100. The population size of 10, 20, 30, 50, 100, and 200 are tested to approach the near-optimal solution.

10.4 Case Study

Based on the energy model and energy consumption optimization problem introduced in previous sections, comparisons between the baseline case (i.e. Scenario I) and an optimal case (i.e. Scenario II) of the cellulosic biofuel manufacturing system are illustrated in this section.

- Scenario I: A cellulosic biofuel production design reported by National Renewable Energy Laboratory in the United States.
- Scenario II: A cellulosic biofuel production design obtained by solving the energy consumption optimization problem.

Table 10.4 Energy-related parameters.

Parameter	Value	Unit	Reference
C_k	40.55	–	Section 10.2
H_L	2257	kJ/kg	[12]
R_{in}	1.3	M	[7]
R_{out}	1.315	M	[7]
k_p	13.8	W/(m °C)	[13]
h_{out}	45	W/(m² °C)	Section 10.2
A^p	47.78	m²	[7]
A^h	250	m²	[7]
A^f	200	m²	[7]
R	8.314	–	[a]

[a]Determined based on common knowledge.

10.4.1 Case Settings

The input feedstock is assumed to be 1 kg under two scenarios. The energy-related parameters are summarized in Table 10.4, and the chemical reaction parameters are listed in Table 10.5.

10.4.2 Energy Analysis of Baseline Case

10.4.2.1 Energy Consumption Breakdown

In the baseline case as Scenario I, the total energy consumption is 2470.57 kJ, and the ethanol yield from 1 kg feedstock is 0.241 kg. The energy breakdown for each process is shown in Figure 10.3. The breakdown result indicates that pretreatment process contributes nearly half of the total energy consumption, even though its duration is much shorter in comparison with the hydrolysis and fermentation processes. The relatively higher energy consumption can be mainly attributed to the requirement of high reaction temperature during a pretreatment process. Therefore, it can be expected that the development of new energy-efficient pretreatment technologies would greatly benefit the sustainability of cellulosic biofuel production.

The heating energy breakdown is shown in Figure 10.4, where the energy consumed for heating water and steam accounts for more than 90% of the total heating energy. A possible explanation is that the specific heat capacity of water is much larger than that of biomass. In addition, since the biomass loading is considered constant, the water/biomass mass ratio becomes the main influencing factor in heating energy consumption.

Table 10.5 Chemical reaction-dependent parameters.

Arrhenius reaction rate parameters			
Reaction rate	**A (1/s)**	**E_a (kJ/mol)**	
k_{xo}	1.0×10^{15}	110	[19]
k_{xl}	8.0×10^{17}	130	[19]
k_g	2.0×10^{22}	180	[20]
k_{h2}	2.0×10^{7}	5570	[5]
Molecular weight (g/mol)		Value	
M_x		132	[7]
M_{xo}		450	[7]
M_{xl}		150	[7]
M_g		162	[7]
M_{gl}		180	[7]
Inhibition constants (g/l)		Value	
K_g		0.1	[17]
K_{xl}		0.1	[17]
Other parameters		Value	
ρ_w (kg/m^3)		1	–
q_n		0.213	[21]
D_e (m^2/s)		0.68×10^{-10}	[12]
Overall substrate utilization rates (g/l)		Value	
$q_{emax,g}$		10.9	[18]
$q_{emax,x}$		3.27	[18]
$q_{smax,g}$		5.12	[18]
$q_{smax,x}$		1.59	[18]
Substrates inhibitory concentration (g/l)		Value	
$K_{i,g}$		600	[18]
$K_{i,x}$		186	[18]
Threshold ethanol concentration (g/l)		Value	
$P_{i,g}$		57.4	[18]
$P_{i,x}$		81.2	[18]
$P_{m,g}$		57.4	[18]
$P_{m,x}$		81.2	[18]

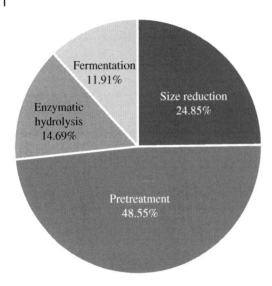

Figure 10.3 Distribution of energy consumption by process.

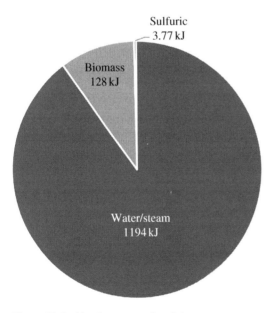

Figure 10.4 Heating energy breakdown.

Table 10.6 The statistical results of the energy consumption optimization problem under six different population sizes (20 independent trial runs for each size).

Population size	Min (kJ)	Mean (kJ)	Max (kJ)	Std. dev.
10	1964.3782	2101.1899	2242.9214	132.74
20	1949.9644	1989.6188	2094.5133	54.9090
30	1949.9644	1964.0309	1987.1477	13.3324
50	1949.3492	1952.6224	1965.2769	4.7360
100	1949.3472	1951.3232	1963.6914	4.4136
200	1949.3468	1949.6940	1950.9635	0.5325

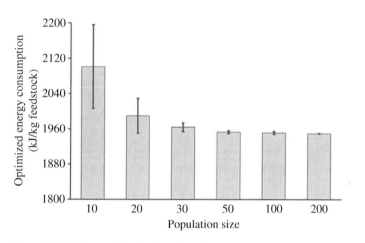

Figure 10.5 Averaged optimal results of energy consumption with 95% confidence interval. *Source:* [3]/With permission of Elsevier.

10.4.3 Energy Analysis of Optimal Results

By solving the proposed energy consumption optimization problem using PSO, the statistical results after 100 iterations under six population sizes are shown in Table 10.6.

Figure 10.5 shows the average values in 20 independent trail runs with the 95% confidence interval of the optimized energy consumption under six population sizes. It can be observed that as the population size increases, the optimal solution

Table 10.7 Comparison of the near-optimal solution and baseline case.

Decision variables	Baseline case	Near-optimal solution	Unit
Feedstock particle diameter, L_2	6.67	20.32	mm
Water loading (per kg feedstock), m_w	2.2	1.62	kg
Diluted acid loading, C_{acid}	1.0	2.49	w/w %
Pretreatment temperature, T_p	158	168.21	°C
Pretreatment time, t_p	5	4	min
Enzymatic hydrolysis temperature, T_h	48	44	°C
Enzymatic hydrolysis time, t_h	84	93.68	h
Fermentation temperature, T_f	32	28	°C
Fermentation time, t_f	36	42	h
Ethanol yield, ce_f	0.2411	0.2416	kg/kg feedstock
Total energy consumption, E_{total}	2470.57	1949.347	kJ/kg feedstock

gradually converges to 1950 kJ, and the 95% confidence interval is narrowed to (−0.38, 0.39). Therefore, the solution with the lowest energy consumption is selected as the near-optimal solution.

The result comparisons between the baseline case (Scenario I) and the optimal solution (Scenario II) are shown in Table 10.7. Using the cellulosic biofuel production design obtained from the optimization problem, the total energy consumption of a biofuel production system can be reduced by 21.09% compared with the baseline case while maintaining the same level of ethanol yield.

Problems

10.1 What are the advantages of cellulosic biofuels?

10.2 Demonstrate the major processes involved in cellulosic biofuel manufacturing.

10.3 Based on the energy consumption model of cellulosic biofuel production, which process has the largest impact on the total energy consumption?

10.4 In addition to PSO [22, 23], please use other intelligent optimization algorithm, e.g. Ant Colony Algorithm [24], Artificial Bee Colony algorithm [25, 26], beetle antennae search optimizer [27], cuckoo search [28–30], Dandelion Algorithm [31], Differential Evolution [32–34], gravitational search algorithm [35, 36], Jaya Algorithm [37], and Scatter Search [38–41] to solve the problem in (10.21)–(10.31).

References

1 Lynd, L.R. (2017). The grand challenge of cellulosic biofuels. *Nat. Biotechnol.* 35 (10): 912–915. https://doi.org/10.1038/nbt.3976.

2 Farrell, A.E. (2006). Ethanol can contribute to energy and environmental goals. *Science (80-.).* 311 (5760): 506–508. https://doi.org/10.1126/science.1121416.

3 Ge, Y. and Li, L. (2018). System-level energy consumption modeling and optimization for cellulosic biofuel production. *Appl. Energy* 226 (June): 935–946. https://doi.org/10.1016/j.apenergy.2018.06.020.

4 Sitaraman, H., Kuhn, E.M., Nag, A. et al. (2015). Multiphysics modeling and simulation of high-solids dilute-acid pretreatment of corn stover in a steam-explosion reactor. *Chem. Eng. J.* 268: 47–59. https://doi.org/10.1016/j.cej.2015.01.020.

5 Kadam, K.L., Rydholm, E.C., and McMillan, J.D. (2004). Development and validation of a kinetic model for enzymatic saccharification of lignocellulosic biomass. *Biotechnol. Prog.* 20 (3): 698–705. https://doi.org/10.1021/bp034316x.

6 Jeon, Y.J., Svenson, C.J., Joachimsthal, E.L., and Rogers, P.L. (2002). Kinetic analysis of ethanol production by an acetate-resistant strain of recombinant *Zymomonas mobilis. Biotechnol. Lett.* 24 (10): 819–824. https://doi.org/10.1023/A:1015546521000.

7 Humbird, D., Davis, R., Tao, L. et al. (2011). *Process Design and Economics for Conversion of Lignocellulosic Biomass to Ethanol*. Golden, CO: National Renewable Energy Lab. (NREL).

8 Tanaka, T. (1966). Comminution laws. Several probabilities. *Ind. Eng. Chem. Process Des. Dev.* 5 (4): 353–358. https://doi.org/10.1021/i260020a001.

9 Austin, L.G. and Klimpel, R.R. (1964). The theory of grinding operations. *Ind. Eng. Chem.* 56 (11): 18–29. https://doi.org/10.1021/ie50659a004.

10 Bitra, V.S.P., Womac, A.R., Igathinathane, C. et al. (2009). Direct measures of mechanical energy for knife mill size reduction of switchgrass, wheat straw, and corn stover. *Bioresour. Technol.* 100 (24): 6578–6585. https://doi.org/10.1016/j.biotech.2009.07.069.

11 Zheng, Y., Pan, Z., and Zhang, R. (2009). Overview of biomass pretreatment for cellulosic ethanol production. *Int. J. Agric. Biol. Eng.* 2 (3): 51–68. https://doi.org/10.3965/j.issn.1934-6344.2009.03.051-068.

12 Doble, M. (2007). *Perry's Chemical Engineers' Handbook*, 8e. New York: McGraw-Hill.

13 Mafe, O.A.T., Davies, S.M., Hancock, J., and Du, C. (2015). Development of an estimation model for the evaluation of the energy requirement of dilute acid pretreatments of biomass. *Biomass Bioenergy* 72: 28–38. https://doi.org/10.1016/j.biombioe.2014.11.024.

14 Li, L. and Ge, Y. (2017). System-level cost evaluation for economic viability of cellulosic biofuel manufacturing. *Appl. Energy* 203: 711–722. https://doi.org/10.1016/j.apenergy.2017.06.074.

15 Mellor, J.W. (1909). *Chemical Statics and Dynamics. Longmans.* Longmans, Green, and Company.

16 Williams, M.M.R. (1977). The mathematics of diffusion. *Ann. Nucl. Energy* 4 (4–5): 205–206. https://doi.org/10.1016/0306-4549(77)90072-X.

17 Zhang, Y.-H.P. and Lynd, L.R. (2004). Toward an aggregated understanding of enzymatic hydrolysis of cellulose: noncomplexed cellulase systems. *Biotechnol. Bioeng.* 88 (7): 797–824. https://doi.org/10.1002/bit.20282.

18 Leksawasdi, N., Joachimsthal, E.L., and Rogers, P.L. (2001). Mathematical modelling of ethanol production from glucose/xylose mixtures by recombinant *Zymomonas mobilis. Biotechnol. Lett.* 23 (13): 1087–1093. https://doi.org/10.1023/A:1010599530577.

19 Esteghlalian, A., Hashimoto, A.G., Fenske, J.J., and Penner, M.H. (1997). Modeling and optimization of the dilute-sulfuric-acid pretreatment of corn stover, poplar and switchgrass. *Bioresour. Technol.* 59 (2–3): 129–136. https://doi.org/10.1016/S0960-8524(97)81606-9.

20 Wolfrum, E.J. and Sluiter, A.D. (2009). Improved multivariate calibration models for corn stover feedstock and dilute-acid pretreated corn stover. *Cellulose* 16 (4): 567–576. https://doi.org/10.1007/s10570-009-9320-2.

21 Patzek, T.W. (2004). Thermodynamics of the corn-ethanol biofuel cycle. *CRC. Crit. Rev. Plant Sci.* 23 (6): 519–567. https://doi.org/10.1080/07352680490886905.

22 Yun, L., Ma, S., Li, L., and Liu, Y. (2022). CPS-enabled and knowledge-aided demand response strategy for sustainable manufacturing. *Adv. Eng. Informatics* 52: 101534. https://doi.org/10.1016/j.aei.2022.101534.

23 Yun, L., Li, L., and Ma, S. (2022). Demand response for manufacturing systems considering the implications of fast-charging battery powered material handling equipment. *Appl. Energy* 310: 118550. https://doi.org/10.1016/j.apenergy.2022.118550.

24 Feng, Y., Zhou, M.C., Tian, G. et al. (2019). Target disassembly sequencing and scheme evaluation for CNC machine tools using improved multiobjective ant colony algorithm and fuzzy integral. *IEEE Trans. Syst. Man, Cybern. Syst.* 49 (12): 2438–2451. https://doi.org/10.1109/TSMC.2018.2847448.

25 Tian, G., Ren, Y., Feng, Y. et al. (2019). Modeling and planning for dual-objective selective disassembly using and/or graph and discrete artificial bee colony. *IEEE Trans. Ind. Informatics* 15 (4): 2456–2468. https://doi.org/10.1109/TII.2018.2884845.

26 Jia, H., Miao, H., Tian, G. et al. (2019). Multiobjective bike repositioning in bike-sharing systems via a modified artificial bee colony algorithm. *IEEE Trans. Autom. Sci. Eng.* 17 (2): 909–920. https://doi.org/10.1109/TASE.2019.2950964.

27 Khan, A.H., Cao, X., Li, S. et al. (2020). BAS-ADAM: an ADAM based approach to improve the performance of beetle antennae search optimizer. *IEEE/CAA J. Autom. Sin.* 7 (2): 461–471. https://doi.org/10.1109/JAS.2020.1003048.

28 Zhao, J., Liu, S., Zhou, M. et al. (2018). Modified cuckoo search algorithm to solve economic power dispatch optimization problems. *IEEE/CAA J. Autom. Sin.* 5 (4): 794–806. https://doi.org/10.1109/JAS.2018.7511138.

29 Cao, Z., Lin, C., Zhou, M., and Huang, R. (2019). Scheduling semiconductor testing facility by using cuckoo search algorithm with reinforcement learning and surrogate modeling. *IEEE Trans. Autom. Sci. Eng.* 16 (2): 825–837. https://doi.org/10.1109/TASE.2018.2862380.

30 Cao, Z., Lin, C., and Zhou, M. (2021). A knowledge-based cuckoo search algorithm to schedule a flexible job shop with sequencing flexibility. *IEEE Trans. Autom. Sci. Eng.* 18 (1): 56–69. https://doi.org/10.1109/TASE.2019.2945717.

31 Zhu, H., Liu, G., Zhou, M. et al. (2019). Dandelion algorithm with probability-based mutation. *IEEE Access* 7: 97974–97985. https://doi.org/10.1109/ACCESS.2019.2927846.

32 Gao, S., Yu, Y., Wang, Y. et al. (2021). Chaotic local search-based differential evolution algorithms for optimization. *IEEE Trans. Syst. Man, Cybern. Syst.* 51 (6): 3954–3967. https://doi.org/10.1109/TSMC.2019.2956121.

33 Sun, J., Gao, S., Dai, H. et al. (2020). Bi-objective elite differential evolution algorithm for multivalued logic networks. *IEEE Trans. Cybern.* 50 (1): 233–246. https://doi.org/10.1109/TCYB.2018.2868493.

34 Tian, G., Ren, Y., and Zhou, M. (2016). Dual-objective scheduling of rescue vehicles to distinguish forest fires via differential evolution and particle swarm optimization combined algorithm. *IEEE Trans. Intell. Transp. Syst.* 17 (11): 3009–3021. https://doi.org/10.1109/TITS.2015.2505323.

35 Wang, Y., Gao, S., Zhou, M., and Yu, Y. (2021). A multi-layered gravitational search algorithm for function optimization and real-world problems. *IEEE/CAA J. Autom. Sin.* 8 (1): 94–109. https://doi.org/10.1109/JAS.2020.1003462.

36 Ji, J., Gao, S., Wang, S. et al. (2017). Self-adaptive gravitational search algorithm with a modified chaotic local search. *IEEE Access* 5: 17881–17895. https://doi.org/10.1109/ACCESS.2017.2748957.

37 Gao, K., Yang, F., Zhou, M. et al. (2019). Flexible job-shop rescheduling for new job insertion by using discrete Jaya Algorithm. *IEEE Trans. Cybern.* 49 (5): 1944–1955. https://doi.org/10.1109/TCYB.2018.2817240.

38 Guo, X., Zhou, M., Liu, S., and Qi, L. (2020). Lexicographic multiobjective scatter search for the optimization of sequence-dependent selective disassembly subject to multiresource constraints. *IEEE Trans. Cybern.* 50 (7): 3307–3317. https://doi.org/10.1109/TCYB.2019.2901834.

39 Guo, X., Liu, S., Zhou, M., and Tian, G. (2018). Dual-objective program and scatter search for the optimization of disassembly sequences subject to multiresource constraints. *IEEE Trans. Autom. Sci. Eng.* 15 (3): 1091–1103. https://doi.org/10.1109/TASE.2017.2731981.

40 Guo, X., Liu, S., Zhou, M., and Tian, G. (2016). Disassembly sequence optimization for large-scale products with multiresource constraints using scatter search and petri nets. *IEEE Trans. Cybern.* 46 (11): 2435–2446. https://doi.org/10.1109/TCYB.2015.2478486.

41 Tan, Y., Zhou, M., Zhang, Y. et al. (2020). Hybrid scatter search algorithm for optimal and energy-efficient steelmaking-continuous casting. *IEEE Trans. Autom. Sci. Eng.* 17 (4): 1814–1828. https://doi.org/10.1109/TASE.2020.2979079.

11

Energy-consumption Minimized Scheduling of Flexible Manufacturing Systems

In Chapter 3, Petri nets (PNs) have been introduced as a modeling tool for manufacturing systems. In other chapters, a production line consisting of multiple machines is mainly used. Motivated by the need to minimize the energy consumption of complex manufacturing systems, the flexible manufacturing systems (FMS) is taken as example in this chapter to illustrate how to achieve various production plans and goals. FMS exhibit the high degree of resource sharing and product route flexibility, which face deadlock issues that must be well resolved. In this chapter, the dynamic program of FMS is formulated to find a deadlock-free schedule that aims to minimize energy consumption while meeting the productivity requirements. For small-size scheduling problems, their optimal solutions can be obtained by the conventional dynamic programming (DP) algorithm. However, the conventional DP fails to solve large-scale problems due to exponentially increased number of explored states. Therefore, a modified DP (MDP) algorithm is presented to obtain an optimal or suboptimal schedule in an acceptable time. MDP adopts two novel ways in which only the most promising states are explored. One is to keep only one transition sequence for each state through an evaluation function. Another is to select the most promising states in each stage for further exploration through a heuristic function. To guarantee that the generated states are safe, a deadlock controller is applied in MDP. Experimental results on manufacturing systems and comparisons with existing methods are provided to show its effectiveness. Specifically, Section 11.1 gives an introduction and reviews some existing and related work. Section 11.2 introduces the formulation of a place-timed PN model for FMS. Section 11.3 presents two different resource status modes and their corresponding energy consumption functions. Section 11.4 shows how to perform DP for FMS scheduling. Section 11.5 introduces MDP. Section 11.6 adopts several examples to show the effectiveness of MDP. Section 11.7 summarizes this chapter.

Sustainable Manufacturing Systems, First Edition. Lin Li and MengChu Zhou.
© 2023 The Institute of Electrical and Electronics Engineers, Inc.
Published 2023 by John Wiley & Sons, Inc.

11.1 Introduction

Energy consumption becomes a major issue in our society nowadays as stated in the previous chapters. Because of the increasing energy prices and environmental pressure, reducing energy consumption has become a critical issue for manufacturing industry and has attracted increasing attention in recent years [1].

Most energy consumption optimization problems concentrate on single machines [2–6], job shops [7–13], and flow shop production systems [14–18]. In various shop floor environments and single machine systems studied in these references, it is often assumed that unlimited buffers exist between resources. In such systems, every job can leave its resource immediately after its processing is finished [19]; hence, deadlock will never occur. The deadlock refers to the situation where the system or a part remains indefinitely blocked and cannot terminate its intended task. However, in practical situations, the number of buffers, machines, robots, and other tools are usually limited as illustrated in the previous chapters. Deadlock could occur without proper control when various parts enter the system and compete for limited resources.

This chapter focuses on the scheduling problem of minimizing the total energy consumption of all production processes in deadlock-prone FMS. Two energy consumption functions, which are based on two resource status modes, are considered in this chapter. In the first resource status mode, it is assumed that a resource has two possible states, i.e. busy or idle. While in the second resource status mode, resources can be in one of three states, i.e. working, occupied, or idle. In addition, the reachability graph (RG) of a place-timed PN model is constructed by taking all the reachable states as its nodes. DP is performed on RG to guarantee that only the optimal transition sequence is kept for each node. Based on the place-timed PN scheduling model, the Bellman equation is established to optimize the total energy consumption of FMS.

Since the number of generated states increases exponentially with the FMS size, it is computationally infeasible to perform DP on large scheduling problems. To obtain an optimal or suboptimal solution in an acceptable time, DP is revised and an MDP is introduced. In MDP, fewer states are generated and evaluated through two functions: (i) the transition sequence evaluation function, aiming at keeping one transition sequence for each marking, and (ii) the heuristic function to select the most "promising" states in each stage. Correspondingly, only a part of RG is generated and searched. To guarantee that the generated states in the searching process are safe, deadlock controllers are used separately [20]. Experimental tests on FMSs and comparison with the existing work [21] are conducted to show the effectiveness of this scheduling algorithm.

11.2 Construction of Place-timed PN for FMS Scheduling

A PN is an efficient mathematical tool that can model, simulate, and analyze various kinds of systems, such as the stamping system [22], semiconductor manufacturing systems [23], [24], and crude oil operations [25]. This section introduces how to build up a scheduling model for FMS.

Recall from Chapter 3 that a PN $Z = (P, T, I, O, M)$:

1) P is a set of places.
2) T is a set of transitions.
3) I is a $|P| \times |T|$ matrix, called an input function, which represents the arcs connecting places to transitions. Its value at the ith row and jth column, denoted as $I(p_i, t_j)$, represents the number of arcs directly connecting place $p_i \in P$ to transition $t_j \in T$. If there is no direct connection from place p_i to transition t_j, then $I(p_i, t_j) = 0$. Otherwise, if $I(p_i, t_j) > 0$, p_i is called an input place of t_j.
4) O is a $|P| \times |T|$ matrix, called an output function, which represents the arcs connecting transitions to places. Similarly, the value of $O(p_i, t_j)$ represents the number of arcs that directly connect transition t_j to place p_i. If $O(p_i, t_j) > 0$, p_i is called an output place of t_j.
5) M is a vector of non-negative integers and referred to as a marking, where M_0 represents the initial marking.

11.2.1 Basic Definitions of PN

For $x \in P \cup T$, its pre-set is defined as

$$^{\bullet}x = \{y \in P \cup T \mid I(y,x) > 0 \text{ or } O(x,y) > 0\} \tag{11.1}$$

The post-set of x is defined as

$$x^{\bullet} = \{y \in P \cup T \mid I(x, y) > 0 \text{ or } O(y, x) > 0\} \tag{11.2}$$

Let $N^+ = \{0, 1, 2, 3, ...\}$ and $N^k = \{1, 2, 3, ..., k\}$ for positive integer k. Given a place $p \in P$ and a marking M, $M(p)$ denotes the number of tokens in p under M, and p is called "marked" if $M(p) \neq 0$. Let $S \subseteq P$ be a set of places, the sum of tokens in all places of S under M is denoted by $M(S)$, i.e. $M(S) = \sum_{p \in S} M(p)$.

A transition $t \in T$ is enabled at M, denoted by $M[t>$, if $\forall\, p \in {}^{\bullet}t$, $M(p) > 0$. An enabled transition t at M can fire, resulting in a new marking M', denoted by $M[t > M'$, where

$$M'(p) = M(p) - I(p, t) + O(p, t), \quad \forall p \in P \tag{11.3}$$

A sequence of transitions $\beta = t_1, t_2, ..., t_k$ is feasible from M if $M_i[t_i > M_i + 1, i \in N_k$, where $M_1 = M$.

11.2.2 Place-timed PN Scheduling Models of FMS

An FMS consists of m types of resources and is able to process n types of parts. The set of resource types is denoted as $R = \{r_i, i \in N_m\}$ and the capacity of type r_i resource, denoted as $\chi(r_i)$, is the maximum number of parts that resource r_i can simultaneously manage. The set of part types is denoted as $Q = \{q_i, i \in N_m\}$. The number of type q_i parts to be processed is u_i. Then the total number of parts to be processed is $u = \sum_{i \in N_n} u_i$, and all parts can be indexed as integers from 1 to u. For example, parts 1 to u_1 are of type q_1, parts $u_1 + 1$ to u_2 are of type q_2. A processing route of a part is an ordered sequence of operations to be processed on resources, and each operation requires only one predetermined resource for processing. Each part may have more than one route to process, and the same type of parts has the same processing routes.

This FMS model is called a place-timed PNs for scheduling (PNS) [26], in which operations and resources are modeled by operation and resource places, respectively. Let $\omega_i = o_{i1}o_{i2}... o_{il}$ denote a processing route of a type q_k part, where o_{ij} is the jth operation in ω_i and l is the length of ω_i. In the PNS model, ω_i is modeled by a path of transitions and operation places, which is called an operation path and is formulated as

$$\rho(\omega_i) = p_{k0}t_{i1}p_{i1}t_{i2}... p_{il}t_{i(l + 1)}p_{kf} \tag{11.4}$$

where p_{ij} is an operation place representing operation o_{ij}; places p_{k0} and p_{kf} are used for storing raw and finished type q_k parts, respectively; t_{ij} is a transition, whose firing implies the start of operation o_{ij}. If the operation represented by place $p \in P$ is processed on resource r, denoted by $R(p) = r$, then there are arcs from r to each input transition of p and from each output transition of p to r, representing the request and release of r, respectively. Then the PNS model can be formulated as

$$(N, M_0) = (P \cup P_0 \cup P_f \cup P_R, T, F, M_0) \tag{11.5}$$

where P is the set of operation places, $P_0 = \{p_{i0}|i \in N_n\}$ and $P_f = \{p_{if}| i \in N_n\}$. P_R is the set of all resource places. The initial marking M_0 is defined as

$$M_0(p_{is}) = u_i, \quad \forall p_{is} \in P_0$$
$$M_0(p) = 0, \quad \forall p \in P \cup P_f \tag{11.6}$$
$$M_0(r) = \chi(r), \quad \forall r \in P_R$$

Let $d(o_{ij})$ be the time needed for processing operation o_{ij}. Note that o_{ij} is modeled by operation place p_{ij}, then there is a delay of at least $d(o_{ij})$ associated with p_{ij}. Let $d(p_{ij})$ denote this delay, and $d(p_{ij}) = d(o_{ij})$. After a token arrives at p_{ij}, it needs $d(p_{ij})$ to be able to enable p_{ij}'s output transitions. Clearly, $d(p) = 0, \forall p \in P_0 \cup P_f \cup P_R$. When operations of all parts are processed, PNS reaches its final marking, denoted as M_f, where

$$M_f(p) = 0, \qquad \forall p \in P \cup P_0$$

$$M_f(p_{if}) = M_0(p_{i0}), \quad \forall p_{if} \in P_f \qquad\qquad (11.7)$$

$$M_f(r) = \chi(r), \qquad \forall r \in P_R$$

A feasible sequence of transitions α is complete if $M_0[\alpha > M_f$. Then the scheduling problem of FMS is to find a complete sequence of transitions such that a predefined scheduling objective function is optimized.

Example 11.1 Consider an FMS that has three machines $\{m_1, m_2, m_3\}$ and two robots $\{r_1, r_2\}$ and can process two types of parts q_1 and q_2. Each machine can process two parts concurrently, while each robot can handle one part at a time, i.e. $\chi(m_i) = 2, i \in \{1, 2, 3\}$ and $\chi(r_j) = 1, j \in \{1, 2\}$. The numbers of type q_1 and q_2 raw parts to be processed, denoted as u_1 and u_2, respectively, are two, i.e. $u_1 = u_2 = 2$. Suppose the FMS has the following two assumptions:

1) Part q_1 requires robot r_1 to upload (five seconds). Its first operation can be performed by machine m_1 (30 seconds) or m_3 (20 seconds), while second operation can be done by m_2 only (44 seconds). The processed part is unloaded by robot r_1 (five seconds).
2) Part q_2 requires robot r_2 to upload (five seconds). Once the operation is performed by machine m_3 (38 seconds), it is unloaded by robot r_1 (five seconds).

Please formulate the PNS for this system.

Solution

The PNS can be formulated following two steps. The first step focuses on the route of all operations for each part type; while the second one adds the resources, initial marking, and time requirements of the model.

Step 1:

For part q_1, first, let p_{10} denote the available raw part, and let p_{11} denote the available parts for robot r_1 to load. Then, design two alternative operational routes, i.e. (i) p_{12} and p_{13} represent the first route in which q_1's operations are performed by machine m_1 and then m_2; and (ii) p_{22} and p_{23} represent the second route in which q_1's operations are performed by machine m_3 and then m_2. Next, let P_{14} indicate the finished part that is ready to be unloaded by robot r_2, and let p_{1f} indicate the finished part is in storage, respectively. The transitions among all the places, which represent start and/or end of their corresponding operations, are denoted by $t_{11}, t_{12}, t_{13}, t_{14}, t_{15}, t_{22}, t_{23}, t_{24}$, and t_{15}. Follow the same strategy, the

(a)   (b)

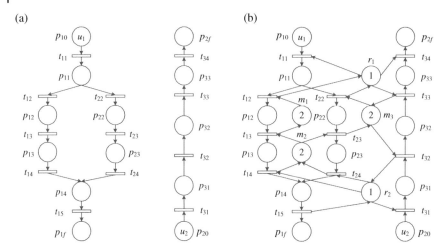

Figure 11.1 The PNS model of an FMS in Example 11.1: (a) PNS formulation step one, and (b) PNS formulation step two.

PN can be designed for part q_2. The PNs, as the results of first step PNS formulation, are represented in Figure 11.1a.

Step 2:

Let m_1, m_2, m_3, r_1, r_2 denote five resource places for all resources (three machines and two robots). According to whether a particular operation requires a part or resource, directed arcs are added between resource places and transitions. For example, operation in p_{11} requires robot r_1, and thus a directed arc from r_1 to t_{11} is added. Next, mark the places with a proper number of tokens as the initial marking for this system with the following relationships.

$$u_1 = u_2 = 2 \rightarrow M_0(p_{10}) = M_0(p_{20}) = 2.$$
$$\chi(mi) = 2 \rightarrow M_0(mi) = 2$$
$$\chi(rj) = 1 \rightarrow M_0(rj) = 1$$

Finally, note that the processing time of operations in $p_{11}, p_{12}, p_{13}, p_{14}, p_{22}, p_{23},$ $p_{31}, p_{32},$ and p_{33} is associated with time delay of 5, 30, 44, 5, 20, 34, 5, 38, and 5, respectively. The final PNS is shown in Figure 11.1b.

11.3 Energy Consumption Functions

Two energy consumption functions are considered based on two different resource status modes. Since the resource state transformation is related to the firing time of transitions, before introducing the energy consumption functions, a method to calculate the earliest firing time of transitions in a feasible transition sequence is first discussed. In addition, all parts entering an operation place for processing follow the first-in-first-out rule in a scheduling process.

11.3.1 Calculating the Earliest Firing Time of Transitions

Assume $\pi_n = t_1 t_2 ... t_n$ is a feasible transition sequence starting from M_0. Let $t_k[o_{ij}]$ denote the firing of t_k, which starts operation o_{ij}. Let $f(t_k[o_{ij}])$ be the earliest firing time of t_k in π_n, which is also the start time of operation o_{ij}. Considering the order of transition firings in π_n, $t_k[o_{ij}]$ can fire only after $t_{k-1}[o_{uv}]$ fires by assuming that the firing of t_{k-1} corresponds to the start of operation o_{uv}, i.e. $f(t_k[o_{ij}]) \geq f(t_{k-1}[o_{uv}])$. Besides, $t_k[o_{ij}]$ can be fired only after the preceding operation of o_{ij}, say $o_{i(j-1)}$, is finished. Suppose that the firing of t_s corresponds to the start of operation $o_{i(j-1)}$. Then, $f(t_k[o_{ij}])$ can be calculated by

$$f\left(t_k\left[o_{ij}\right]\right) = \max\left\{f\left(t_s\left[o_{i(j-1)}\right]\right) + d\left(o_{i(j-1)}\right), f\left(t_{k-1}[o_{uv}]\right)\right\} \tag{11.8}$$

Denote the ith type q_1 part entering the system as part i, the kth type q_2 part entering the system as part $u_1 + k$, etc. For a type q_1 part, assume that its jth operation, denoted as o_{1j}, has started under π_n (the start of o_{1j} corresponds to the firing of t_{1j}. a_{ij} is set as t_{1j}, i.e. $a_{ij} = t_{1j}$. Then $A_i(\pi_n) = a_{i1}, a_{i2}, ..., a_{ij}$ is the sequence of fired transitions on the operation path of part i under π_n. Let $B_i(\pi_n) = b_{i1} b_{i2} ... b_{ij}$ be the sequence of the earliest firing time of transitions in $A_i(\pi_n)$. Assume that a_{ij} corresponds to the lth transition of π_n, $b_{ij} = f(t_l[o_{1j}])$. Let $\varpi(\pi_n) = (\varpi_1, \varpi_2, ..., \varpi_u)$ be the route vector, where ϖ_i records the processing route of part i.

Assume that the number of processing routes of type q_i parts is s_i. The processing routes can be denoted as $\omega_1 - \omega_{s1}$ for type q_1 parts, $\omega_{s1+1} - \omega_{s1+s2}$ for type q_2 parts, etc. For a type q_1 part, part j, if the transitions in $A_j(\pi_n)$ are only on $\rho(\omega_{s1})$, then $\varpi_j = \omega_{s1}$. Before any transition fires, the route vector is initialized as

$$\begin{aligned}\varpi(\varepsilon) &= (\varpi_1, \varpi_2, ..., \varpi_u) \\ &= (\omega_1, ..., \omega_1, \omega_{s1+1}, ..., \omega_{s1+1}, \omega_{s1+s2+1}, ..., \omega_{s1+s2+1}, ...)\end{aligned} \tag{11.9}$$

where ε is an empty string, and $\varpi_i = \omega_1, i \in \{1, 2, ..., u_1\}$, $\varpi_i = \omega_{s1+1}, i \in \{u_1 + 1, u_1 + 2, ..., u_1 + u_2\}$, etc.

Example 11.2 Find the fired transitions, the earliest firing time of transitions, and the process route for Example 11.1.

Solution

In Example 11.1, there are two parts of type q_1 (i.e. parts 1 and 2) and two parts of type q_2 (i.e. parts 3 and 4) to be processed. There are two processing routes for type

q_1 parts, denoted as ϖ_1 and ϖ_2, and one processing route for type q_2 parts, denoted as ϖ_3. The initial route vector is

$$\varpi(\varepsilon) = (\varpi_1, \varpi_2, \varpi_3, \varpi_4) = (\omega_1, \omega_1, \omega_3, \omega_3)$$

In PNS, the numbers of transitions on the operation path of type q_1 and q_2 parts are 5 and 4, respectively. Define π_{11} as

$$\pi_{11} = t_{31}t_{11}t_{32}t_{31}t_{22}t_{11}t_{23}t_{32}t_{33}t_{34}t_{24}$$

It can be checked that it is feasible from M_0. For π_{11}, the fired transitions and the earliest firing time of transitions can be represented as

$$A_1(\pi_{11}) = a_{11}, a_{12}, a_{13}, a_{14} = t_{11}, t_{22}, t_{23}, t_{24}$$
$$A_2(\pi_{11}) = a_{21} = t_{11}$$
$$A_3(\pi_{11}) = a_{31}, a_{32}, a_{33}, a_{34} = t_{31}, t_{32}, t_{33}, t_{34}$$
$$A_4(\pi_{11}) = a_{41}, a_{42} = t_{31}, t_{32}$$

$$B_1(\pi_{11}) = 0, 5, 25, 59$$

$$B_2(\pi_{11}) = 5$$

$$B_3(\pi_{11}) = 0, 5, 43, 48$$

$$B_4(\pi_{11}) = 5, 10$$

The transitions on $\rho(\omega_1)$ and $\rho(\omega_2)$ are $t_{11}, t_{12}, t_{13}, t_{14}, t_{15}$ and $t_{11}, t_{22}, t_{23}, t_{24}, t_{15}$, respectively. For part 1 with $A_1(\pi_{11})$, its processing route is ω_2. While for part 2, $A_2(\pi_{11}) = t_{11}$, its processing route is not known and $\varpi_2 = \omega_1$. Clearly, the processing routes of parts 3 and 4 are ω_3. The route vector is $\varpi(\pi_{11}) = (\omega_2, \omega_1, \omega_3, \omega_3)$.

Let $\pi_{12} = \pi_{11}t_{22}$, where t_{22} is feasible under π_{11}. Then there exist

$$A_1(\pi_{12}) = t_{11}, t_{22}, t_{23}, t_{24}$$
$$A_2(\pi_{12}) = t_{11}, t_{22}$$
$$A_3(\pi_{12}) = t_{31}, t_{32}, t_{33}, t_{34}$$
$$A_4(\pi_{12}) = t_{31}, t_{32}$$

The processing route of part 2 can be determined and is ω_2. The route vector becomes $\varpi(\pi_{12}) = (\omega_2, \omega_2, \omega_3, \omega_3)$.

11.3.2 Two Energy Consumption Functions

Energy consumption in FMS is decided by the states of all equipment or resources during the whole manufacturing process. In this chapter, two resource status modes are considered. In the first mode, a resource has two possible states, busy and idle. Specifically, a resource is in busy state when it processes a part; otherwise, it is idle. In the second mode, resources can be in one of the three states: working, occupied but not working or occupied for short, and idle. Specifically, a resource is processing a part, it is in working state; while it finishes the processing of a part, but the part on it has to wait for its next resource for processing, the resource is in an occupied state. A resource is idle when there is no part on it.

Note that in PNS, a token in a resource place represents a unit of idle resources. A token in operation place p represents that a part is on a unit of resource $R(p)$. In the first mode, such a resource is in a busy state; while in the second mode, it is in either working or occupied state. In the second mode, if a token enters operation place p at time t, then the resource is in a working state before time $t + d(p)$; it is in an occupied state from time $t + d(p)$ until the token leaves p.

When a resource is in different states, the corresponding energy consumption rate (ECR), i.e. energy consumption per unit time, is different. For resource r, let $c(r)$ denote its ECR in the idle state. For a resource in p, in the first mode, its ECR is denoted as $c_b(p)$. In the second mode, the ECRs are denoted as $c_w(p)$ and $c_a(p)$ for its working and occupied states, respectively. In this chapter, it is assumed that $c_w(p) = c_b(p)$.

Let $\pi_n = t_1 t_2 ... t_n$ be a feasible sequence of transitions and $M_0[\pi_n > M_n$. Let π_l be a prefix with l transitions of π_n, i.e. $\pi_l = t_1 t_2 ... t_l$, $M_0[\pi_l > M_l$, and $l \leq n$. For simplicity, let τ_l denote the earliest firing time of transition t_l in π_n. Suppose that the firing of t_l corresponds to the start of operation o_{xy}, then $\tau_l = f(t_l[o_{xy}])$. Two energy consumption functions of the system under π_n are defined as follows.

11.3.2.1 Energy Consumption Function E_1

Let $E_1(\pi_n)$ denote the energy consumption of the system from M_0 to M_n, i.e. from the beginning to moment τ_n, in the first mode. Then it can be calculated by the following recursive formula.

$$E_1(\pi_1) = 0 \tag{11.10}$$

$$E_1(\pi_l) = E_1(\pi_{l-1}) + B(\pi_l) + I(\pi_l), \quad 1 < l \leq n \tag{11.11}$$

where $B(\pi_l) = \sum_{p \in P}(\tau_l - \tau_{l-1})c_b(p)M_{l-1}(p)$ and $I(\pi_l) = \sum_{r \in PR}(\tau_l - \tau_{l-1})c(r)$ $M_{l-1}(r)$ are the total energy consumptions of resources in busy and idle states in time interval $[\tau_{l-1}, \tau_l]$, respectively.

11.3.2.2 Energy Consumption Function E_2

Let $E_2(\pi_n)$ denote the energy consumption of the system from M_0 to M_n, i.e. from the beginning to moment τ_n, in the second mode, which can be calculated by the following recursive formula.

$$E_2(\pi_1) = 0 \tag{11.12}$$

$$E_2(\pi_l) = E_2(\pi_{l-1}) + E_p(\pi_l) + I(\pi_l), \quad 1 < l \le n \tag{11.13}$$

where $E_p(\pi_l)$ is the total energy consumption of resources that have parts on them, i.e. resources in operation places during $[\tau_{l-1}, \tau_l]$. In this period, resources can be in an occupied, working then occupied, or working states. $I(\pi_l)$ is the total energy consumption of idle resources during $[\tau_{l-1}, \tau_l]$, and its expression is the same as the one in $E_1(\pi_l)$.

For a resource $R(p)$ in operation place p, let $E(p, \tau_{l-1}, \tau_l)$ denote its energy consumption during $[\tau_{l-1}, \tau_l]$. To calculate $E(p, \tau_{l-1}, \tau_l)$, the resource state during $[\tau_{l-1}, \tau_l]$ should be considered in detail. Assume that part i is on $R(p)$ during $[\tau_{l-1}, \tau_l]$, and let $h(p, i)$ denote its processing completion time in p. In time interval $[\tau_{l-1}, \tau_l]$, the state of $R(p)$ could be in one of the following three cases depending on the relationship among $h(p, i)$, τ_{l-1}, and τ_l (Figure 11.2).

Let λ be a binary indicator defined as

$$\lambda(x) = \begin{cases} 1, & x > 0 \\ 0, & \text{otherwise} \end{cases} \tag{11.14}$$

Then $E(p, \tau_{l-1}, \tau_l)$ can be expressed as

$$\begin{aligned}
E(p, \tau_{l-1}, \tau_l) = & (\tau_l - h(p,i))c_b(p)\lambda(h(p,i) - \tau_l) \\
& + (h(p,i) - \tau_{l-1})c_b(p)\lambda(h(p,i) - \tau_{l-1}) \\
& + (\tau_l - h(p,i))c_a(p)\lambda(\tau_l - h(p,i)) \\
& + (h(p,i) - \tau_{l-1})c_a(p)\lambda(\tau_{l-1} - h(p,i))
\end{aligned} \tag{11.15}$$

Before the expression of $E_p(\pi_l)$ is given, let $\Theta(\pi_{l-1}, i)$ represent a marked place holding part i under π_{l-1}. $\Theta(\pi_{l-1}, i)$ can be part i in p_{k0}, or p_{kf}, or with resource R (p) in operation place p. Let a_{iy} denote the last element in $A_i(\pi_{l-1})$. Then part i is in place $(a_{iy})^{\bullet}$ under π_{l-1} and $\Theta(\pi_{l-1}, i) = (a_{iy})^{\bullet}$. To calculate $E_p(\pi_l)$, all the marked operation places under π_{l-1}, i.e. the resources on parts, should be considered. Hence, $E_p(\pi_l)$ can be expressed as

$$E_p(\pi_l) = \sum_{i \in Z_u, \, p = \Theta(\pi_{l-1}, i) \in P} E(p, \tau_{l-1}, \tau_l) \tag{11.16}$$

For $\Theta(\pi_{l-1}, i) = (a_{iy})^{\bullet} \in P$, the processing of part i in $p = (a_{iy})^{\bullet}$ starts at b_{iy}, i.e. $R(p)$ on part i starts its working state at b_{iy}. The completion time of the processing is $b_{iy} + d(p)$. Hence, $h(p, i) = b_{iy} + d(p)$, $i \in Z_u$, $p \in P$.

Example 11.3 Consider the transition sequence in Example 11.1. Find the processing completion time for two parts.

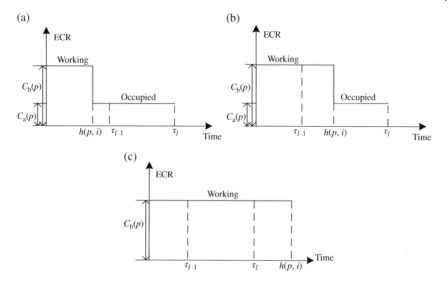

Figure 11.2 (a) $R(p)$ is in occupied state during $[\tau_{l-1}, \tau_l]$; (b) $R(p)$ is in working state during $[\tau_{l-1}, h(p, i)]$ and occupied state in $[h(p, i), \tau_l]$; and (c) $R(p)$ is in working state during $[\tau_{l-1}, \tau_l]$.

Solution

In the transition sequence in Example 11.1:

$$A_1(\pi_{11}) = a_{11}, a_{12}, a_{13}, a_{14} = t_{11}, t_{22}, t_{23}, t_{24}$$

Therefore,

$$\Theta(\pi_{11}, 1) = (a_{14})^{\bullet} = (t_{24})^{\bullet} = p_{14}$$
$$\Theta(\pi_{11}, 2) = p_{11}$$
$$\Theta(\pi_{11}, 3) = p_{2f}$$
$$\Theta(\pi_{11}, 4) = p_{32}$$

For part 1, the processing in p_{14} can be finished at

$$h(p_{14}, 1) = b_{14} + d(p_{14}) = 59 + 5 = 64$$

For part 2, the processing in p_{11} can be finished at

$$h(p_{11}, 2) = 10$$

For part 4, the processing in p_{32} can be finished at

$$h(p_{32}, 4) = 48$$

11.4 Dynamic Programming for Scheduling FMS

DP is a powerful mathematical optimization method for making a sequence of interrelated decisions. The term DP is formalized by Bellman and is used to solve the optimization problem of a multistage decision process [27]. In this section, DP for optimizing the total energy consumption in FMS scheduling is presented.

11.4.1 Formulation of DP for FMSs

The optimization problem aims to find a solution, i.e. a transition sequence that can lead the system PNS from initial marking M_0 to its final marking M_f, to minimize energy consumption. To formulate DP for FMS scheduling, the following definitions and notations are introduced first.

11.4.1.1 States and Stages

Given PNS (N, M_0), let π_l be a feasible transition sequence satisfying $M_0[\pi_l > M$. For part k, let j_k be the start processing time of its latest operation (being processed or completed) under π_l. Let $J(\pi_l) = (j_1, j_2, ..., j_u)$ refer to the start processing time vector of π_l. Then $j_1 = b_1 y_1$, $j_2 = b_2 y_2$, ..., $j_u = b_u y_u$, where y_i is the number of elements in $B_i(\pi_l)$, $i \in Z_u$. For π_0 (empty string), $J(\pi_0) = (0, 0, ..., 0)$. In the DP algorithm, state S is formed as

$$S = (M, J(\pi_l), \varpi(\pi_l), \pi_l) \tag{11.17}$$

where S is the system state after the firing of the last transition of π_l. The initial state is

$$S_0 = (M_0, J(\pi_0), \varpi(\pi_0), \pi_0).$$

Given two states $S_1 = (M_1, J(\beta_1), \varpi(\beta_1), \beta_1)$ and $S_2 = (M_2, J(\beta_2), \varpi(\beta_2), \beta_2)$, they are the same if $(M_1, J(\beta_1), \varpi(\beta_1)) = (M_2, J(\beta_2), \varpi(\beta_2))$. This means that two different transition sequences may lead to the same marking, start processing time vector, and route vector.

According to the length of the firing transition sequences in states, the state space can be divided into stages. A state $S = (M, J(\pi), \varpi(\pi), \pi)$ is in stage k if the length of π is k.

11.4.1.2 State Transition Equation

To avoid deadlock situations in FMS scheduling, deadlock controllers proposed in [20, 28], and [29] are used separately in this chapter. For more detailed descriptions, readers can refer to these publications. Note that with these controllers, PNS and its live-controlled PNS (denoted as CPNS) have the same transition set, and the scheduling algorithm discussed here is based on such a live CPNS.

Let $S_l = (M_l, J(\pi_l), \varpi(\pi_l), \pi_l)$ be a state in stage l and $\Xi(M_l)$ be the set of all transitions enabled at M_l. The $\Xi(M_l)$ can be expressed as $\Xi(S_l)$ if M_l is the marking of S_l. Then the state transition equation can be stated as follows:

$$S_{l+1} = \phi(S_l, t), \quad t \in \Xi(S_l) \tag{11.18}$$

where $S_{l+1} = (M_{l+1}, J(\pi_{l+1}), \varpi(\pi_{l+1}), \pi_{l+1})$, $\pi_{l+1} = \pi_l t$, and $M_l[t > M_l + 1$. Once π_{l+1} is obtained, $J(\pi_{l+1})$ and $\varpi(\pi_{l+1})$ can be determined.

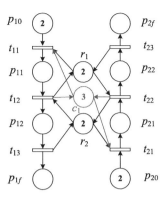

Figure 11.3 The controlled PNS.

11.4.1.3 Bellman Equation

In FMSs, a state could be reached through different transition sequences, and may lead to different energy consumption. For the multiple transition sequences to the same state, according to the Bellman Optimality Principle, only the one with the smallest energy consumption value should be kept.

Let $E^*(S_l)$ denote the optimal energy consumption from S_0 to S_l. The forward recursive formula for the minimal energy consumption can be calculated as

$$E^*(S_{l+1}) = \min \{E^*(S_l) + E(S_l, t) \mid t \in \Xi(S_l)\} \tag{11.19}$$

where $E(S_l, t)$ is the energy consumed in the period from state S_l to S_{l+1}.

11.4.2 Reachability Graph of PNS

For an FMS, its legal (or liveness) behavior can be completely tracked by the reachability graph (RG) of its CPNS. RG takes all reachable states of CPNS as its nodes or vertices. A directed arc from node S_1 to S_2 labeled with transition t if the firing of t leads to S_2 from S_1. For $S_0 = (M, J(\pi), \varpi(\pi), \pi)$ in RG, its energy consumption is $E^*(S)$, i.e. at present, the best energy consumption from S_0 to S is $E^*(S) = E(\pi)$. If another transition sequence α with energy consumption $E(\alpha) < E^*(S)$ is generated for S, then π is pruned and α is added for S and $E^*(S) = E(\alpha)$. Otherwise, α is not added for S and $E^*(S) = E(\pi)$. An example is given to show how DP is performed on RG to guarantee that only the optimal transition path of a node is kept.

Example 11.4 Consider the PNS model in Figure 11.3, where the optimal deadlock controller with control place c_1 is added according to the deadlock control policy in [29]. The processing time of operations in p_{11}, p_{12}, p_{21}, and p_{22} is 30, 10, 20, and 10, respectively. The ECRs of resources in their busy state are $c_b(p_{11}) = c_b(p_{22}) = 1$ and $c_b(p_{12}) = c_b(p_{21}) = 2$, respectively. While the ECRs of r_1 and r_2 in their idle state are 0.1 and 0.2, respectively.

Let $S_{i,j}$ denote the jth node in stage i. The enabled transitions under M_0 are t_{11} and t_{21}. Let t_{11} and t_{21} fire, resulting in new nodes $S_{1,1}$ and $S_{1,2}$ with markings

$M_{1,1} = p_{1s} + 2p_{2s} + p_{11} + r_1 + 2r_2 + 2c_1$ and $M_{1,2} = 2p_{1s} + p_{2s} + p_{21} + 2r_1 + r_2 + 2c_1$, respectively. Note that the above marking expression is often used to save the space. For example, given $P = \{p_1, p_2, ..., p_5\}$, $M = 2p_2 + 3p_5$ can be represented by $M = (0\ 2\ 0\ 0\ 5)^r$. Repeat this process for the newly generated nodes, the whole RG can be constructed. Partial RG for the model is shown in Figure 11.4.

For the first generated node $S_{2,3}$ with marking $M_{2,3} = p_{1s} + p_{2s} + p_{11} + p_{21} + r_1 + r_2 + c_1$ and transition sequence $\pi_{2,3} = t_{11}t_{21}$, its energy consumption is $E_1^*(S_{2,3}) = 0$. $\alpha_{2,3} = t_{21}t_{11}$ is a different transition sequence resulting in $S_{2,3}$ and $E_1(S_{2,3}) = 0$. Since $E_1(\alpha_{2,3}) = E_1^*(S_{2,3})$, the arc from $S_{1,2}$ to $S_{2,3}$ labeled with t_{11} (marked by the dashed line in Figure 11.4) is not included in RG. Similarly, in stage 3, $E_1(\alpha_{3,5}) = E_1^*(S_{3,5}) = 0$ for $\pi_{3,5} = t_{11}t_{11}t_{21}$ and $\alpha_{3,5} = t_{11}t_{21}t_{11}$, and $E_1(\alpha_{3,8}) = E_1^*(S_{3,8}) = 0$ for $\pi_{3,8} = t_{11}t_{21}t_{21}$ and $\alpha_{3,8} = t_{21}t_{21}t_{11}$.

The arc from $S_{2,3}$ to $S_{3,5}$ marked by t_{11} and from $S_{2,4}$ to $S_{3,8}$ labeled with t_{11} are not included on RG. For node $S_{3,6}$ with $\pi_{3,6} = t_{11}t_{21}t_{12}$, $\tau_1 = 0$, $\tau_2 = 0$, and $\tau_3 = 30$. Its energy consumption can be calculated as

$$
\begin{aligned}
E_1^*(S_{3,6}) &= E_1^*(S_{2,3}) + B(\pi_{2,3}) + I(\pi_{2,3}) \\
&= 0 + \{(\tau_3 - \tau_2)[M_{2,3}(p_{11})c_b(p_{11}) + M_{2,3}(p_{21})c_b(p_{21})]\} \\
&\quad + \{(\tau_3 - \tau_2)[M_{2,3}(r_1)c(r_1) + M_{2,3}(r_2)c(r_2)]\} \\
&= 99
\end{aligned}
$$

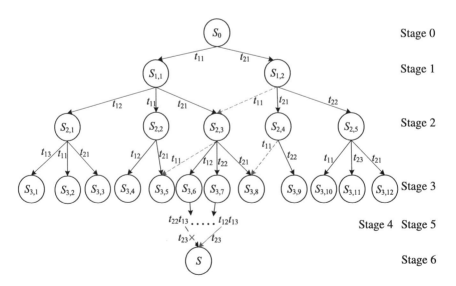

Figure 11.4 Partial RG of the model in Figure 11.3.

For a node S in stage 6 with marking $M = p_{1s} + p_{2s} + 2r_1 + 2r_2 + 3c_1 + p_{1e} + p_{2e}$ and transition sequence $\pi = t_{11}t_{21}t_{12}t_{22}t_{13}t_{23}$, its energy consumption $E_1^*(S) = 132$. $\alpha = t_{11}t_{21}t_{22}t_{12}t_{13}t_{23}$ is a different transition sequence resulting in S and $E_1(\alpha) = 123$. Since $E_1^*(S) > E_1(\alpha)$, the arc to S labeled with t_{23} is pruned as shown in Figure 11.4 and another arc to S also labeled with t_{23} is added for S and $E_1^*(S) = E_1(\alpha)$.

11.4.3 DP Implementation for Scheduling FMS

DP for optimizing the total energy consumption in FMS is realized in Algorithm 11.1. An example is presented to show its execution.

Algorithm 11.1 DP for FMS Scheduling

Input: A live controlled PNS(N, M_0);
Output: Finial state S_f and $E^*(S_f)$;
Begin
Set OPEN $= \emptyset$, CLOSED $= \{S_0\}$;
Let S_f denote a final state, and at beginning, set it a fictional value with $E^*(S_f) = \infty$;
While (CLOSED $\neq \emptyset$) **do{**
For($S \in$ CLOSED){ /*$S = (M, J(\pi)$, $\varpi(\pi)$, π)*/
Compute $\Xi(S)$;/*In the live controlled PNS, all the reachable states are safe. There exists at least one enabled transition under a state, hence $\Xi(S)$ $\neq \emptyset$.*/
Delete S from CLOSED;
For($t \in \Xi(S)$){
Let $\alpha = \pi t$, $M[t > M_1$, and $S_1 = (M_1, J(\alpha)$, $\varpi(\alpha)$, $\alpha)$;
Calculate the energy consumption to S_1: $E(S_1) = E^*(S) + E(S, t)$;
Delete t from $\Xi(S)$;
If($M_1 = M_f$){
If($E(S_1) < E^*(S_f)$){
$S_f := S_1$; $E^*(S_f) = E(S_1)$;}}
Else{
If(there exists state S_2 in OPEN such that $S_2 = S_1$){/*$S_2 = (M_2, J(\beta)$, $\varpi(\beta)$, $\beta)$*/
If($E(S_1)$ $E^*(S_2)$){
$S_2 := S_1$; $E^*(S_2) = E(S_1)$;}}
Else{
Add S_1 into OPEN;
$E^*(S_1) = E(S_1)$;}}
} /*End **For**($t \in \Xi(S)$)*/

} /*End **For**$(S \in \text{CLOSED})$*/
CLOSED := OPEN; OPEN $= \emptyset$;
} /*End **While***/
Output S_f and $E^*(S_f)$; /*The optimal schedule and energy consumption.*/
End

Example 11.5 Consider the model in Figure 11.1. The ECRs of resources in busy and idle states are listed in Table 11.1. Two cases are discussed here: in the first case, both the numbers of type q_1 and q_2 parts are 2; in the second case, the numbers of type q_1 and q_2 parts are 2 and 3, respectively. For the two cases, no deadlock state exists, and no deadlock controller is needed.

In the first case, the optimal energy consumption is 641.4. The number of explored states is 304 206, and its corresponding transition sequence is

$$t_{11}t_{22}t_{11}t_{22}t_{23}t_{31}t_{23}t_{32}t_{31}t_{32}t_{24}t_{15}t_{24}t_{33}t_{15}t_{34}t_{33}t_{34}$$

The optimal energy consumption for the second case is 797.64. The number of explored states is 4 016 308, and its corresponding transition sequence is

$$t_{11}t_{31}t_{22}t_{32}t_{23}t_{11}t_{22}t_{33}t_{31}t_{32}t_{34}t_{31}t_{23}t_{32}t_{24}t_{15}t_{24}t_{33}t_{15}t_{34}t_{33}t_{34}$$

The computation time of the two cases is 2 087s and 366 296s, respectively. Hence, with the number of type q_2 part increasing from 2 to 3, the number of explored states increases 12 times and the computation time increases 174 times. Such computation makes it difficult for DP to meet industrial application needs given a sizable FMS. Hence, an MDP to shorten the computational time is introduced in the next section.

11.5 Modified Dynamic Programming for Scheduling FMS

As shown in the previous section, although the optimal schedule for FMS can be found by DP, the number of explored states increases exponentially with the size of the system. Hence, DP is computationally infeasible for sizable industrial

Table 11.1 ECRs of resources in busy and idle states.

State	m_1	m_2	m_3	r_1	r_2
Busy	2.3	1.9	3.4	1	1
Idle	0.23	0.19	0.34	0.1	0.1

scheduling problems. To obtain an optimal or suboptimal schedule within acceptable computation time, an MDP method is introduced in this section.

11.5.1 Evaluation Function of Transition Sequences

To alleviate the state explosion problem in FMS scheduling, the condensed state space method and other methods are used in [30–36]. In this method, all the states that have the same marking can be considered equivalent even if they differ on start processing time and route information. For states with the same marking, they are compared through heuristic functions and only the one with the best performance is kept. Through this method, the number of states need to be searched is greatly reduced. Similarly, in MDP, only the states determined by the marking and feasible sequence of transitions are considered. In this method, a state can be expressed as $S = (M, \pi)$, where M is a marking and π is a feasible sequence of transitions such that $M_0[\pi > M$. The initial state is represented by (M_0, π_0).

All feasible transition sequences from M_0 to M are evaluated by the evaluation function and only one with the best function value, say π, is kept in the state representation. The evaluation function is denoted as $g(\pi) = E_1(\pi) + \zeta(\pi)$ based on the energy consumption function E_1 or $g(\pi) = E_2(\pi) + \zeta(\pi)$ based on E_2. $\zeta(\pi)$ is the energy consumption needed to complete the processing of all operations, which have been started under M and $M_0[\pi > M$. $\zeta(\pi)$ can be calculated as

$$\zeta(\pi) = \sum_{i \in Z_u, \, p = \Theta(\pi_{l-1}, \, i) \in P} \phi(h(p, i) - \tau)c_b(p) \tag{11.20}$$

where τ is the earliest firing time of the last transition in π. $\phi(x)$ is defined as

$$\phi(x) = \begin{cases} x, & x > 0 \\ 0, & \text{otherwise} \end{cases} \tag{11.21}$$

If a marking M can be reached by respective transition sequences β_1 and β_2, and $g(\beta_1) > g(\beta_2)$, then in the search process, only node (M, β_2) is kept. The entire search space RG becomes the reduced RG (RRG).

Example 11.6   Reconsider the controlled PNS in Figure 11.3. Under the energy consumption function E_1, its RG in Figure 11.4 is compressed into RRG shown in Figure 11.5.

Consider $S_{3,3}$ and $S_{3,6}$ in RG in Figure 11.4. Their markings are the same, i.e. markings $M = p_{1s} + p_{2s} + p_{12} + p_{21} + 2r_1 + 2c_1$, and the two transition sequences

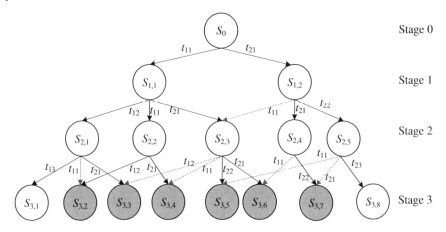

Figure 11.5 The first four stages of RRG.

reaching M are $\alpha = t_{11}t_{12}t_{21}$ and $\beta = t_{11}t_{21}t_{12}$, respectively. For α, the earliest firing time of transitions are $\tau_1 = 0$, $\tau_2 = 30$, and $\tau_3 = 30$, and its energy consumption is $E_1(\alpha) = 45$. Based on $h(p_{12}, 1) = 40$ and $h(p_{21}, 3) = 50$, the following results can be obtained:

$$\zeta(\alpha) = \phi(h(p_{12}, 1) - \tau_3) \times c_b(p_{12}) + \phi(h(p_{21}, 3) - \tau_3) \times c_b(p_{21})$$
$$= 10 \times 2 + 20 \times 2$$
$$= 60$$

$$g(\alpha) = E_1(\alpha) + \zeta(\alpha) = 105.$$

Similarly, $E_1(\beta) = 99$, $\zeta(\beta) = 20$, and $g(\beta) = 119$. Since $g(\alpha) < g(\beta)$, only α is kept for M and the state is denoted as $S_{3,3} = (M, \alpha)$. By applying this method, nodes $S_{3,3}$ and $S_{3,6}$ in RG are compressed into one node, represented as the gray node $S_{3,3}$ in RRG. The last transition in β labeled with t_{12} (marked by the dashed line) is not included in RRG, as shown in Figure 11.5. Similarly, the nodes with the same marking in Figure 11.4 are compressed into one and the correspondence of nodes between RG and RRG in stage 3 is given in Table 11.2.

11.5.2 Heuristic Function

To solve scheduling problems with DP in reasonable time, the maximum number of nodes in each stage is limited to a given number, denoted as $\hat{W}$. A heuristic function $f(M, \pi)$ is applied to maintain the most promising nodes.

For a given place $p \in P \cup P_s$, let $\gamma = pt_1p_1t_2...t_np_nt_{n+1}p_e$ be an operation path from p to the end place p_e, where p_i and t_i are the ith operation place and transition,

Table 11.2 The correspondence of nodes between RG and RRG in stage 3.

RG in Figure 11.4	RRG in Figure 11.5
$S_{3,1}$	$S_{3,1}$
$S_{3,2}$ and $S_{3,4}$	$S_{3,2}$
$S_{3,3}$ and $S_{3,6}$	$S_{3,3}$
$S_{3,5}$	$S_{3,4}$
$S_{3,7}$ and $S_{3,10}$	$S_{3,5}$
$S_{3,8}$	$S_{3,6}$
$S_{3,9}$ and $S_{3,12}$	$S_{3,7}$
$S_{3,11}$	$S_{3,8}$

respectively. Then $\eta(\gamma) = \sum_{i \in N_n} c_b(p_i) d(p_i)$ is the energy consumption of moving a part along path γ from p to p_e. Note that $\eta(\gamma)$ does not include energy consumption in p. In addition, denote $X(p)$ as

$$X(p) = \min \{\eta(\gamma) \mid \gamma \text{ is an operation path from } p \text{ to } p_e\} \quad (11.22)$$

The heuristic function is a mapping from the state space to a non-negative real number set defined as $f(M, \pi) = E_1(\pi) + h(\pi)$ based on energy consumption function E_1, or $f(M, \pi) = E_2(\pi) + h(\pi)$ based on E_2. $h(\pi) = \zeta(\pi) + \xi(\pi)$ is a "heuristic estimate" of the energy consumption from state (M, π) to the final state. $\xi(\pi) = \sum_{p \in P \cup P_s} X(p) M(p)$ is a minimum estimate of energy consumption needed to complete all not-yet-started operations of all parts.

As the energy consumption of resources in idle and occupied states is not considered in $h(\pi)$ since $h(\pi) \leq h^*(\pi)$, where $h^*(\pi)$ is the actual minimal energy consumption needed to complete all the parts from the current (M, π). Hence, $h(\pi)$ is admissible.

11.5.3 MDP Algorithm for FMS Scheduling

In the MDP algorithm, two kinds of lists are used. OPEN[i] is used for storing the states generated in stage i, and the states saved are different from each other. CLOSED[i] stores the states selected from OPEN[i] for further exploration through the heuristic function. For the states in CLOSED[i], MDP generates different states to constitute OPEN[$i + 1$] and the most promising states are put into CLOSED [$i + 1$]. Repeat this process for CLOSED[$i + 1$], CLOSED[$i + 2$], and so on. If there is no state generated, the optimal or suboptimal schedule is obtained. The corresponding transition sequence and energy consumption result are the outputs. MDP is realized in Algorithm 11.2.

Algorithm 11.2   MDP Algorithm for Optimizing E_1 in FMS

Input: A live controlled PNS(N, M_0);
Output: Finial state S_f and $E^*(S_f)$;
Begin
Set OPEN[i] = $\emptyset$, CLOSED[i] = $\emptyset$, CLOSED[0] = $\{S_0\}$, $E^*(S_0)$ = 0;/* $i \in Z$ */
Let S_f denote a final state, and at beginning, set it a fictional energy consumption value with $E^*(S_f) = \infty$;
While(CLOSED[i] $\neq \emptyset$) **do**{
  For($S \in$ CLOSED[i]) { /*$S = (M, \pi)$*/
  Delete S from CLOSED;
  Compute $\Xi(S)$; /*The set of all enabled transitions at S, i.e. under M. */
    For($t \in \Xi(S)$){
    Delete t from $\Xi(S)$;
    Let $\alpha = \pi t$, $M[t > M_1$, and $S_1 = (M_1, \alpha)$;
    Calculate the energy consumption to S_1: $E(S_1) = E^*(S) + E(S, t)$; /*Only one transition sequence is kept for S, it is treated as the optimal transition sequence from S_0 to S. The corresponding energy consumption is denoted as $E^*(S)$, and $E^*(S) = E(\pi)$. */
    If($M_1 = M_f$){
      If($E(S_1)$ $E^*(S_f)$){
      S_f := S_1; $E^*(S_f) = E(S_1)$;}}
    Else{
      If(there exists state S_2 in OPEN[i+1] such that $S_2 = S_1$)
  {/*$S_2 = (M_2, \beta)$*/
      If($g(\alpha) < g(\beta)$){
      $E^*(S_2) = E(S_1)$, and $S_2 = (M_2, \alpha)$;}}
    Else{
      Add S_1 into OPEN[i+1];
      $E^*(S_1) = E(S_1)$;}}
    }/*End **For**($t \in \Xi(S)$)*/
  }/*End **For**($S \in$ CLOSED[i])*/
  Sorting the states in OPEN in an ascending order based on their $f(M, \pi)$ values;
  Put the front $\hat{W}$ nodes in CLOSED[i+1]; /* If the number of nodes in OPEN[i+1] is less than $\hat{W}$, put all of them into CLOSED[i+1].*/
  i := i+1;
}/*End **While***/
Output S_f and $E^*(S_f)$;
End

11.6 Case Study

Consider the PNS from [21] and [26] as shown in Figure 11.6. It is the PN model of an FMS that consists of four machines and three robots and can process three types of parts q_1-q_3. The numbers of various parts to be scheduled are u_1, u_2, and u_3, respectively. The processing time of operations are listed in Table 11.3. In Table 11.4, the ECRs of resources in busy, occupied, and idle states are presented. According to the numbers of parts to be processed and resources in the system, 10 instances are considered.

In Instances 1–5, the capacities of resources are $\chi(m_i) = 2$, $i \in Z_4$, $\chi(r_1) = \chi(r_3) = 1$, and $\chi(r_2) = 2$. In Instances 6–10, the capacities of resources are $\chi(m_i) = 2$, $i \in Z_4$, $\chi(r_i) = 1$, $i \in Z_3$. While the numbers of parts are

Instances 1 and 6: $u_1 = 4$, $u_2 = 8$, and $u_3 = 4$.
Instances 2 and 7: $u_1 = 5$, $u_2 = 9$, and $u_3 = 5$.
Instances 3 and 8: $u_1 = 6$, $u_2 = 10$, and $u_3 = 6$.

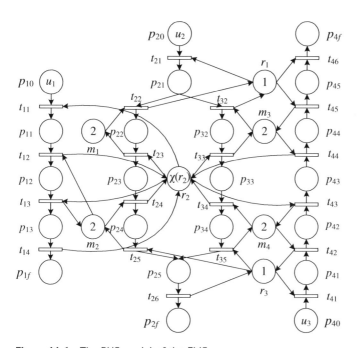

Figure 11.6 The PNS model of the FMS.

Table 11.3   Processing time of operations.

q_1	q_2		q_3
ω_1	ω_2	ω_3	ω_4
o_{11}: 8	o_{21}: 4	o_{21}: 4	o_{41}: 5
o_{12}: 34	o_{22}: 32	o_{32}: 23	o_{42}: 22
o_{13}: 5	o_{23}: 8	o_{33}: 6	o_{43}: 4
	o_{24}: 38	o_{34}: 20	o_{44}: 17
	o_{25}: 5	o_{25}: 5	o_{45}: 6

Table 11.4   ECRs of resources in different states.

State	m_1	m_2	m_3	m_4	r_1	r_2	r_3
Busy	2.3	1.9	3	3.7	1.2	1.1	1.3
Occupied	0.92	0.76	1.2	1.48	0.48	0.44	0.42
Idle	0.23	0.19	0.3	0.37	0.12	0.11	0.13

Instances 4 and 9: $u_1 = 7$, $u_2 = 11$, and $u_3 = 7$.
Instances 5 and 10: $u_1 = 8$, $u_2 = 12$, and $u_3 = 8$.

The values of $\hat{W}$ are randomly set as 10^2, 10^3, and 10^4, respectively. The simulation results for instances 1–10 under E_1 and E_2 are given in Tables 11.5 and 11.6, where H-, P-, and X-controllers are deadlock controller proposed in [20], [28], and [29], respectively. As there is no available deadlock controller for instances 1–5 in [20] and [28], calculation for instances 1–5 under H- and P-controllers cannot be performed and results are not available. The columns denoted as 10^2, 10^3, and 10^4 correspond to the results obtained under $\hat{W} = 10^2$, 10^3, and 10^4, respectively.

As shown in Tables 11.5 and 11.6, when $\hat{W}$ is set to 10^2, the MDP algorithm with X-controller performs best in optimizing both E_1 and E_2. When $\hat{W}$ is set to 10^3, MDP with X-controller performs best in minimizing E_1, while MDP with H-controller performs best in minimizing E_2. In the case of $\hat{W} = 10^4$, there are no significant differences between the results of the different controllers from the point of minimizing the total energy consumption of the FMS in this example.

Table 11.5 Simulation results under E_1.

Instance	X-controller			H-controller			P-controller		
	$\hat{W} = 10^2$	$\hat{W} = 10^3$	$\hat{W} = 10^4$	$\hat{W} = 10^2$	$\hat{W} = 10^3$	$\hat{W} = 10^4$	$\hat{W} = 10^2$	$\hat{W} = 10^3$	$\hat{W} = 10^4$
1	2459.95	2458.13	2458.13	/	/	/	/	/	/
2	2910.94	2876.07	2873.4	/	/	/	/	/	/
3	3324.11	3293.59	3301.82	/	/	/	/	/	/
4	3731.21	3745.78	3705.29	/	/	/	/	/	/
5	4163.79	4171.53	4126.23	/	/	/	/	/	/
6	2583.56	2532.97	2536.53	2573.35	2576.14	2517.75	2587.28	2550.67	2508.69
7	2967.03	2961.77	2969.83	3056.9	2997.39	2942.03	3075.6	3012.53	2959.76
8	3420.48	3395.16	3395.16	3483.54	3424.25	3415.89	3493.64	3419.94	3379.27
9	3835.29	3794.27	3794.27	3935.34	3862.67	3789.12	3965.14	3889.74	3781.28
10	4312.84	4274.37	4212.71	4409.56	4317.67	4293.59	4470.08	4333.42	4205.3

Table 11.6 Simulation results under E_2.

Instance	X-controller			H-controller			P-controller		
	$\hat{W} = 10^2$	$\hat{W} = 10^3$	$\hat{W} = 10^4$	$\hat{W} = 10^2$	$\hat{W} = 10^3$	$\hat{W} = 10^4$	$\hat{W} = 10^2$	$\hat{W} = 10^3$	$\hat{W} = 10^4$
1	2398.55	2398.55	2398.55	/	/	/	/	/	/
2	2843.46	2843.46	2843.46	/	/	/	/	/	/
3	3277.09	3238.88	3240.59	/	/	/	/	/	/
4	3675.87	3662.36	3667.48	/	/	/	/	/	/
5	4110.32	4083.64	4108.64	/	/	/	/	/	/
6	2504.43	2504.43	2500.32	2561.89	2509.31	2492.41	2559.97	2534.77	2487.72
7	2918.55	2916.2	2907.47	2987.68	2951.53	2916.01	3009.52	2950.85	2916.01
8	3350.03	3370.73	3325.73	3404.5	3324.91	3324.91	3377.83	3411.32	3323.18
9	3798.94	3763.32	3748.66	3900.88	3711.97	3711.97	3815.79	3851.36	3751.74
10	4322.8	4185.57	4182.11	4379.07	4164.77	4160.39	4238.73	4278.97	4143.61

Table 11.7 Computation time (second) under E_1 and E_2 under $\hat{W} = 10^2$, 10^3, and 10^4.

| | E_1 | | | | | | | | | E_2 | | | | | | | | |
| | X-controller | | | H-controller | | | P-controller | | | X-controller | | | H-controller | | | P-controller | | |
	10^2	10^3	10^4	10^2	10^3	10^4	10^2	10^3	10^4	10^2	10^3	10^4	10^2	10^3	10^4	10^2	10^3	10^4
1	4	274	20 873	/	/	/	/	/	/	3	254	14 052	/	/	/	/	/	/
2	4	380	33 957	/	/	/	/	/	/	5	359	23 645	/	/	/	/	/	/
3	5	495	47 207	/	/	/	/	/	/	7	433	31 965	/	/	/	/	/	/
4	7	591	58 968	/	/	/	/	/	/	8	574	39 240	/	/	/	/	/	/
5	9	695	63 129	/	/	/	/	/	/	10	728	44 253	/	/	/	/	/	/
6	2	127	6 891	2	109	8 664	3	114	6 422	2	114	7 492	2	120	7 611	2	121	9 451
7	3	168	10 804	3	159	12 800	3	157	9 531	3	155	11 316	3	171	12 706	3	161	13 224
8	3	206	20 913	3	205	19 076	3	204	17 419	3	205	14 745	4	233	15 861	4	221	15 477
9	5	257	25 476	4	256	23 720	4	250	21 675	4	254	17 953	4	301	20 933	4	276	20 555
10	5	293	32 382	5	325	28 155	5	283	27 705	5	306	19 647	5	357	25 976	5	346	26 255

In addition, for different instances under X-, P-, and H-controllers, most of the best results are obtained under $\hat{W} = 10^4$. For most instances, the results become better as $\hat{W}$ increases. From results in Tables 11.5 and 11.6, the biggest improvement percentage of energy consumption from $\hat{W} = 10^2$ to 10^3 and from $\hat{W} = 10^3$ to 10^4 are 4.89% (Instance 10 under H-controller in Table 11.6) and 2.96% (Instance 10 under P-controller in Table 11.5), respectively.

In Table 11.7, the computation time is recorded for all 10 instances under E_1 and E_2. As $\hat{W}$ increases from 10^2 to 10^3 and from 10^3 to 10^4, the computation time increases several tens of times.

To summarize, for this FMS example, as $\hat{W}$ increases from 10^2 to 10^3 and from 10^3 to 10^4, the computation time increases tens of times. However, the improvement percentage of objective value is rather limited. Hence, by restricting the maximum number of states in each stage to between 10^2 and 10^3, the MDP algorithm can successfully find a feasible solution in an acceptable computation time.

11.7 Summary

With the PN models of FMS and DP, this chapter presents the total energy consumption optimization problem for deadlock-prone FMS. To calculate the total energy consumption, two energy consumption functions are proposed. For small-size energy optimization problems, DP is formulated and the optimal solutions can be found. As for large-scale scheduling problems, an MDP method is introduced. In MDP, through the transition sequence evaluation function and heuristic function, only the most promising states are explored such that the searched state space is largely reduced. By restricting the maximum number of nodes in each stage to a suitable value, MDP can optimize the total energy consumption and return an optimal or suboptimal schedule in a short time. Deadlock controllers are used to guarantee that MDP is performed on a live PN in which no deadlock occurs.

Problems

11.1 Please build up the Petri net model of an FMS consisting of machine m_1–m_4 with their capacity being 1, 2, 3, and 4 and three robots r_1–r_3 with capacity being 1 such that it can process u_A parts of Type A and u_B parts of Type B. A Type-A part has two routes, (i) loading via r_3, processing by m_1 and m_3 and then unloaded by r_2, and (ii) loading via r_2, processing by m_4 and m_3 and then unloaded by r_3. A Type-B part has two routes: (i) loading via r_3, processing by m_4 and m_3, and then unloaded by r_1; and (ii) loading via r_1, processing by m_4 and m_2 and then unloaded by r_1.

11.2 Assuming $u_A = u_B = 1$, 2, and 3, please estimate the number of possible transition firing sequences.

11.3 Please realize other intelligent optimization methods to find the schedules for the Petri net shown in Figure 11.6 and compare the performance in terms of makespan, energy consumption, and computation time.

11.4 Please design an algorithm that minimizes both makespan and energy consumption when scheduling FMS.

References

1 Li, X., Xing, K., Zhou, M. et al. (2019). Modified dynamic programming algorithm for optimization of total energy consumption in flexible manufacturing systems. *IEEE Trans. Autom. Sci. Eng.* 16 (2): 691–705. https://doi.org/10.1109/TASE. 2018.2852722.

2 Mouzon, G. and Yildirim, M.B. (2008). A framework to minimise total energy consumption and total tardiness on a single machine. *Int. J. Sustain. Eng.* 1 (2): 105–116.

3 Yildirim, M.B. and Mouzon, G. (2012). Single-machine sustainable production planning to minimize total energy consumption and total completion time using a multiple objective genetic algorithm. *IEEE Trans. Eng. Manag.* 59 (4): 585–597. https://doi.org/10.1109/TEM.2011.2171055.

4 Che, A., Wu, X., Peng, J., and Yan, P. (2017). Energy-efficient bi-objective single-machine scheduling with power-down mechanism. *Comput. Oper. Res.* 85: 172–183. https://doi.org/10.1016/j.cor.2017.04.004.

5 Cheng, J., Chu, F., Liu, M. et al. (2017). Bi-criteria single-machine batch scheduling with machine on/off switching under time-of-use tariffs. *Comput. Ind. Eng.* 112: 721–734. https://doi.org/10.1016/j.cie.2017.04.026.

6 Chen, X., Li, C., Yang, Q. et al. (2022). Toward energy footprint reduction of a machining process. *IEEE Trans. Autom. Sci. Eng.* 19 (2): 772–787. https://doi.org/ 10.1109/TASE.2021.3062648.

7 Han, Z., Zhao, J., and Wang, W. (2017). An optimized oxygen system scheduling with electricity cost consideration in steel industry. *IEEE/CAA J. Autom. Sin.* 4 (2): 216–222. https://doi.org/10.1109/JAS.2017.7510439.

8 Liu, Y., Dong, H., Lohse, N. et al. (2014). An investigation into minimising total energy consumption and total weighted tardiness in job shops. *J. Clean. Prod.* 65: 87–96. https://doi.org/10.1016/j.jclepro.2013.07.060.

9 Mokhtari, H. and Hasani, A. (2017). An energy-efficient multi-objective optimization for flexible job-shop scheduling problem. *Comput. Chem. Eng.* 104: 339–352. https://doi.org/10.1016/j.compchemeng.2017.05.004.

10 Zhang, R. and Chiong, R. (2016). Solving the energy-efficient job shop scheduling problem: a multi-objective genetic algorithm with enhanced local search for minimizing the total weighted tardiness and total energy consumption. *J. Clean. Prod.* 112: 3361–3375. https://doi.org/10.1016/j.jclepro.2015.09.097.

11 Salido, M.A., Escamilla, J., Giret, A., and Barber, F. (2016). A genetic algorithm for energy-efficiency in job-shop scheduling. *Int. J. Adv. Manuf. Technol.* 85 (5–8): 1303–1314. https://doi.org/10.1007/s00170-015-7987-0.

12 Zhang, L., Tang, Q., Wu, Z., and Wang, F. (2017). Mathematical modeling and evolutionary generation of rule sets for energy-efficient flexible job shops. *Energy* 138: 210–227. https://doi.org/10.1016/j.energy.2017.07.005.

13 Lv, Y., Li, C., Tang, Y., and Kou, Y. (2021). Toward energy-efficient rescheduling decision mechanisms for flexible job shop with dynamic events and alternative process plans. *IEEE Trans. Autom. Sci. Eng.* 1–17. https://doi.org/10.1109/TASE.2021.3115821.

14 Fang, K., Uhan, N., Zhao, F., and Sutherland, J.W. (2011). A new approach to scheduling in manufacturing for power consumption and carbon footprint reduction. *J. Manuf. Syst.* 30 (4): 234–240.

15 Mashaei, M. and Lennartson, B. (2013). Energy reduction in a pallet-constrained flow shop through on–off control of idle machines. *IEEE Trans. Autom. Sci. Eng.* 10 (1): 45–56. https://doi.org/10.1109/TASE.2012.2225426.

16 Dai, M., Tang, D., Giret, A. et al. (2013). Energy-efficient scheduling for a flexible flow shop using an improved genetic-simulated annealing algorithm. *Robot. Comput. Integr. Manuf.* 29 (5): 418–429. https://doi.org/10.1016/j.rcim.2013.04.001.

17 Tang, D., Dai, M., Salido, M.A., and Giret, A. (2016). Energy-efficient dynamic scheduling for a flexible flow shop using an improved particle swarm optimization. *Comput. Ind.* 81: 82–95. https://doi.org/10.1016/j.compind.2015.10.001.

18 Lu, C., Gao, L., Li, X. et al. (2017). Energy-efficient permutation flow shop scheduling problem using a hybrid multi-objective backtracking search algorithm. *J. Clean. Prod.* 144: 228–238. https://doi.org/10.1016/j.jclepro.2017.01.011.

19 Dashora, Y., Kumar, S., Tiwari, M.K., and Newman, S.T. (2007). Deadlock-free scheduling of an automated manufacturing system using an enhanced colored time resource Petri-net model-based evolutionary endosymbiotic learning automata approach. *Int. J. Flex. Manuf. Syst.* 19 (4): 486–515. https://doi.org/10.1007/s10696-008-9046-8.

20 Huang, Y., Jeng, M., Xie, X., and Chung, S. (2001). Deadlock prevention policy based on Petri nets and siphons. *Int. J. Prod. Res.* 39 (2): 283–305. https://doi.org/10.1080/00207540010002405.

21 Li, X., Xing, K., Wu, Y. et al. (2017). Total energy consumption optimization via genetic algorithm in flexible manufacturing systems. *Comput. Ind. Eng.* 104: 188–200. https://doi.org/10.1016/j.cie.2016.12.008.

22 Pang, C.K. and Le, C.V. (2014). Optimization of total energy consumption in flexible manufacturing systems using weighted P-timed Petri nets and dynamic programming. *IEEE Trans. Autom. Sci. Eng.* 11 (4): 1083–1096. https://doi.org/10.1109/TASE.2013.2265917.

23 Xiong, H.H. and Zhou, M. (1998). Scheduling of semiconductor test facility via Petri nets and hybrid heuristic search. *IEEE Trans. Semicond. Manuf.* 11 (3): 384–393. https://doi.org/10.1109/66.705373.

24 Wu, N. and Zhou, M. (2012). Modeling, analysis and control of dual-arm cluster tools with residency time constraint and activity time variation based on Petri nets. *IEEE Trans. Autom. Sci. Eng.* 9 (2): 446–454. https://doi.org/10.1109/TASE.2011.2178023.

25 Wu, N., Zhou, M., and Li, Z. (2015). Short-term scheduling of crude-oil operations: enhancement of crude-oil operations scheduling using a Petri net-based control-theoretic approach. *IEEE Robot. Autom. Mag.* 22 (2): 64–76. https://doi.org/10.1109/MRA.2015.2415047.

26 Xing, K., Han, L., Zhou, M., and Wang, F. (2012). Deadlock-free genetic scheduling algorithm for automated manufacturing systems based on deadlock control policy. *IEEE Trans. Syst. Man, Cybern. Part B* 42 (3): 603–615. https://doi.org/10.1109/TSMCB.2011.2170678.

27 Bellman, R. (1966). Dynamic programming. *Science (80-.).* 153 (3731): 34–37.

28 Piroddi, L., Cordone, R., and Fumagalli, I. (2008). Selective siphon control for deadlock prevention in Petri nets. *IEEE Trans. Syst. Man, Cybern. Part A Syst. Humans* 38 (6): 1337–1348. https://doi.org/10.1109/TSMCA.2008.2003535.

29 Xing, K., Zhou, M.C., Liu, H., and Tian, F. (2009). Optimal Petri-net-based polynomial-complexity deadlock-avoidance policies for automated manufacturing systems. *IEEE Trans. Syst. Man, Cybern. Part A Syst. Humans* 39 (1): 188–189. https://doi.org/10.1109/TSMCA.2008.2007947.

30 Christensen, S., Kristensen, L.M., and Mailund, T. (2001). Condensed state spaces for timed Petri nets. *International Conference on Application and Theory of Petri Nets*, Newcastle upon Tyne, UK (25–29 June 2001), 101–120.

31 Luo, J., Xing, K., Zhou, M., and Wang, X. (2015). Deadlock-free scheduling of automated manufacturing systems using Petri nets and hybrid heuristic search. *IEEE Trans. Syst. Man, Cybern. Syst.* 45 (3): 530–541. https://doi.org/10.1109/TSMC.2014.2351375.

32 Lei, H., Xing, K., Han, L. et al. (2014). Deadlock-free scheduling for flexible manufacturing systems using Petri nets and heuristic search. *Comput. Ind. Eng.* 72: 297–305. https://doi.org/10.1016/j.cie.2014.04.002.

33 Baruwa, O.T., Piera, M.A., and Guasch, A. (2015). Deadlock-free scheduling method for flexible manufacturing systems based on timed colored Petri nets and anytime heuristic search. *IEEE Trans. Syst. Man, Cybern. Syst.* 45 (5): 831–846. https://doi.org/10.1109/TSMC.2014.2376471.

34 Luo, J., Zhou, M., and Wang, J.-Q. (2021). AB&B: an anytime branch and bound algorithm for scheduling of deadlock-prone flexible manufacturing systems. *IEEE Trans. Auto. Sci. Eng.* 18 (4): 2011–2021.

35 Huang, B., Zhou, M., Abusorrah, A., and Sedraoui, K. (2022). Scheduling robotic cellular manufacturing systems with timed Petri Net, A* search, and admissible heuristic function. *IEEE Trans. Auto. Sci. Eng.* 19 (1): 243–250.

36 Zhao, Z., Liu, S., Zhou, M. et al. (2022). Heuristic scheduling of batch production processes based on Petri nets and iterated Greedy Algorithms. *IEEE Trans. Auto. Sci. Eng.* 19 (1): 251–261.

Part V

Summaries and Conclusions

12

Research Trends and Future Directions in Sustainable Industrial Development

In the previous chapters of this book, the fundamentals of sustainable manufacturing and energy consumption of industrial manufacturing systems have been introduced, followed by detailed discussions on representative sustainable manufacturing practices and energy efficiency in advanced manufacturing processes and systems. More precisely, in the first part of the book, the fundamental knowledge about the manufacturing system as well as the associated technical issues with energy efficiency and energy saving are demonstrated to deliver a better understanding of manufacturing sustainability. In the second part, mathematical tools and modeling approaches for investigating manufacturing activity-related energy consumption and energy efficiency management are demonstrated. In the third and fourth parts, representative manufacturing systems and processes are used as practical examples to illustrate the real-world problem-solving processes using the methodologies presented in previous chapters. In the past decades, numerous efforts were made to address the challenges of sustainable manufacturing and a variety of methodologies for the improvement of energy efficiency in manufacturing processes and systems were proposed. Due to the increased competition on resources and persistent demands for environmentally responsible practices, there is a growing need to incorporate the sustainability requirements in a product life cycle, which go beyond the performance evaluation for sustainable manufacturing. In this chapter, the research trends and future directions in sustainable industrial development are discussed, which provides a holistic view of industrial and environmental sustainability.

12.1 Insights into Sustainable Industrial Development

Numerous researchers have studied the interrelationships among energy consumption, environmental impacts, and resource conservation toward sustainable industrial development. The scopes of the research questions and the seminal

Sustainable Manufacturing Systems, First Edition. Lin Li and MengChu Zhou.
© 2023 The Institute of Electrical and Electronics Engineers, Inc.
Published 2023 by John Wiley & Sons, Inc.

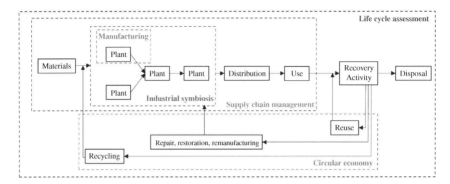

Figure 12.1 The life cycle of a product and associated research topics.

contributions in existing literature are diverse, which cannot be introduced in detail within the scope of a single book. This chapter places particular emphasis on some major research topics in the context of sustainable industrial development. More specifically, the sustainability methodologies and techniques are aligned with different stages during a product lifecycle. Figure 12.1 depicts the major stages in the life cycle of a product and the associated material flows. The research topics and potential problems toward sustainable manufacturing are discussed from the following five aspects: energy and resource efficiency in manufacturing, industrial symbiosis, supply chain management, circular economy, and life cycle assessment.

12.2 Energy and Resource Efficiency in Manufacturing

12.2.1 Equipment Design

The existing study has demonstrated that machine tools could account for 8–36% of the life cycle energy impacts, depending on the types of machines and production environment [1]. Therefore, energy efficiency improvements at the machine level have immense potential for reducing energy consumption during manufacturing processes. The major challenge for machine tool manufacturers, i.e. original equipment manufacturer (OEM), to deliver continuous improvement in the energy and resource efficiency of their products without compromising the designated functions [2]. The flexibility in component design and/or specification during the design stage of machining equipment can provide great potential for improvement in energy efficiency [3].

 The use of more efficient components can effectively improve the energy efficiency of machining equipment. The energy efficiency of equipment components can be improved through the perspectives of functional fit, dimensioning, and technological changes [4]. In particular, the functional fit indicates that machine components can be selected according to functional expectations to maximize the efficiency of the machines. For example, the adoption of speed control in a cooling lubricant pump instead of a valve can fulfill variable demands in machine tool cooling, which can help manufacturers with high-precision grinding machines reduce approximately 20% of the total energy consumption [4]. Apart from functional fit, over-dimensioning issues in machine tool design can lead to high energy expenditure during the equipment use stage due to the excessive energy consumption with respect to what is necessary to fulfill a given function. This issue can be primarily attributed to the lack of information about the actual customer needs and the controllability of auxiliaries related to the designated machining process [5]. Over-dimensioning can be solved by machine tool redesign and reconfiguration, which usually involve activities of compiling a set of predefined components while taking into account restrictions for machine tool assembly [6]. The efficiency of equipment can also be reduced through technological changes, which are achieved by machine tool upgrading with the implementation of innovative technologies [7]. For example, the existing study's results indicate that by replacing the traditional CO_2 laser source with a fiber laser source in the laser cutting machine, an annual energy saving of 16.6 MWh can be achieved [8].

12.2.2 Smart Manufacturing

Smart manufacturing takes advantage of emerging technologies such as sensor networks [9–12], big data analysis [13–19], cloud computing [20–25], additive manufacturing [26–30], Internet of Things (IoT) [31–37], cyber-physical systems (CPS) [38–43], and digital twins [44–46] to improve the overall sustainability performance [47–56], as shown in Figure 12.2.

Several opportunities for sustainable manufacturing can be identified by considering intelligent cross-linking and process digitalization among different value creation modules with respect to machines, production lines, manufacturing cells, and entire manufacturing plant. Some of the representative opportunities are discussed as follows [57].

- Equipment. Intelligent equipment is the foundation of smart manufacturing, which enables real-time data collection and massive data analysis through direct data processing or data uploading to the cloud for operation adjustment toward sustainable manufacturing [58]. For example, by adding smart sensors on ball mills and analyzing the acquired data, the overall energy consumption can be reduced by 3% in a pulp workshop, which corresponds to a 4% reduction of associated energy costs [19]. Despite the benefits of implementing smart equipment, the brand-new manufacturing equipment with embedded sensors, actuators, and/or telecommunication modules may require relatively higher

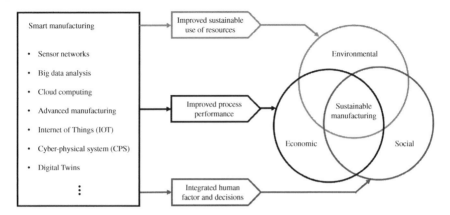

Figure 12.2 Smart manufacturing towards enhanced sustainability.

capital investment, which may not be an economically attractive option to manufacturers. In this case, existing equipment retrofitting is also considered as a less costly alternative equipment upgrading solution in sustainable manufacturing practices [59].

- Process. Advanced manufacturing technologies such as additive manufacturing are often envisioned to bring potential economic benefits by reducing the time delay between design and manufacturing. Compared with conventional manufacturing, additive manufacturing can offer additional manufacturing complexity and improved material efficiency [60, 61]. Despite the continuous development in advanced manufacturing, the associated environmental implications and impacts on industrial sustainability have not been fully elucidated [60]. To date, only a few sustainability performance evaluation studies have been conducted on a small number of advanced manufacturing processes. Most of these studies only consider either limited environmental impact categories or suffer from very limited or low-quality data [62]. In addition, the occupational health and workplace safety risks associated with advanced manufacturing processes are often not that dissimilar to the counterpart in traditional industrial settings, such as operating process-related physical injury, and harmful emissions. However, the use of unconventional materials in advanced manufacturing may cause emerging risks that need to be further characterized [63, 64].

- Organization. The link between work organization and smart manufacturing technologies has drawn increasing attention in recent literature. From the social-technical perspective, smart manufacturing can affect the work organization at both the micro- and macro-levels. At the micro-level, smart manufacturing can affect the work design with respect to the job breadth, cognitive demand of operator activities, and social interaction. Attributable to the development of sensing and communication technologies, manufacturing processes can be managed by CPS, which functions as the center of the value chain in a manufacturing system. Although the routine jobs may be limited to CPS monitoring, there also exist new workforce needs in the design, implementation, and training of CPS, leading to the recruitment of both the highly skilled personnel and operators with average technical qualifications [65]. At the macro-level, process digitalization and automation facilitate the information exchange processes, which allows the transition to decentralized decision-making in the plant. The sustainability-oriented decentralized work organization can improve allocation efficiency for resources such as materials, energy, water, and products [66].

12.3 Industrial Symbiosis

Apart from the energy and resource management within one manufacturing site, new opportunities and challenges toward sustainable manufacturing would appear when the interactions among different industrial facilities are taken into consideration. The interactions may occur between independent companies through the material and/or energy exchange toward mutual benefits [67]. This mutualism in the cross-industrial interactions leads to the concept of "industrial symbiosis." A commonly accepted definition of industrial symbiosis is "Industrial symbiosis engages traditionally separate industries in a collective approach to competitive advantage involving physical exchange of materials, energy, water, and/or by-products. The keys to industrial symbiosis are collaboration and the synergistic possibilities offered by geographic proximity [68]."

One fundamental notion of industrial symbiosis is that it treats the "waste" of one industrial facility as the input resource or energy for other facilities. In Figure 12.3, the material exchange among a cellulosic ethanol plant, a microalgae biodiesel plant, and a biogas plant is provided as an example of industrial symbiosis. As shown in Figure 12.3, the waste gas and wastewater of the cellulosic ethanol plant can be used as input resources in the microalgae biodiesel plant. In particular, carbon dioxide is the major constituent in waste gas, and wastewater mainly consists of low-molecular-weight organic compounds and ammonia ions. These chemical compounds are essential in microalgae cultivation. In addition, after the lipid extraction from microalgae, the residual biomass, which contains fermentable carbohydrates, from the microalgae biodiesel plant can be used as the raw material in cellulosic ethanol production. Furthermore, the organic solid waste from the cellulosic ethanol plant can be transported to the biogas plant as the input resource for biogas production. By employing this collective and multi-industrial collaboration, the resource cost and waste generation associated with three biofuel production plants can be effectively reduced.

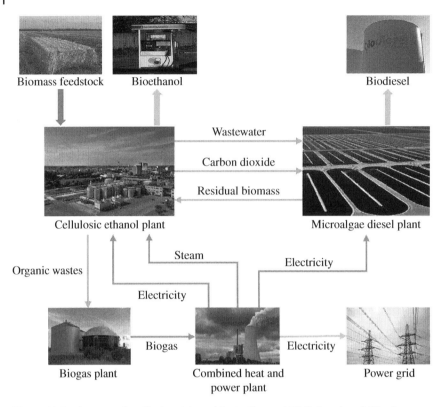

Figure 12.3 An example of industrial symbiosis. *Source:* [69]/University of Illinois.

To make industrial symbiosis more economically attractive, the physical distance between different industrial plants should be short enough to favor cross-facility networking. To address this issue, the concept of an eco-industrial park is proposed. The U.S. EPA defines the eco-industrial park as "a community of manufacturing and service businesses seeking enhanced environmental and economic performance through collaboration in managing environmental and resource issues including energy, water, and materials. By working together, the community of businesses seeks a collective benefit that is greater than the sum of the individual benefits each company would realize if it optimized its individual performance only" [70]. The concept of eco-industrial parks has now been adopted worldwide. For example, in Guiyang, China, municipal solid waste is used as an energy source for the local steel and iron industry, and in return, the waste heat from the steel and iron industry is used for residential heating. It is estimated that this partnership can help the related stakeholders reduce the urban carbon footprint by about one million metric tons per year [71]. In Iran, the application of the eco-industrial park is evaluated in a pilot study in Mourcheh Khort, which is an existing industrial park in the north of the city of Isfahan. A pilot investigation of the wastewater flows of different industries within the representative site is conducted to identify the wastewater reuse potentials, and three linking scenarios are proposed to reduce the consumption of freshwater and reuse water within the industrial park. A prominent example is the implementation of linkage between the polyamide fiber factory and the dyeing shop; 48.5% of the wastewater generated from polyamide fiber production is only slightly contaminated and can be reused as process water in the dyeing shop. Through the connection between these two industries, the overall freshwater consumption can be reduced by 48.5 m^3/day, which corresponds to approximately 33% of the total water demand and 53% of the total wastewater reduction [72].

Despite the benefits of industrial symbiosis, some challenging barriers need to be overcome. One challenge is how to prove the economic benefits of a specific industrial symbiosis plan, as most benefits of existing industrial symbiosis system may be considered confidential and commercially sensitive [73]. As an alternative evaluation method, the costs and benefits associated with the implementation of industrial symbiosis are usually compared with the attributes of stand-alone industrial facilities. However, this reference selection can be questionable, especially when the plants generate a vast amount of waste or by-products. In this case, it is unlikely that the manufacturer would bear the high disposal costs without seeking alternative solutions to improve waste management and/or waste treatment efficiency. Another barrier to increasing industrial symbiosis activities is matching the "waste" outputs with "resource" inputs across diverse industries. To address this issue, the semantic approach [74] and big data analytics approach [75] have been employed to facilitate the industrial symbiosis input/output matching issue. In addition, an open-source database of company feedstocks and wastes has also been established [76], where a database engine is designed to provide information of technology required to make the waste-to-resource conversion happen [77] as well as the use of "green" social networking to stimulate social connections and reveal material flow compatibilities [78].

12.4 Supply Chain Management

A supply chain can be defined as a set of three or more entities (suppliers, manufacturers, and customers) directly involved in the upstream and downstream flows of products, services, finances, and/or information [79]. Based on this definition, sustainable supply chain management mainly focuses on three stages: premanufacturing, manufacturing, and use. In contrast to the industrial symbiosis, which primarily considers the manufacturing stage, sustainable supply chain management emphasizes the value of the product related to the environmental and social burdens incurred through the entire system, from a source to a customer. For example, the supply chain of bioethanol involves biomass feedstock suppliers, biorefinery plants, warehouses, gas stations, and logistics network (i.e. connections among the components), as shown in Figure 12.4. Even if the biorefinery process exerts a nonsignificant environmental impact, the supply chain may not be considered sustainable due to nonrenewable resource consumption or high transport emissions.

The efforts that contribute to sustainable supply chain management can be categorized into the following four aspects [81]:

- Business strategy. Large companies are asked to consider the environmental and social problems involved in their entire supply chain, especially for the focal companies who can rule or govern the supply chain. For these firms, the objectives of the business strategies involve not only the profits or economic performance but also the environmental and social metrics [82]. It has been reported that if a firm adopts a more environmentally friendly strategy and moves toward a sustainable supply chain, it can ultimately lead to economic benefits [83, 84].
- Operational level. Research efforts on supply chain design issues at the operational level mainly involve the performance measurement in sustainable supply chain management [85], logistics network design [86], and strategic supply chain modeling and planning [80, 87].
- Decisions at functional interfaces. The function can be viewed as a series of activities involved in a supply chain that can be implemented individually or within the cross-functional alignment. These activities include but are not limited to supplier selection [88], low carbon production [89], logistics provider selection [90], and green shipping [91].
- Energy perspective. The research efforts on the employment of energy-related performance measures in supply chain management involve renewable energy supply chain [92], energy efficiency evaluation and improvement [93, 94], and energy sources selection [95].

With the increased complexity in supply chains, new challenges emerge from sustainable supply chain development. The complex supply chain may lead to more complexity in measuring the associated economic, environmental, and social impacts, which would pose dilemmas in the choice of suppliers, fuel, logistics route, production setup, etc. These challenges elicit a critical need for exploring and developing more holistic models for solving complex supply chain problems. Moreover, uncertainty may lead to the need for risk and resilience management in a sustainable supply chain. The uncertainty can originate from product demand, products or raw materials prices, and resource availability. The presence of uncertainty highlights the need for identifying the risk and resilience metrics, which

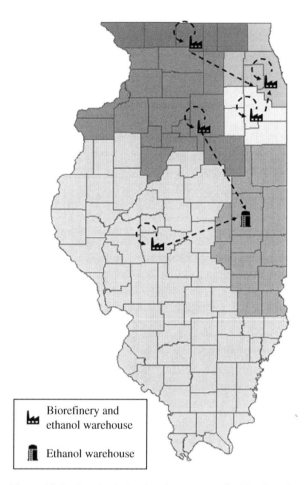

Biorefinery and
ethanol warehouse

Ethanol warehouse

Figure 12.4 Supply chain planning strategy for bioethanol production in the state of Illinois in the United States. *Source:* Adapted from [80].

should be incorporated into sustainable supply chain management to ensure the robustness and validity of management decisions in long-term development.

12.5 Circular Economy

The manufacturing system plays a significant role in boosting economic growth and promoting technological innovations. In accordance with the report from National Association Manufacturing (NAW) in the United States, the manufacturing value-added output was USD 2.57 trillion in the third quarter of 2021, and the manufacturing gross output was USD 6.44 trillion [96]. Traditionally, the primary objectives of many manufacturing systems are to enhance productivity with special considerations on performance metrics such as throughput and machine utilization rate. Due to the global resource scarcity and environmental degradation, significant concerns are also arising with respect to the heavy dependency of virgin materials and energy in manufacturing systems and the vast amount of end-of-life (EOL) product disposed of in landfill sites, which often contains valuable components or materials. A prominent example of the fast-growing waste stream is the waste of electrical and electronic equipment (WEEE). According to the 2020 report released by the United Nations University (UNU), it is estimated that the global generation of WEEE reached a striking 53.6 million metric tons in 2019, with the fate of 82.6% of e-waste being uncertain [97].

Owing to the growing concerns for the increased competition for resource and manufacturing environmental footprints, the circular economy has been proposed as a precondition for sustainable development in manufacturing industries. In particular, as an environmentally responsible alternative to the traditional linear "take-make-dispose" model, the circular economy is considered as an economic model, where the resource acquisition, procurement process, product manufacturing, and EOL reprocessing are systematically managed to minimize the waste generation and maximize the resource efficiency in the early production and distribution stages of a product, as well as the EOL residual value recovery through extended lifecycles [98]. According to the report released by the United Nations Environment Programme (UNEP) [99], circular economy builds connections among economic development, environmental protection, and resource conservation.

The efforts that promote the transition to circular economy mainly involve the following three central aspects: reduction, reuse, and recycling of materials and energy (which are often referred to as "3R" principles).

- Reduction aims to improve the resource utilization efficiency in product manufacturing (e.g. the reduction in feedstock material consumption and energy intensity, as well as the displacement of virgin material intake with by-products) and reduce the waste generation and dispersion of toxic substances.
- Reuse aims to enhance the circularity of products or core components that are discarded but still in good condition and fulfill their designed functions.
- Recycling aims to recover materials (often referred to as secondary materials) from the EOL products to obtain the same (i.e. high grade) or lower (low grade) quality.

Attributed to the close association of circular economy with the three pillars of sustainable manufacturing, i.e. economic, environmental, and social (referring to Section 1.1), the strategies involved in the circular economy has been further categorized into nine subcategories (as referred to as "9R" framework), as demonstrated in Figure 12.5 [101].

As illustrated in Figure 12.5, the varieties of original "3R" principles are classified with a hierarchical structure, where the first R is prioritized over the second R, and so forth. Particularly, in terms of smarter product use and manufacture, the *refuse* principle suggests the avoidance of redundant products that can ultimately deliver the same or equivalent functionalities, whereas the *rethink* principle indicates more intensive product use through product sharing or the manufacturing of multifunctional products. As the efforts to extend the product lifecycle, strategies such as repair, refurbish, and remanufacturing are proposed to improve the circularity of products or components that are defective, damaged, or outdated. If the reuse or remanufacturing options are not possible, the incineration of materials or waste with energy recovery needs to be considered.

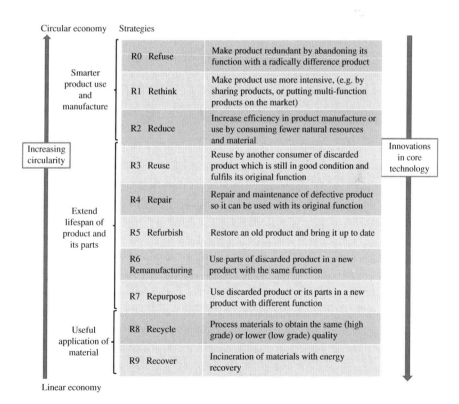

Figure 12.5 Illustration of strategies involved in the circular economy. *Source:* Adapted from [100].

To date, circular economy has gained considerable momentum as a new sustainability paradigm in national governments and intergovernmental agencies. The European Commission launched a standardization mandate to assess material efficiency aspects of products [102], and the product design for sustainability has been identified as one of the critical enabling strategies. In Singapore, a data library of materials and recyclability attributes has been established to determine the suitability of a material for recycling purposes [103]. In China, the first "Circular Economy Promotion Law" took effect in January 2009, and the EOL waste recycling rate was improved from 7.2% in 1995 to 17% in 2015 [104]. In Japan, leading companies such as Hitachi and Toshiba have been developing design software that can be used to help practitioners perform more sustainable product designs, which involves resource consumption in product manufacturing and EOL recycling criteria in design processes [105]. In addition, the U.S. EPA hosted the first America Recycles Day Summit in November 2018, which brought together stakeholders from across the US recycling systems to identify specific action areas in addressing the challenges that the US recycling system was facing. As a product of this summit, the National Framework for Advancing the US recycling system has been proposed. In this framework, education and outreach plans for enhancing public awareness of EOL product management, upgrading of outdated material management and recycling infrastructures, developing the new-end market for secondary materials, and improving recycling metrics and measures have been proposed as the priority actions [106].

12.6 Life Cycle Assessment

Based on the information introduced in the previous section, it is understandable that the environmental impacts of a product are related to all stages in its lifespan, ranging from raw material processing, production, logistics to end-use, recycle, and disposal. For example, the fluorescent light bulb has a longer service life and consumes less energy than traditional incandescent bulbs, which makes it more environmentally friendly and cost-effective in the use stage. However, more feedstock materials are required in fluorescent light bulb manufacturing, and the heavy metals can be contained during the process, which requires additional treatments before disposal [107]. Therefore, to address sustainability problems throughout the lifecycle of products, life cycle assessment (LCA) has emerged as a holistic assessment tool. According to the International Organization for Standardization (ISO), LCA is defined as a method for analyzing and determining the environmental impact along the production chain [108]. Figure 12.6 illustrates the major lifecycle stages involved in LCA.

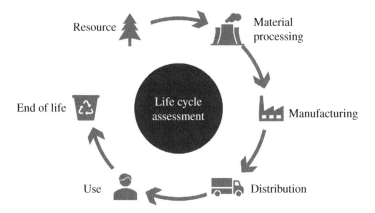

Figure 12.6 Illustration of lifecycle stages involved in the LCA.

Different strategies for sustainable product design using the LCA approach are summarized as follows [109]:

- Selection of environmentally low-impact materials. Material selection is of vital importance in sustainable product design as it determines the consumption of resources and energy during the production and use stages of products, as well as the associated environmental impacts. The environmentally low-impact materials include but are not limited to nontoxic materials, renewable materials, low-energy content materials, recycled materials, or recyclable materials [110].
- Reduction in material flows. A direct way to reduce material flows is to dematerialize a product. For example, the concept of light-weighting design has been proposed to prioritize the use of advanced materials and engineering methods in product design, and the goal is to ensure the satisfaction of product quality or even enhanced performance with less material use [111].
- Design for production. At the production level, factors such as the environmental performance of production techniques, number of production steps, and potential strategies for improving the production energy efficiency need to be incorporated in the design stage [112].
- Design for distribution. Sustainability-focused product distribution strategies include using less/cleaner/reusable packaging, selecting more energy-efficient transport modes, and choosing local suppliers to avoid long-distance transport [113].
- Design for "green" use. The goal of sustainable use is to identify, evaluate, and apply sustainable product design approached to change consumer use behavior toward reduced energy consumption and waste generation during the use stage [114].

- Design for long service life. The strategies for the design of long-life products include improving the product quality and making the most vulnerable components easy to be repaired or replaced, which can lead to a slowdown of the resource flow [115].
- End-of-life design. The strategies in this category highlight the importance of end-of-life product management, material recycling, valuable component remanufacturing, etc. The design strategies for improving resource efficiency include the design for disassembly [116], design for remanufacturing [117], and design for recycling [118].

Considering different sustainable product design strategies, extensive effort has been devoted to quantitatively making trade-offs across different life cycle stages. For example, the material efficiency during the processing stage of raw materials is related to various technical factors and can be influenced by the demand of material from the market and the extent of recycling [119].

Even though LCA has become one of the most widely used tools for sustainability evaluation and can provide a wealth of information to support sustainable product development, the data-intensive nature of LCA limits its application in product design. In the early conceptual product design stage, the ideas are diverse and numerous, but the details of different ideas are scarce. However, these early ideas could largely determine the environmental impacts of the final product. Therefore, many research efforts have been made to address the data scarcity issue. For example, artificial neural networks (ANN) are trained using the known characteristics of existing products to estimate the environmental impact of a new product during the concept design without new models [120]. In addition, further development of knowledge-based ANN can facilitate the extraction of general product attributes and link them with environmental characteristics [121].

References

1 Diaz, N., Helu, M., Jayanathan, S. et al. (2010). Environmental analysis of milling machine tool use in various manufacturing environments. *Proceedings of the 2010 IEEE International Symposium on Sustainable Systems and Technology*, (17–19 May 2010), 1–6.

2 CECIMO (2009). Concept Description for CECIMO's Self-Regulatory Initiative (SRI) for the Sector Specific Implementation of the Directive 2005/32/EC (EuP Directive).

3 Dornfeld, D. and Lee, D.E. (2007). *Precision Manufacturing*. New York: Springer.

4 Züst, S., Züst, R., Schudeleit, T., and Wegener, K. (2016). Development and application of an eco-design tool for machine tools. *Procedia CIRP* 48: 431–436.

5 Brecher, C., Triebs, J., Heyers, C., and Jasper, D. (2012). *Effizienzsteigerung von Werkzeugmaschinen durch Optimierung der Technologien zum Komponentenbetrieb-EWOTeK: Verbundprojekt im Rahmenkonzept "Forschung für die Produktion von morgen", "Ressourceneffizienz in der Produktion" des Bundesministeriums für Bildu*. Apprimus.

6 Mittal, S. and Frayman, F. (1989). Towards a generic model of configuration tasks. *Proceedings of the 11th International Joint Conference on Artificial Intelligence,* Detroit, Michigan, USA (20–25 August 1989)

7 Haapala, K.R., Zhao, F., Camelio, J. et al. (2013). A review of engineering research in sustainable manufacturing. *J. Manuf. Sci. Eng. Trans. ASME* 135 (4): 1–16.

8 Duflou, J.R., Kellens, K., Devoldere, T. et al. (2010). Energy related environmental impact reduction opportunities in machine design: case study of a laser cutting machine. *Int. J. Sustain. Manuf.* 2 (1): 80.

9 Duan, Y., Li, W., Fu, X. et al. (2018). A methodology for reliability of WSN based on software defined network in adaptive industrial environment. *IEEE/CAA J. Autom. Sin.* 5 (1): 74–82.

10 Yan, J., Zhou, M., and Ding, Z. (2016). Recent advances in energy-efficient routing protocols for wireless sensor networks: a review. *IEEE Access* 4: 5673–5686.

11 Zhu, X., Li, J., and Zhou, M. (2019). Optimal deployment of energy-harvesting directional sensor networks for target coverage. *IEEE Syst. J.* 13 (1): 377–388.

12 Zhu, X., Li, J., and Zhou, M. (2019). Target coverage-oriented deployment of rechargeable directional sensor networks with a mobile charger. *IEEE Internet Things J.* 6 (3): 5196–5208.

13 Ding, K. and Jiang, P. (2018). RFID-based production data analysis in an IoT-enabled smart job-shop. *IEEE/CAA J. Autom. Sin.* 5 (1): 128–138.

14 Luo, X., Chen, M., Wu, H. et al. (2021). Adjusting learning depth in nonnegative latent factorization of tensors for accurately modeling temporal patterns in dynamic QoS data. *IEEE Trans. Autom. Sci. Eng.* 18 (4): 2142–2155.

15 Luo, X., Zhou, M., Li, S., and Shang, M. (2018). An inherently nonnegative latent factor model for high-dimensional and sparse matrices from industrial applications. *IEEE Trans. Ind. Informatics* 14 (5): 2011–2022.

16 Luo, X., Zhou, M., Li, S. et al. (2020). Non-negativity constrained missing data estimation for high-dimensional and sparse matrices from industrial applications. *IEEE Trans. Cybern.* 50 (5): 1844–1855.

17 Shang, M., Luo, X., Liu, Z. et al. (2019). Randomized latent factor model for high-dimensional and sparse matrices from industrial applications. *IEEE/CAA J. Autom. Sin.* 6 (1): 131–141.

18 Ma, S., Zhang, Y., Lv, J. et al. (2020). Big data driven predictive production planning for energy-intensive manufacturing industries. *Energy* 211: 118320.

19 Zhang, Y., Ma, S., Yang, H. et al. (2018). A big data driven analytical framework for energy-intensive manufacturing industries. *J. Clean. Prod.* 197: 57–72.

20 Bi, J., Yuan, H., and Zhou, M. (2019). Temporal prediction of multiapplication consolidated workloads in distributed clouds. *IEEE Trans. Autom. Sci. Eng.* 16 (4): 1763–1773.

21 Ghahramani, M.H., Zhou, M., and Hon, C.T. (2017). Toward cloud computing QoS architecture: analysis of cloud systems and cloud services. *IEEE/CAA J. Autom. Sin.* 4 (1): 6–18.

22 Yuan, H., Liu, H., Bi, J., and Zhou, M. (2021). Revenue and energy cost-optimized biobjective task scheduling for green cloud data centers. *IEEE Trans. Autom. Sci. Eng.* 18 (2): 817–830.

23 Yuan, H., Zhou, M., Liu, Q., and Abusorrah, A. (2020). Fine-grained and arbitrary task scheduling for heterogeneous applications in distributed green clouds. *IEEE/CAA J. Autom. Sin.* 1–13.

24 Zhang, P., Kong, Y., and Zhou, M. (2018). A domain partition-based trust model for unreliable clouds. *IEEE Trans. Inf. Forensics Secur.* 13 (9): 2167–2178.

25 Zhang, P. and Zhou, M. (2018). Dynamic cloud task scheduling based on a two-stage strategy. *IEEE Trans. Autom. Sci. Eng.* 15 (2): 772–783.

26 Han, M., Yang, Y., and Li, L. (2021). Techno-economic modeling of 4D printing with thermo-responsive materials towards desired shape memory performance. *IISE Trans.* 1–13.

27 Han, M., Yang, Y., and Li, L. (2020). Energy consumption modeling of 4D printing thermal-responsive polymers with integrated compositional design for material. *Addit. Manuf.* 34: 101223.

28 Zhao, J., Han, M., and Li, L. (2021). Modeling and characterization of shape memory properties and decays for 4D printed parts using stereolithography. *Mater. Des.* 203: 109617.

29 Peng, T., Zhu, Y., Leu, M., and Bourell, D. (2020). Additive manufacturing-enabled design, manufacturing, and lifecycle performance. *Addit. Manuf.* 36: 101646.

30 Peng, T., Lv, J., Majeed, A., and Liang, X. (2021). An experimental investigation on energy-effective additive manufacturing of aluminum parts via process parameter selection. *J. Clean. Prod.* 279: 123609.

31 Bekiroglu, K., Srinivasan, S., Png, E. et al. (2020). Recursive approximation of complex behaviours with IoT-data imperfections. *IEEE/CAA J. Autom. Sin.* 7 (3): 656–667.

32 Fortino, G., Russo, W., Savaglio, C. et al. (2018). Agent-oriented cooperative smart objects: from IoT system design to implementation. *IEEE Trans. Syst. Man, Cybern. Syst.* 48 (11): 1939–1956.

33 Fortino, G., Messina, F., Rosaci, D., and Sarne, G.M.L. (2020). ResIoT: an IoT social framework resilient to malicious activities. *IEEE/CAA J. Autom. Sin.* 7 (5): 1263–1278.

34 Fortino, G., Savaglio, C., Spezzano, G., and Zhou, M. (2021). Internet of Things as system of systems: a review of methodologies, frameworks, platforms, and tools. *IEEE Trans. Syst. Man, Cybern. Syst.* 51 (1): 223–236.

35 Peng, T., He, Q., Zhang, Z. et al. (2021). Industrial Internet-enabled resilient manufacturing strategy in the wake of COVID-19 pandemic: a conceptual framework and implementations in China. *Chinese J. Mech. Eng.* 34 (1): 48.

36 Peng, C., Peng, T., Liu, Y. et al. (2021). Industrial Internet of Things enabled supply-side energy modelling for refined energy management in aluminium extrusions manufacturing. *J. Clean. Prod.* 301: 126882.

37 Liu, W., Peng, T., Tang, R. et al. (2020). An Internet of Things-enabled model-based approach to improving the energy efficiency of aluminum die casting processes. *Energy* 202: 117716.

38 Guo, Y., Hu, X., Hu, B. et al. (2018). Mobile cyber physical systems: current challenges and future networking applications. *IEEE Access* 6: 12360–12368.

39 Liu, T., Tian, B., Ai, Y., and Wang, F.-Y. (2020). Parallel reinforcement learning-based energy efficiency improvement for a cyber-physical system. *IEEE/CAA J. Autom. Sin.* 7 (2): 617–626.

40 White, A., Karimoddini, A., and Karimadini, M. (2020). Resilient fault diagnosis under imperfect observations - a need for Industry 4.0 era. *IEEE/CAA J. Autom. Sin.* 7 (5): 1279–1288.

41 Liu, Y., Peng, Y., Wang, B. et al. (2017). Review on cyber-physical systems. *IEEE/CAA J. Autom. Sin.* 4 (1): 27–40.

42 Yun, L., Ma, S., Li, L., and Liu, Y. (2022). CPS-enabled and knowledge-aided demand response strategy for sustainable manufacturing. *Adv. Eng. Informatics* 52: 101534.

43 Ma, S., Zhang, Y., Lv, J. et al. (2019). Energy-cyber-physical system enabled management for energy-intensive manufacturing industries. *J. Clean. Prod.* 226: 892–903.

44 Wang, Q., Jiao, W., Wang, P., and Zhang, Y. (2021). Digital twin for human-robot interactive welding and welder behavior analysis. *IEEE/CAA J. Autom. Sin.* 8 (2): 334–343.

45 Wei, Q., Li, H., and Wang, F.-Y. (2020). Parallel control for continuous-time linear systems: a case study. *IEEE/CAA J. Autom. Sin.* 7 (4): 919–928.

46 Tao, F., Zhang, H., Liu, A., and Nee, A.Y.C. (2019). Digital twin in industry: state-of-the-art. *IEEE Trans. Ind. Informatics* 15 (4): 2405–2415.

47 Chavarría-Barrientos, D., Batres, R., Wright, P.K., and Molina, A. (2018). A methodology to create a sensing, smart and sustainable manufacturing enterprise. *Int. J. Prod. Res.* 56 (1–2): 584–603.

48 Hessami, A.G., Hsu, F., and Jahankhani, H. (2009). A systems framework for sustainability. *International Conference on Global Security, Safety, and Sustainability*, London, UK (1–2 September 2009), 76–94.

49 Guo, X., Zhou, M., Abusorrah, A. et al. (2021). Disassembly sequence planning: a survey. *IEEE/CAA J. Autom. Sin.* 8 (7): 1308–1324.

50 Yun, L., Li, L., and Ma, S. (2022). Demand response for manufacturing systems considering the implications of fast-charging battery powered material handling equipment. *Appl. Energy* 310: 118550.

51 Ma, S., Zhang, Y., Ren, S. et al. (2021). A case-practice-theory-based method of implementing energy management in a manufacturing factory. *Int. J. Comput. Integr. Manuf.* 34 (7–8): 829–843.

52 Ma, S., Zhang, Y., Liu, Y. et al. (2020). Data-driven sustainable intelligent manufacturing based on demand response for energy-intensive industries. *J. Clean. Prod.* 274: 123155.

53 Li, J., Zhang, Y., Qian, C. et al. (2020). Research on recommendation and interaction strategies based on resource similarity in the manufacturing ecosystem. *Adv. Eng. Informatics* 46: 101183.

54 Lin, Q., Zhang, Y., Yang, S. et al. (2020). A self-learning and self-optimizing framework for the fault diagnosis knowledge base in a workshop. *Robot. Comput. Integr. Manuf.* 65: 101975.

55 Lv, J., Wang, Z., and Ma, S. (2020). Calculation method and its application for energy consumption of ball mills in ceramic industry based on power feature deployment. *Adv. Appl. Ceram.* 119 (4): 183–194.

56 Lv, J., Peng, T., Zhang, Y., and Wang, Y. (2021). A novel method to forecast energy consumption of selective laser melting processes. *Int. J. Prod. Res.* 59 (8): 2375–2391.

57 Stock, T. and Seliger, G. (2016). Opportunities of sustainable manufacturing in Industry 4.0. *Procedia CIRP, 2016*, Binh Duong New City, Vietnam (16–18 September 2015).

58 Zheng, P., Sang, Z., Zhong, R.Y. et al. (2018). Smart manufacturing systems for Industry 4.0: conceptual framework, scenarios, and future perspectives. *Front. Mech. Eng.* 13 (2): 137–150.

59 Guerreiro, B.V., Lins, R.G., Sun, J., and Schmitt, R. (2018). Definition of smart retrofitting: first steps for a company to deploy aspects of Industry 4.0. In: *Advances in Manufacturing* (ed. A. Hamrol, O. Ciszak, S. Legutko and M. Jurczyk), 161–170. Springer.

60 Ford, S. and Despeisse, M. (2016). Additive manufacturing and sustainability: an exploratory study of the advantages and challenges. *J. Clean. Prod.* 137: 1573–1587.

61 Hagel, J. III, Brown, J.S., Kulasooriya, D. et al. (2015). The future of manufacturing-making things in a changing world. *Futur. Bus. Landsc.* 4–18.

62 Jin, M., Tang, R., Ji, Y. et al. (2017). Impact of advanced manufacturing on sustainability: an overview of the special volume on advanced manufacturing for sustainability and low fossil carbon emissions. *J. Clean. Prod.* 161: 69–74.

63 Rejeski, D., Zhao, F., and Huang, Y. (2018). Research needs and recommendations on environmental implications of additive manufacturing. *Addit. Manuf.* 19: 21–28.

64 Yang, Y. and Li, L. (2018). Total volatile organic compound emission evaluation and control for stereolithography additive manufacturing process. *J. Clean. Prod.* 170: 1268–1278.

65 Cagliano, R., Canterino, F., Longoni, A., and Bartezzaghi, E. (2019). The interplay between smart manufacturing technologies and work organization. *Int. J. Oper. Prod. Manag.* 39 (6/7/8): 913–934.

66 Carvalho, N., Chaim, O., Cazarini, E., and Gerolamo, M. (2018). Manufacturing in the fourth industrial revolution: a positive prospect in sustainable manufacturing. *Procedia Manuf.* 21: 671–678.

67 Brehm, C. and Layton, A. (2021). Nestedness of eco-industrial networks: exploring linkage distribution to promote sustainable industrial growth. *J. Ind. Ecol.* 25 (1): 205–218.

68 Chertow, M.R. (2000). Industrial symbiosis: literature and taxonomy. *Annu. Rev. energy Environ.* 25 (1): 313–337.

69 Ge, Y. (2020). *System-Level Sustainability Assessment for Economic Viability of Cellulosic Biofuel Manufacturing*. University of Illinois at Chicago.

70 Lowe, E.A., Moran, S.R., and Holmes, D.B. (1995). Fieldbook for the Development of Eco-industrial Parks. Draft report. Oakland, CA Indigo Dev. Co.

71 Fang, K., Dong, L., Ren, J. et al. (2017). Carbon footprints of urban transition: tracking circular economy promotions in Guiyang, China. *Ecol. Modell.* 365: 30–44.

72 J. von Koerber, W. Raber, and P. Schneider, "Nexus-oriented approach for sharing water resources: development of eco-industrial parks in the catchment of Zayandeh Rud River, Iran," in Stephan Hülsmann, Mahesh Jampani *A Nexus Approach for Sustainable Development*, Cham: Springer International Publishing, 2021, pp. 203–221.

73 Duflou, J.R., Sutherland, J.W., Dornfeld, D. et al. (2012). Towards energy and resource efficient manufacturing: a processes and systems approach. *CIRP Ann.* 61 (2): 587–609.

74 Trokanas, N., Cecelja, F., and Raafat, T. (2014). Semantic input/output matching for waste processing in industrial symbiosis. *Comput. Chem. Eng.* 66: 259–268.

75 Bin, S., Zhiquan, Y., Jonathan, L.S.C. et al. (2015). A big data analytics approach to develop industrial symbioses in large cities. *Procedia CIRP* 29: 450–455.

76 Ben Zhu, G.K. and Davis, C.B. (2014). Information synergy of industrial symbiosis. http://uest.ntua.gr/conference2014/pdf/zhu.pdf (accessed 19 March 2021).

77 Low, J.S.C., Tjandra, T.B., Yunus, F. et al. (2018). A collaboration platform for enabling industrial symbiosis: application of the database engine for waste-to-resource matching. *Procedia CIRP* 69: 849–854.

78 Ghali, M.R., Frayret, J.-M., and Robert, J.-M. (2016). Green social networking: concept and potential applications to initiate industrial synergies. *J. Clean. Prod.* 115: 23–35.

79 Mentzer, J.T. et al. (2001). Defining supply chain management. *J. Bus. Logist.* 22 (2): 1–25.

80 Ge, Y., Li, L., and Yun, L. (2021). Modeling and economic optimization of cellulosic biofuel supply chain considering multiple conversion pathways. *Appl. Energy* 281: 116059.

81 Ghadimi, P., Wang, C., and Lim, M.K. (2019). Sustainable supply chain modeling and analysis: past debate, present problems and future challenges. *Resour. Conserv. Recycl.* 140: 72–84.

82 Mani, V., Gunasekaran, A., Papadopoulos, T. et al. (2016). Supply chain social sustainability for developing nations: evidence from India. *Resour. Conserv. Recycl.* 111: 42–52.

83 Tseng, M.-L., Lim, M., Wu, K.-J. et al. (2018). A novel approach for enhancing green supply chain management using converged interval-valued triangular fuzzy numbers-grey relation analysis. *Resour. Conserv. Recycl.* 128: 122–133.

84 Eltayeb, T.K., Zailani, S., and Ramayah, T. (2011). Green supply chain initiatives among certified companies in Malaysia and environmental sustainability: investigating the outcomes. *Resour. Conserv. Recycl.* 55 (5): 495–506.

85 Ahmadi, H.B., Kusi-Sarpong, S., and Rezaei, J. (2017). Assessing the social sustainability of supply chains using Best Worst Method. *Resour. Conserv. Recycl.* 126: 99–106.

86 Lee, D.-H., Dong, M., and Bian, W. (2010). The design of sustainable logistics network under uncertainty. *Int. J. Prod. Econ.* 128 (1): 159–166.

87 Chaabane, A., Ramudhin, A., and Paquet, M. (2012). Design of sustainable supply chains under the emission trading scheme. *Int. J. Prod. Econ.* 135 (1): 37–49.

88 Kumar, D.T., Palaniappan, M., Kannan, D., and Shankar, K.M. (2014). Analyzing the CSR issues behind the supplier selection process using ISM approach. *Resour. Conserv. Recycl.* 92: 268–278.

89 Seo, S., Kim, J., Yum, K.-K., and McGregor, J. (2015). Embodied carbon of building products during their supply chains: case study of aluminium window in Australia. *Resour. Conserv. Recycl.* 105: 160–166.

90 Prakash, C. and Barua, M.K. (2016). An analysis of integrated robust hybrid model for third-party reverse logistics partner selection under fuzzy environment. *Resour. Conserv. Recycl.* 108: 63–81.

91 Lai, K.-H., Lun, V.Y.H., Wong, C.W.Y., and Cheng, T.C.E. (2011). Green shipping practices in the shipping industry: conceptualization, adoption, and implications. *Resour. Conserv. Recycl.* 55 (6): 631–638.

92 Ye, F., Li, Y., and Yang, Q. (2018). Designing coordination contract for biofuel supply chain in China. *Resour. Conserv. Recycl.* 128: 306–314.

93 Zhang, Q., Tang, W., and Zhang, J. (2018). Who should determine energy efficiency level in a green cost-sharing supply chain with learning effect? *Comput. Ind. Eng.* 115: 226–239.

94 Marchi, B. and Zanoni, S. (2017). Supply chain management for improved energy efficiency: review and opportunities. *Energies* 10 (10): 1618.

95 Xie, G., Yue, W., and Wang, S. (2017). Energy efficiency decision and selection of main engines in a sustainable shipbuilding supply chain. *Transp. Res. Part D Transp. Environ.* 53: 290–305.

96 Bureau of Economic Analysis. Industry economic accounts. https://www.bea.gov/data/economic-accounts/industry (accessed 20 March 2021).

97 Forti, V., Baldé, C., Kuehr, R., and Bel, G. (2020). The global e-waste monitor 2020. *United Nations Univ. (UNU), The Global E-waste Monitor 2020: Quantities, flows and the circular economy potential*, Bonn, Geneva and Rotterdam (2 July 2020).

98 Stahel, W.R. (2016). The circular economy. *Nature* 531 (7595): 435–438.

99 U. C. Economy (2006). An alternative for economic development. *UNEP DTIE Paris, Fr.*

100 Potting, J., Hekkert, M., Worrell, E., and Hanemaaijer, A. (2017). *Circular Economy: Measuring Innovation in the Product Chain*. The Hague, Netherlands: PBL Netherlands Environmental Assessment Agency.

101 Kirchherr, J., Reike, D., and Hekkert, M. (2017). Conceptualizing the circular economy: an analysis of 114 definitions. *Resour. Conserv. Recycl.* 127: 221–232.

102 Talens Peiró, L., Ardente, F., and Mathieux, F. (2017). Design for disassembly criteria in EU product policies for a more circular economy: a method for analyzing battery packs in PC-tablets and subnotebooks. *J. Ind. Ecol.* 21 (3): 731–741.

103 Akbarnezhad, A., Ong, K.C.G., and Chandra, L.R. (2014). Economic and environmental assessment of deconstruction strategies using building information modeling. *Autom. Constr.* 37: 131–144.

104 Wang, H., Schandl, H., Wang, X. et al. (2020). Measuring progress of China's circular economy. *Resour. Conserv. Recycl.* 163: 105070.

105 Bogue, R. (2007). Design for disassembly: a critical twenty-first century discipline. *Assem. Autom.* 27 (4): 285–289.

106 EPA (U.S. Environmental Protection Agency) (2019). National Framework for Advancing the U.S. Recycling System.

107 Guinee, J.B., Heijungs, R., Huppes, G. et al. (2011). *Life Cycle Assessment: Past, Present, and Future*. ACS Publications.

108 International Organization for Standardization (2006). *ISO 14040:2006, Environmental Management—Life Cycle Assessment–Principles and Framework*. Geneva, Switzerland: ISO.

109 Zbicinski, I. (2006). *Product Design and Life Cycle Assessment*, vol. 3. Baltic University Press.

110 Ljungberg, L.Y. (2007). Materials selection and design for development of sustainable products. *Mater. Des.* 28 (2): 466–479.

111 Zhu, L., Li, N., and Childs, P.R.N. (2018). Light-weighting in aerospace component and system design. *Propuls. Power Res.* 7 (2): 103–119.

112 Kluczek, A. (2019). An energy-led sustainability assessment of production systems – an approach for improving energy efficiency performance. *Int. J. Prod. Econ.* 216: 190–203.

113 Coelho, P.M., Corona, B., ten Klooster, R., and Worrell, E. (2020). Sustainability of reusable packaging–current situation and trends. *Resour. Conserv. Recycl. X* 6: 100037.

114 Bhamra, T., Lilley, D., and Tang, T. (2011). Design for sustainable behaviour: using products to change consumer behaviour. *Des. J.* 14 (4): 427–445.

115 Bocken, N.M.P., de Pauw, I., Bakker, C., and van der Grinten, B. (2016). Product design and business model strategies for a circular economy. *J. Ind. Prod. Eng.* 33 (5): 308–320.

116 Abuzied, H., Senbel, H., Awad, M., and Abbas, A. (2020). A review of advances in design for disassembly with active disassembly applications. *Eng. Sci. Technol. Int. J.* 23 (3): 618–624.

117 Boorsma, N., Balkenende, R., Bakker, C. et al. (2020). Incorporating design for remanufacturing in the early design stage: a design management perspective. *J. Remanufacturing* 11 (1): 25–48.

118 Kishawy, H., Hegab, H., and Saad, E. (2018). Design for sustainable manufacturing: approach, implementation, and assessment. *Sustainability* 10 (10): 3604.

119 Allwood, J.M., Ashby, M.F., Gutowski, T.G., and Worrell, E. (2011). Material efficiency: a white paper. *Resour. Conserv. Recycl.* 55 (3): 362–381.

120 Sousa, I., Wallace, D., and Eisenhard, J.L. (2000). Approximate life-cycle assessment of product concepts using learning systems. *J. Ind. Ecol.* 4 (4): 61–81.

121 Park, J.-H. and Seo, K.-K. (2006). A knowledge-based approximate life cycle assessment system for evaluating environmental impacts of product design alternatives in a collaborative design environment. *Adv. Eng. Informatics* 20 (2): 147–154.

Glossary

α	The ratio between the manufacturing power and the total power
δ	Calorific value of the natural gas (kWh/Nm^3)
ϵ	The electricity to heat ratio of the CHP system
η_1	Light efficiency for a specific UV source
η_2	Ratio of effective wavelength over the total wavelength for a specific UV source
η_3	Material absorptivity for a specific UV source
η_A	Efficiency of the auxiliary boiler (%)
η_C	Efficiency of the CHP system (%)
$\eta_i(t)$	Production rate of machine M_i
$\eta_{SYS}(t)$	Production rate of the manufacturing system
$\bar{\eta}$	Average cumulative production of the system
λ_i	Efficiency of machine M_i
ω	Inertial weight in PSO
ω_i	A processing route for part i
$\omega(\pi_n)$	Route vector under π_n
π_n	A transition sequence starting from the initial marking with length n
σ	Penalty coefficient
τ	Duration of each time step
A	The surface area of biomass reactor
$A_i(\pi_n)$	Fired transition sequence on the operation path of part i under π_n
B_i	The i-th buffer in produciton line
$B_i(\pi_n)$	Firing time of the transitions in $A_i(\pi_n)$
c_1	Cognitive parameter in PSO
c_2	Social parameter in PSO
$c_a(p)$	Energy consumption rate of the resource in an occupied state in p
C_A^0	Startup costs for the auxiliary boiler

Sustainable Manufacturing Systems, First Edition. Lin Li and MengChu Zhou.
© 2023 The Institute of Electrical and Electronics Engineers, Inc.
Published 2023 by John Wiley & Sons, Inc.

C_A^{gas}	Natural gas costs of the auxiliary boiler
C_C	The operational costs of the CHP system and auxiliary boiler
C_C^0	Startup costs for the CHP system
C_C^{gas}	Natural gas costs of the CHP system
C_D	Electricity demand cost
c_D	Unit electricity demand cost
C_E	Electricity consumption cost
c_E	Unit electricity consumption cost
C_F	Fixed electricity cost
C_G	The cost of electricity purchased from the power grid
c_g	Glucose concentration
$c_h(p)$	Energy consumption rate of the resource in working/busy state in p
$c_i(t)$	Buffer content in buffer B_i at time t
$\hat{C}_i$	Maximum capacity of buffer B_i
C_k	Kick's constant
C_{p_b}	Heat capacity of biomass
C_{p_s}	Heat capacity of steam
C_{p_w}	Heat capacity of water
c_{xl}	Xylose concentration
c_{xo}	Xylose oligomers concentration
$c(r)$	Energy consumption rate of resource r in idle state
d	Layer thickness of a 3D-printed part
D_p	The peak demand of the combined manufacturing and HVAC system during the demand response event
d_T	Peak demand
$E_{cooling}$	Energy consumption of cooling system
E_{curing}	Energy consumption of the UV curing process
E_i	Energy consumption of machine M_i during the entire production planning horizon
$e_i(t)$	Energy consumption of machine M_i at time t
e_K	UV curing energy consumption for the K-th layer of a 3D-printed part
$E_{platform}$	Energy consumption of building platform movement
$E(t)$	The electricity consumption of production line at time t
h	Total height of a 3D-printed part
h_{out}	Convection coefficient at the outside surface of a biomass reactor
$I_{measure}$	Measured current
k	A positive constant of Newton's law of cooling
K	Number of layers of a 3D-printed part
k_1	Heat capacity of the entire manufacturing plant

k_g	Reaction rates of the conversion from glucan to glucose
K_g	Substrate inhibition coefficient of glucose
k_{xl}	Reaction rates of the conversion from xylan to xylose
K_{xl}	Substrate inhibition coefficient of xylose
k_{xo}	Reaction rates of the conversion from xylan to xylose oligomers
M_i	The i-th machine in production line
$P_{cooling}$	Power output of a cooling fan
p_g	Glucan mass fraction in biomass
P_i	The rated power of machine M_i
p_i	Reliability of machine M_i in Bernoulli distribution
$P_i^B(t)$	The probability that machine M_i is blocked at time t
$P_i^S(t)$	The probability that machine M_i is starved at time t
P_m	Power output of the stepper motor
P_{UV}	The output power of the UV light source
p_x	Xylan mass fraction in biomass
$\hat{p}$	The output power capacity of the CHP system
$q_C(t)$	The instantaneous convective heat transferred to the surroundings at time t
$q_G(t)$	Rate of the thermal energy generated from a manufacturing system at time t
$Q_i(t)$	State transition matrix for buffer B_i
$Q_{i,(j_2|j_1)}(t)$	Transition probability that the state of B_i changes from j_1 to j_2 at time t
$q_{i,k}(t)$	The probability that buffer B_i contains k parts at time t
$q_R(t)$	The total radiant heat transferred to the surroundings at time t
$q(t)$	Rate of the thermal energy transferred from the manufacturing system to the plant indoor environment at time t
R_{in}	The inner radiuses of a biomass reactor
R_{out}	The outer radiuses of a biomass reactor
R_p	Committed limitation of power demand during a demand response event
$s_i(t)$	A binary decision variable decides whether machine M_i is turned on or off at time t
T	Total time steps in production planning horizon
t_D	The time interval defined by utility to calculate the highest average power during the billing period
t_k	The curing time for the k-th layer of a 3D-printed part
T_L	Lower bounds of acceptable indoor temperature
T_{out}	Outdoor temperature
T_U	Upper bound of acceptable indoor temperature

u	Total number of all types of parts
u_i	Total number of the i-th type parts
U_{measure}	Measured voltage
$U(t)$	Consecutive on/off time of CHP system at time
Y	Production yield of the manufacturing system during the production horizon
Y^o	Predefined production target
$Z_{\text{A}}(t)$	The operation status of the auxiliary boiler at time t

Acronyms

AM	Additive manufacturing
ANN	Artificial neural networks
BAS	Blocked after service
BBS	Blocked before service
CAD	Computer-aided design
CDF	Cumulative distribution function
CFL	Compact fluorescent lamp
CHP	Combined heat and power
CPP	Critical peak pricing
CPS	Cyber-physical systems
DLC	Direct load control
DMD	Digital micromirror device
DOE	Design of experiments
DP	Dynamic programming
DSM	Demand-side management
EOL	End of life
ERC	Energy consumption rate
FMS	Flexible manufacturing systems
GA	Genetic algorithm
GDP	Gross domestic product
GHG	Greenhouse gas
HGLs	Hydrocarbon gas liquids
HVAC	Heating, ventilation, and air conditioning
IOT	Internet of things
LCA	Life cycle assessment
LED	Light-emitting diode
MDP	Modified dynamic programming
MHS	Material handling system

Sustainable Manufacturing Systems, First Edition. Lin Li and MengChu Zhou.
© 2023 The Institute of Electrical and Electronics Engineers, Inc.
Published 2023 by John Wiley & Sons, Inc.

MIP	Mask image projection
MTBF	Mean time between failure
MTTR	Mean time to repair
MVA	Manufacturing value added
PDF	Probability density function
PMF	Probability mass function
PN	Petri nets
PNS	Petri net for scheduling
PSO	Particle swarm optimization
RG	Reachability graph
RTP	Real-time pricing
SI	International System of Units
SL	Stereolithography
TOU	Time-of-use
UV	Ultraviolet
WIP	Work in process

Index

Sustainable Manufacturing Systems, First Edition. Lin Li and MengChu Zhou.
© 2023 The Institute of Electrical and Electronics Engineers, Inc.
Published 2023 by John Wiley & Sons, Inc.

IEEE Press Series on Systems Science and Engineering

Editor: Mengchu Zhou, *New Jersey Institute of Technology*

Co-Editors: Han-Xiong Li, *City University of Hong-Kong*
Margot Weijnen, *Delft University of Technology*

The focus of this series is to introduce the advances in theory and applications of systems science and engineering to industrial practitioners, researchers, and students. This series seeks to foster system-of-systems multidisciplinary theory and tools to satisfy the needs of the industrial and academic areas to model, analyze, design, optimize, and operate increasingly complex man-made systems ranging from control systems, computer systems, discrete event systems, information systems, networked systems, production systems, robotic systems, service systems, and transportation systems to Internet, sensor networks, smart grid, social network, sustainable infrastructure, and systems biology.

Printed and bound by CPI Group (UK) Ltd, Croydon, CR0 4YY

15/11/2022

03163228-0001